Principles of Soil Science

Principles of Soil Science

Edited by **Henry Wang**

Callisto Reference
New York

Published by Callisto Reference,
106 Park Avenue, Suite 200,
New York, NY 10016, USA
www.callistoreference.com

Principles of Soil Science
Edited by Henry Wang

International Standard Book Number: 978-1-63239-703-4 (Hardback)

Printed in China.

Contents

Preface

The main aim of this book is to educate learners and enhance their research focus by presenting diverse topics covering this vast field. This is an advanced book which compiles significant studies by distinguished experts. This book addresses successive solutions to the challenges arising in the area of application, along with it; the book provides scope for future developments.

Soil is one of the many natural resources present in abundance and variety. Soil science deals with the structure, composition, mapping and classification of soils all over the globe. The environmentalists and soil scientists are trying hard to preserve the quality of soils and arable lands. This book provides an extensive analysis on a wide array of topics such as nutrient management, water management, wetlands, sensitive and unstable soils, manure, contamination and soil conservation, etc. This book is compiled in such a manner, that it will provide in-depth knowledge about the theory and practice of soil science. A number of latest researches have been included to keep the readers up-to-date with the global concepts in this area of study. It will serve as a reference guide for soil scientists, geologists, environmentalists, ecologists, researchers, professionals and students involved with the field of soil science at various levels.

It was a great honour to edit this book, though there were challenges, as it involved a lot of communication and networking between me and the editorial team. However, the end result was this all-inclusive book covering diverse themes in the field.

Finally, it is important to acknowledge the efforts of the contributors for their excellent chapters, through which a wide variety of issues have been addressed. I would also like to thank my colleagues for their valuable feedback during the making of this book.

Editor

1

Greenhouse Gas Emission from Cauliflower Grown under Different Nitrogen Rates and Mulches

A. A. Farag[1] and Shaimaa H. Abd-Elrahman[2*]

[1]Central Laboratory for Agricultural Climate, Agricultural Research Center, Egypt.
[2]Department of Soils and Water, Faculty of Agriculture, Ain Shams University, Egypt.

Authors' contributions

This work was carried out in collaboration between the two authors. Both authors designed the study, wrote the protocol, managed the literature searches and managed the experimental process. Both authors read and approved the final manuscript.

Editor(s):
(1) Junhong Bai, School of Environment, Beijing Normal University, Beijing, China.
Reviewers:
(1) Aneeza Soobadar, Mauritius Sugarcane Industry Research Institute, Mauritius.
(2) P. K. Wahome, University of Swaziland, Swaziland.
(3) Mrityunjoy Biswas, Sylhet Agricultural University, Bangladesh.

ABSTRACT

A field experiment was conducted during the two successive seasons of 2012/2013 and 2013/2014 at El-Bosaily farm, El-Behira governorate, Egypt, to study the effect of different nitrogen rates and mulch treatments on plant growth and yield of cauliflower (*Brassica oleracia* var. *botrytis*). The study includes three mulch treatments (transparent and black polyethylene sheets, and bare soil as a control) and four N rates 30, 45, 60 and 75 kg/feddan (1 feddan = 4200 m^2 and hectare = 2.4 feddan) designed in a split plots with three replicates. The greenhouse gas (GHG) emissions of N_2O and CO_2 from N fertilization were calculated for different treatments. Regarding the polyethylene (PE) mulch, data revealed that transparent PE recorded the highest significant values of cauliflower growth characters during the two studied seasons. Increasing N rate up to 60 kg/feddan enhanced cauliflower growth using different polyethylene mulch treatments; yield was not increased significantly with increasing N rate up to 75 kg/feddan. The highest significant cauliflower yield was obtained by using 60 kg N/feddan combined with transparent mulch, while the lowest yield was obtained in case of 30 kg N/feddan combined with bare soil treatment. Chemical analysis of cauliflower leaves at harvest revealed that the percentage of N, P and K increased

*Corresponding author: E-mail: Shaimaa_Hassan@agr.asu.edu.eg

significantly with increasing N rate up to 75 kg/feddan. Polyethylene mulch treatments increased N, P and K percentages significantly in comparison with bare soil treatment. Moreover, the greenhouse gas emissions (N_2O and CO_2) from fertilization increased but the emission rates (kg CO_2/kg yield) were decreased by increasing N rate due to the highest cauliflower production. The highest yield with low emission was obtained by 60 kg N/feddan.

Keywords: GHG emission; polyethylene mulch; nitrogen rate; cauliflower.

1. INTRODUCTION

Cauliflower (*Brassica oleraceae* var. *botrytis* L.) is a vegetable crop grown for its white curds. It is a cool season crop that demands high levels of nitrogen to maximize yields. The N fertilization affects cauliflower development in terms of yield, quality and aspects of N metabolism. The cauliflower curd yield increased with N applications [1]. Moreover, N fertilization showed a limited effect on storage loss of heads.

In modern agriculture, fertilizers have often been over-applied to maximize crop production. Environmental issues associated with overuse of N fertilizers are of concern. The N nutrient not absorbed by the crop is lost from the soil by leaching or denitrification with NO/N_2O emission. The NO_3^- levels in groundwater close to farms is a major environmental and social concern impacted by N application [2]. Direct sources include N_2O, which is emitted directly to the atmosphere from cultivated fertilized soils [3]. Indirect emissions result from transport of N from agricultural systems into surface and ground waters through surface runoff and drainage as well as emission as ammonia or nitrogen oxides and deposition elsewhere, causing N_2O production. Methodologies for calculating both direct and indirect emissions of N_2O are related to agricultural production. It takes into account anthropogenic N inputs including synthetic fertilizers, animal wastes and other organic fertilizers, biological N fixation by crops, cultivation of organic soils, along with mineralization of crop residues returned to the field [4].

Gases that trap heat in the atmosphere are called greenhouse gases (GHGs); each of these gases (i.e. N_2O, CO_2, CH_4, O_3, etc.) can remain in the atmosphere from a few years to thousands of years. All of these gases remain in the atmosphere long enough to become well mixed, meaning that the amount that is measured in the atmosphere is roughly the same all over the world, regardless of the source of the emission. Global Warming Potential (GWP) has been calculated to reflect how long it remains in the atmosphere, on average, and how strongly it absorbs energy. Gases with a higher GWP absorb more energy than gases with a lower GWP, and thus contribute more to global warming. For example, N_2O has a radiative warming effect which is about 300 times that of CO_2, the dominant GHG [5]. Management practices that can influence emission of N_2O from agricultural soils include N fertilizers (rate, type, timing and application method), crop, tillage, residue management, and irrigation [6]. Decreasing N fertilizer rate may reduce GHG emission [7].

The mulch determines its energy-radiating behavior and its influence on the microclimate around the plant. Today transparent, black, and white mulches dominate in the commercial vegetable production over the world. Transparent and translucent mulches promote a relatively large net radiation at the soil surface, increase soil heat flux and as a consequence, both minimum and maximum soil temperatures are increased [8]. Row covers are a flexible transparent covering which is installed over plants to reduce the water needs and increase air temperature around the crop. Furthermore, their usage has been associated with increased plant growth, higher yields and earliness of harvest [9]. The crop varies in response to polyethylene mulch covers depending on cultivar, materials used and environmental conditions. The beneficial responses of plants to polyethylene mulch, such as higher total yields, earlier production and better fruit quality have been discussed by many authors [10-12]. Mulch can provide additional benefits beyond temperature, moisture and erosion control; a moderate GHG flux is one of such benefits. Application of mulch to cultivated soil increases soil organic carbon (SOC) concentrations [13]. It also has beneficial effects on SOC sequestration and strongly influences the temporal pattern of CO_2 emission from soil [14].

Thus, the aim of this study was to evaluate the effect of different applied rates of N fertilizer and mulch treatments on the growth, yields of cauliflower and also to determine GHG emission from different N fertilization rates.

2. MATERIALS AND METHODS

2.1 The Field Experiment

The experiment was carried out at El-Bosaily farm, El-Behira Governorate, in the North Coast of Egypt. The study included three mulch treatments viz. transparent and black polyethylene sheets, and bare soil as a control, and four N rates viz. 30, 45, 60 and 75 kg/feddan. The recommended N for cauliflower plant was 45 kg/feddan (feddan = 4200 m^2 and hectare = 2.4 feddan).

The permanent wilting point (PWP) and field capacity (FC) of the experimental soil, *Typic Torripsamments*, were determined according to [15]. Some soil physical and chemical properties were investigated as described by [16-18] and the results have presented in Table 1. Soil temperature was measured by soil thermometer at 15 cm depth; it was recorded daily at mid-day during the two studied seasons. The same amount of drip irrigation was applied for all treatments (EC of the water used for irrigation was 0.82 dS m^{-1}).

Five-week old seedlings of healthy cauliflower cv. Amshiry were transplanted on one side of the ridges at a distance of 50 cm on October 18, 2012 and October 15, 2013 for the first and second seasons, respectively. At the end of the two seasons, samples of five plants of each experimental plot were taken to determine growth parameters (plant height, number of leaves, leaf fresh weight, leaf dry weight per plant and stem diameter). The harvesting period was from February 15 to March 5 in the seasons of 2013 and 2014, respectively. At harvesting stage, ten leaves were obtained from five cauliflower plants that were selected from the middle row of each plot. Nutrient contents (N, P and K) were evaluated in the indicated leaves [19].

2.2 Greenhouse Gas Emission from Nitrogen Fertilization in the Experimental Soil

The aerobic microbial oxidation of ammonia to nitrate is called nitrification, while denitrification is the reduction of nitrate to gaseous nitrogen (N_2). Both reactions produce the intermediate gaseous nitrous oxide (N_2O) through microbial activities in the soil and eventually this gas is released to the atmosphere. The emission of N_2O from field was estimated according to [6]; the following equation was adopted:

$$N_2O \text{ emission} = [1.47 + (0.01 \times F)] \times N_2O_{MW} \times N_2O_{GWP}$$

Where:

F	Mass of N applied from synthetic fertilizer, kg N ha^{-1}
N_2O_{MW}	Ratio of molecular weight of N_2O to 2N, kg N_2O (kg N)$^{-1}$
N_2O_{GWP}	Global Warming Potential for N_2O, kg CO_2-e (kg N_2O)$^{-1}$

The GWP value of 298 for N_2O used in the protocol (N_2O_{GWP}) is the 100-year value used in the most recent IPCC fourth assessment report [20]. The CO_2-e equivalent emission for each gas (CO_2, N_2O and CH_4) were summed together to give total CO_2-e.

2.3 Experimental Design and Statistical Analysis

Split plot design with three replicates was used. Nitrogen rates were randomly arranged in the main-plots and mulch treatments were distributed randomly in the sub-plots. Data were statistically analyzed using statistical analysis system (SAS) program [21]. The means that were significant were separated using Duncan's New Multiple Range Test (DNMRT) at $P \leq 0.05$.

Table 1. Some physical and chemical characteristics of the surface layer of the experimental soil (0-30 cm) before crop cultivation

Particle size distribution, %			Texture	SP	FC	PWP	BD	$CaCO_3$	OM
Sand	Silt	Clay	class		%		g cm^{-3}	%	
85.1	5.60	9.30	Sandy	25.1	14.8	4.70	1.60	1.03	0.06
pH	EC_e	Soluble ions, meq L^{-1}							
(1:2.5)	dS m^{-1}	Ca^{2+}	Mg^{2+}	Na^+	K^+	CO_3^{2-}	HCO_3^-	Cl^-	SO_4^{2-}
7.68	1.20	2.95	1.76	6.33	0.96	n.d*	2.98	7.80	1.22

** n.d means not detected*

3. RESULTS AND DISCUSSION

3.1 Soil Temperature

The average soil temperature at 15 cm soil depth in each treatment, in November, December, January and February of 2012/2013 and 2013/2014, is shown in Fig. 1. During the crop cycle, the highest soil temperature was obtained under conditions of the transparent polyethylene mulch (Fig. 1). During the same period, soil temperature under black polyethylene mulch was lower than under the transparent polyethylene mulch by 1-2℃. Soil temperature in control (bare soil) was lower than polyethylene mulch treatments. The obtained results were in agreement with those obtained by [10,22] who observed that the un-mulched plots had the lowest soil temperature (1.0-3.8℃ lower) at different times since planting compared to polyethylene film mulched plots.

3.2 Vegetative Characteristics

The obtained results (Table 2) revealed that mulch and N application rate treatments significantly affected cauliflower vegetative characteristics (plant height, number of leaves, total leaf fresh and dry weight and stem diameter) during the two growing seasons. The highest N rate of 75 kg/feddan produced the highest vegetative characteristics; 60 kg N treatment came in the second order followed by that of 45 kg N; 30 kg N produced the lowest values with significant differences among treatments. Regarding the mulch treatments, data indicated that transparent polyethylene mulch resulted in the highest vegetative characteristics followed by black polyethylene mulch with significant difference between them; the lowest values were obtained in case of control treatment.

The interaction between N rate and mulch treatments was significant for vegetative characteristics during the two growing seasons. The highest values were obtained in cauliflower plants grown using 75 kg N/feddan combined with transparent polyethylene mulch treatment followed by those provided with 60 kg N combined with transparent polyethylene mulch and 75 kg N combined with black polyethylene mulch treatment. The lowest vegetative growth was obtained from plants provided with 30 kg N/feddan treatment along with control. The researchers opined that mulching contributed positively to higher soil temperature and consequently improving growth and yield [23]. While reduced N supply to plants considerably reduced the plant growth [24]. Reductions in growth of N-limited plants correlated with decreases in leaf area, fresh and dry weight along with reallocation of carbon from shoots to roots and a reduction in net assimilation were obtained.

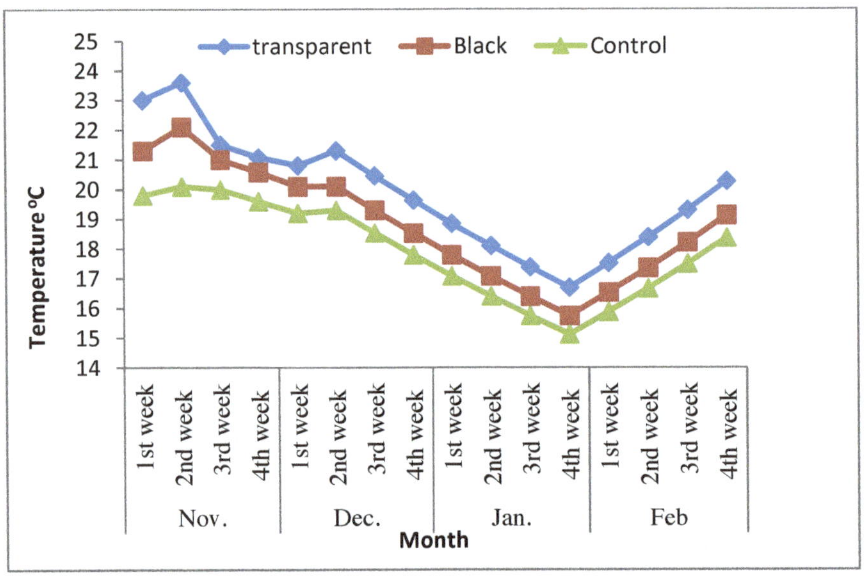

Fig. 1. The average mean soil temperature under transparent, black mulch and control during growing seasons of 2012/2013 and 2013/2014

Table 2. Effect of nitrogen rate and mulch on vegetative characteristics of cauliflower during the studied two seasons of 2012/2013 and 2013/2014

Nitrogen rate, kg/feddan (A)	2012/2013 Polyethylene mulch treatments (B)			Mean (A)	2013/2014 Polyethylene mulch treatments (B)			Mean (A)
	Black	Transparent	Control		Black	Transparent	Control	
Plant height (cm)								
30	61 h	69 f	52 i	**61 D**	64 h	72 f	56 i	**64 D**
45	70 f	77 c	60 h	**69 C**	72 f	78 d	62 h	**71 C**
60	75 d	81 b	66 g	**74 B**	77 d	84 b	67 g	**76 B**
75	79 bc	85 a	72 e	**79 A**	82 c	87 a	75 e	**81 A**
Mean (B)	**71 B**	**78 A**	**63 C**		**74 B**	**80 A**	**65 C**	
No. leaves/plant								
30	12.1 f	12.6 e	10.8 h	**11.8 D**	13.1 f	14.0 e	12.0 h	**13.0 D**
45	12.5 e	13.4 d	11.5 g	**12.4 C**	13.8 e	14.8 d	12.7 g	**13.7 C**
60	13.8 c	14.1 c	12.7 e	**13.5 B**	14.8 d	15.6 c	14.0 e	**14.8 B**
75	15.1 b	16.3 a	13.8 c	**15.1 A**	16.7 b	17.8 a	15.3 c	**16.6 A**
Mean (B)	**13.3 B**	**14.1 A**	**12.2 C**		**14.5 B**	**15.4 A**	**13.4 C**	
Fresh weight of leaves (kg/plant)								
30	1.25 g	1.35 f	1.07 h	**1.22 D**	1.29 g	1.39 f	1.10 h	**1.26 D**
45	1.36 f	1.49 e	1.22 g	**1.36 C**	1.40 f	1.53 e	1.26 g	**1.40 C**
60	1.62 d	1.72 c	1.43 e	**1.59 B**	1.67 d	1.78 c	1.47 e	**1.64 B**
75	1.79 b	1.89 a	1.51 e	**1.73 A**	1.85 b	1.95 a	1.56 e	**1.78 A**
Mean (B)	**1.51 B**	**1.61 A**	**1.31 C**		**1.55 B**	**1.66 A**	**1.35 C**	
Dry weight of leaves (g/plant)								
30	187 h	203 g	161 i	**184 D**	206 h	223 g	177 i	**202 D**
45	204 g	223 ef	184 h	**204 C**	225 g	245 ef	202 h	**224 C**
60	244 d	259 c	215 f	**239 B**	268 d	284 c	236 f	**263 B**
75	269 b	283 a	227 e	**260 A**	296 b	311 a	249 e	**285 A**
Mean (B)	**226 B**	**242 A**	**196 C**		**249 B**	**266 A**	**216 C**	
Stem diameter (cm)								
30	5.41 e	5.51 c	5.19 g	**5.37 C**	5.74 e	5.86 d	5.48 h	**5.69 C**
45	5.46 d	5.53 c	5.23 g	**5.41 C**	5.76 e	5.88 cd	5.53 g	**5.72 C**
60	5.54 c	5.64 b	5.28 f	**5.48 B**	5.90 cd	5.95 b	5.55 g	**5.80 B**
75	5.64 b	5.85 a	5.40 e	**5.63 A**	5.93 bc	6.14 a	5.69 f	**5.92 A**
Mean (B)	**5.51 B**	**5.63 A**	**5.28 C**		**5.83 B**	**5.96 A**	**5.56 C**	

This is a factorial experiment from two factors: nitrogen rates (A), polyethylene mulch treatments (B) in a spilt plot design, letters A B C D among the main factors, letters a b c d ... among the interaction between the two factors (A×B), and different letters means significant

3.3 Yield

Application of 75 kg N/feddan in cauliflower increased the total yield (ton/feddan), curd weight per plant and curd diameter compared to the other N treatments during the two tested seasons (Table 3). The 60 kg N came in the second order without significant differences with 75 kg. These results agree with those who found that maximizing the nitrogen use efficiency of crop production can be achieved by (i) optimizing the supply of N to meet the requirements of crop during growth and development and (ii) growing N-efficient crop genotypes [25]. There is much advantage to be derived from using economic rate that will enhance higher nitrogen use efficiency and maximize production based on the value of increasing N input; the decline was also observed in nitrogen use efficiency in maize plants [26].

Regarding the effect of polyethylene mulch treatments, data revealed that transparent polyethylene mulch recorded the highest values of yield in the two studied seasons followed by black polyethylene mulch (Table 3). The lowest yield was obtained by control treatment with significant difference with other treatments. Polyethylene mulch modified root zone temperature, which has been shown to have an important role in plant growth and yield [23]. Several investigations have demonstrated that transparent and black polyethylene mulches increased yield and improved crop growth of many vegetable crops [11,22].

Referring to the interaction effect between N rate and polyethylene mulch, data indicated that increasing N rate up to 60 kg, led to increase in cauliflower yield under polyethylene mulch; however, increasing N rate more than 60 kg/feddan led to no significant difference in cauliflower yield. The highest yield was obtained in cauliflower plants applied with 75 kg N/feddan with transparent polyethylene mulch followed by 60 kg N with transparent polyethylene mulch without significant differences between them. The lowest yield was obtained in plants in the control treatment with different N rates. Transparent polyethylene mulch gave higher yield and heavier fruit than the other applications which could be explained in the light of beneficial effects of polyethylene mulch which enables retention of soil moisture and prevents soil temperature to rise up at the end of vegetative phase and enables increases in the CO_2 content, which finally results in increasing photosynthesis [27,28].

Of course, the increases in yield might be due to the increased utilization of N fertilizer in stimulating meristimatic activities. The accumulation of synthesized metabolites resulted in high dry matter accumulation and finally high yield. The obtained results were relatively different and it was found that increasing N rate up to 60 kg/feddan had no effect on cauliflower yield [24,29].

3.4 Nutrient Content

The obtained results in Table 4 showed that the N rate and mulch treatments significantly affected the uptake of N, P and K by cauliflower plants during the two studied growing seasons. The highest N, P and K percentages in the cauliflower leaves were obtained by 75 kg N treatment followed by that of 60 kg N while the lowest N, P and K contents were obtained by 30 kg N treatment.

Mulch treatments significantly affected the N, P and K percentages. The transparent polyethylene mulch resulted in the highest average values in cauliflower plants followed by that of black polyethylene mulch. The lowest N, P and K percentages were obtained in case of control treatment during the two studied seasons.

Table 3. Effect of nitrogen rate and mulch on yield characteristics of cauliflower in the two studied seasons of 2012/2013 and 2013/2014

Nitrogen rate, kg/feddan (A)	2012/2013 Polyethylene mulch treatments (B)			2012/2013 Mean (A)	2013/2014 Polyethylene mulch treatments (B)			2013/2014 Mean (A)
	Black	Transparent	Control		Black	Transparent	Control	
Curd weight (kg/plant)								
30	1.15 g	1.22 h	1.07 k	**1.15 C**	1.20 g	1.27 h	1.11 k	**1.19 C**
45	1.27 g	1.31 f	1.19 i	**1.26 B**	1.32 g	1.38 f	1.24 i	**1.32 B**
60	1.41 c	1.45 ab	1.38 ef	**1.41 A**	1.47 c	1.51 ab	1.40 ef	**1.46 A**
75	1.43 bc	1.47 a	1.39 d	**1.43 A**	1.49 bc	1.53 a	1.44 d	**1.49 A**
Mean (B)	**1.31 B**	**1.36 A**	**1.26 C**		**1.37 B**	**1.42 A**	**1.30 C**	
Curd diameter (cm)								
30	24.8 f	25.1 f	22.2 h	**24.1 C**	26.1 f	26.4 f	23.1 h	**25.2 C**
45	25.1 f	26.4 d	24.1 g	**25.2 B**	26.4 f	28.0 d	25.2 g	**26.5 B**
60	27.2 c	28.5 b	25.7 e	**27.1 A**	28.5 c	29.8 b	27.0 e	**28.4 A**
75	27.4 c	28.9 a	26.0 e	**27.4 A**	28.7 c	30.3 a	27.3 e	**28.8 A**
Mean (B)	**26.1 B**	**27.2 A**	**24.5 C**		**27.4 B**	**28.6 A**	**25.7 C**	
Total yield (ton/feddan*)								
30	9.21 g	9.75 f	8.58 h	**9.18 C**	9.61 g	10.2 f	8.86 h	**9.54 C**
45	10.2 e	10.5 d	9.53 f	**10.1 B**	10.6 e	11.1 d	9.94 f	**10.5 B**
60	11.3 b	11.6 a	11.0 c	**11.3 A**	11.8 b	12.1 a	11.5 c	**11.8 A**
75	11.4 b	11.8 a	11.0 c	**11.4 A**	11.9 b	12.3 a	11.5 c	**11.9 A**
Mean (B)	**10.5 B**	**10.9 A**	**10.1 C**		**11.0 B**	**11.4 A**	**10.5 C**	

** 1 feddan = 4200 m^2 and 1 hectare = 2.4 feddan*

This is a factorial experiment from two factors: nitrogen rates (A), polyethylene mulch treatments (B) in a spilt plot design, letters A B C D among the main factors, letters a b c d ... among the interaction between the two factors (A×B), and different letters means significant

Table 4. Effect of nitrogen rate and mulch on cauliflower leaves elemental content of N, P and K in the two studied seasons of 2012/2013 and 2013/2014

Nitrogen rate, kg/feddan (A)	2012/2013 Polyethylene mulch treatments (B)			Mean (A)	2013/2014 Polyethylene mulch treatments (B)			Mean (A)
	Black	Transparent	Control		Black	Transparent	Control	
%N								
30	2.04 hi	2.12 g	1.99 i	**2.05 C**	2.08 i	2.16 g	2.05 i	**2.10 C**
45	2.11 g	2.14 f	2.06 h	**2.10 C**	2.16 g	2.20 fg	2.10 h	**2.15 C**
60	2.25 e	2.34 d	2.21 e	**2.27 B**	2.30 e	2.39 d	2.25 ef	**2.31 B**
75	3.06 b	3.23 a	2.43 c	**2.90 A**	3.09 b	3.26 a	2.45 c	**2.93 A**
Mean (B)	**2.37 B**	**2.46 A**	**2.17 C**		**2.41 B**	**2.51 A**	**2.21 C**	
%P								
30	0.66 h	0.70 f	0.63 i	**0.66 D**	0.67 h	0.69 g	0.65 i	**0.67 D**
45	0.68 g	0.72 e	0.65 h	**0.68 C**	0.68 h	0.75 e	0.67 h	**0.70 C**
60	0.74 d	0.76 c	0.70 f	**0.73 B**	0.74 e	0.81 b	0.71 f	**0.75 B**
75	0.79 b	0.82 a	0.73 de	**0.78 A**	0.79 g	0.86 a	0.76 d	**0.80 A**
Mean (B)	**0.72 B**	**0.75 A**	**0.68 C**		**0.72 B**	**0.78 A**	**0.70 C**	
%K								
30	3.62 f	3.70 e	3.56 g	**3.63 D**	3.55 g	3.63 f	3.46 h	**3.54 D**
45	3.69 e	3.79 c	3.60 f	**3.69 C**	3.58 g	3.70 e	3.49 h	**3.58 C**
60	3.75 d	3.85 b	3.71 e	**3.77 B**	3.68 e	3.77 c	3.64 f	**3.69 B**
75	3.87 b	3.97 a	3.77 cd	**3.87 A**	3.83 b	3.93 a	3.73 d	**3.83 A**
Mean (B)	**3.73 B**	**3.83 A**	**3.66 C**		**3.66 B**	**3.76 A**	**3.58 C**	

This is a factorial experiment from two factors: nitrogen rates (A), polyethylene mulch treatments (B) in a spilt plot design, letters A B C D among the main factors, letters a b c d … among the interaction between the two factors (A×B), and different letters means significant

Regarding the interaction effect between N and mulch treatments, the highest N, P and K contents were obtained in leaves of cauliflower applied with 75 kg N/feddan combined with transparent polyethylene mulch treatment followed by that of 75 kg N/feddan with black polyethylene mulch. The lowest N, P and K contents were obtained in leaves of plant in the control treatment with different N rates during both studied seasons.

Increased yield largely attributed to the increase in soil temperature due to application of mulch covers, which resulted in better soil environment activity around plants roots, led to increasing plant growth, and hence increasing nutrient absorption [10,27]. Optimal root zone temperature conditions allow for adequate root function including proper uptake of water and nutrients. Plant nutrient uptake, plant growth, and yield under mulch fit a quadratic relationship with root zone temperature [24].

3.5 Greenhouse Gas Emission from Nitrogen Fertilization and Impacts on Warming, Pollution and Plant Yield

The results presented in Table 5 and Fig. 2 showed that the N_2O and CO_2 emissions from different applied N rates more increased with increasing N rate from 30 to 75 kg/feddan. Values of N_2O emission varied from 2.90 to 3.78 kg/feddan; those of CO_2 emission were variable between 864 and 1127 kg/feddan. The highest yield (11558 kg/feddan) with low emission (89.9 CO_2 eq g/kg yield) was found with 60 kg N/feddan treatment. Similar results were obtained by other scientists [2].

High N fertilizer application rates resulted in higher N_2O emission and increase air pollution, therefore the vegetable industry potentially contributes a significant proportion of Australia's N_2O emission, with emissions likely to be relatively high on either 'per unit area' or 'per unit production' basis [30]. Changes in the application of fertilizer rate had a lower impact on emissions than on changes in the yield. If the application rate of fertilizer increased by 20%, GHG emission per hectare will increase by 10%. On the other hand, if the application rate of fertilizer decreased by 20%, GHG emission per hectare will decrease by 10% [31]. Numerous field studies conducted on N input gradients in row-crop agriculture and found that emissions of N_2O positively correlated well with fertilizer N rate [32-34]. Nitrogen fertilizer application rate may reduce N_2O and CO_2 emissions substantially where N fertilizer is applied at rates greater than the economic optimum rate. Applying polyethylene mulch to minimize soil wetness can reduce N_2O emissions [11,12,22,27].

Table 5. Emissions for different applied nitrogen rates

N rate kg/feddan*	Total N_2O kg/feddan*	Total CO_2 eq kg/feddan*	Yield kg	CO_2 eq g/kg yield
30	2.90	864	9362	92.3
45	3.19	952	10300	92.4
60	3.49	1040	11558	89.9
75	3.78	1127	11647	96.8

** feddan = 4200 m^2 and hectare = 2.4 feddan*

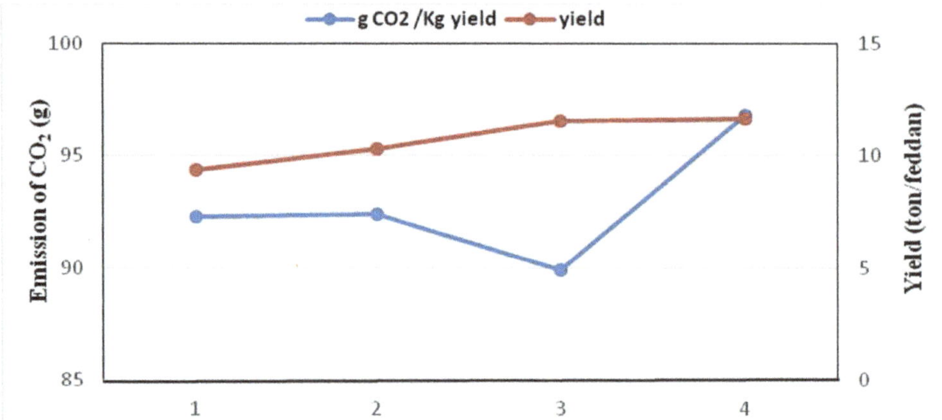

Rate of N fertilization, kg/feddan

Fig. 2. CO_2 emission and total yield under different nitrogen rates (1=30, 2=45, 3=60 and 4=75 kg/feddan)

4. CONCLUSION

The present investigation revealed that, mulch can improve plant growth and yield of cauliflower when compared with bare soil. Using polyethylene mulches, especially the transparent and black, being useful. Increase in the rate of N application up to 60 kg/feddan is recommended for larger and heavy cauliflower curds under field conditions. Use of 60 kg N/feddan with transparent polyethylene mulch is useful for increasing the yield and low N_2O and CO_2 emissions. N_2O is an important GHG that contributes to climate change. Because it has a long atmospheric lifetime (over 100 years) and is about 300 times better at trapping heat than is CO_2, even small emissions of N_2O affect the climate. Improving the management of N fertilizer for field crops can improve nitrogen use efficiency (saving farmers money) and reduce N_2O emission (helping the climate).

ACKNOWLEDGEMENTS

The authors would like to thank Prof. Dr. Mohamed Mostafa, Department of Soils and Water, Faculty of Agriculture, Ain Shams University, for his sincere guidance throughout writing the manuscript.

COMPETING INTERESTS

Authors have declared that no competing interests exist.

REFERENCES

1. Toivonen PMA, Zebarth BJ, Bowen PA. Effect of nitrogen fertilization on head size, vitamin C content and storage life of broccoli (*Brassica oleracea* var. *italica*). Can. J. Plant Sci. 1994;74:607-610.
2. Cui F, Yan G, Zhou Z, Zheng X, Deng J. Annual emissions of nitrous oxide and nitric oxide from a wheat–maize cropping system on a silt loam calcareous soil in the North China plain. Soil Biol. Biochem. 2012;48:10–19.
3. Li H, Huang R, Gordon RJ, Asiedu SK. Nitrogen sink-source relations of three cauliflower cultivars. ASA, CSSA and SSSA annual meeting, 1-5 Nov. Pittsburgh. 2009;102-109.
4. Bouwman AF. Direct emissions of nitrous oxide from agricultural soils. Nut. Cycle. Agroecosys. 1996;46:53-70.

5. Blasing TJ, Smith K. Recent greenhouse gas concentrations. In Trends: A compendium of data on global change. Carbon Dioxide Information Analysis Center, Oak Ridge National Laboratory, US Department of Energy, Oak Ridge, USA; 2011.
6. Millar N, Robertson GP, Hoben JP, Grace PR, Gehl RJ. Nitrogen fertilizer management for nitrous oxide (N_2O) mitigation in intensive corn (maize) production: An emissions reduction protocol for US Midwest agriculture. Mitigation and Adaptation Strategies for Global Change. 2010;15:185-204.
7. IPCC. De Klein CAM, Novoa RSA, Ogle SM, Smith KA, Rochette P, Wirth TC, McConkey BG, Mosier A, Rypdal K. Chapter 11: N_2O emissions from managed soils, and CO_2 emissions from lime and urea application. In IPCC Guidelines for National Greenhouse Gas Inventories; Vol. 4: Agriculture, Forestry and Other Land Use: International Panel on Climate Change; 2006.
8. Lamont WJ. Plastic mulches for the production of vegetable crops. Hort. Tech. 1993;3:35-39.
9. Ibarra L, Flores J, Diaz-Perez JC. Growth and yield of muskmelon in response to plastic mulch and row covers. Sci. Hort. 2001;87:139-145.
10. Fonsecal IC, Klar AE, Goto R, Nevesl CS. Colored polyethylene soil covers and grafting effects on cucumber flowering and yield. Scientia Agricola. 2003;60:643-649.
11. Abdrabbo MAA, Farag AA, Hassanein MK. Irrigation requirements for cucumber under different mulch colors. Egypt. J. Hort. 2009;36:333-346.
12. Shaheen A, Ali S, Stewart BA, Naeem MA, Jilani G. Mulching and synergistic use of organic and chemical fertilizers enhances the yield, nutrient uptake and water use efficiency of sorghum. African J. Agric. Res. 2010;5:2178-2183.
13. Blanco-Canqui H, Lal R. Soil structure and organic carbon relationships following 10 years of wheat straw management in no-till. Soil till. Res. 2007;95:240-254.
14. Jacinthe PA, Lal R, Kimble JM. Carbon budget and seasonal carbon dioxide emission from a central Ohio Luvisol as influenced by wheat residue an amendment. Soil Till. Res. 2002;67:147–157.
15. Israelsen OW, Hansen VE. Irrigation principles and practices. 3rd Ed, John Wiley and Sons Inc., New York, London; 1962.
16. Piper CS. Soil and plant analysis. Inter Science Publisher. Inc., New York, USA; 1950.
17. Jackson ML. Soil chemical analysis. Prentice-Hall. Inc., Englewood Cliffs, New Jersey, USA; 1967.
18. Black CA. Methods of soil analysis. Amer. Soc. Agron. Inc., Madison, USA; 1969.
19. Association of Official Analytical Chemists. Official methods of analysis, 16th Ed. AOAC, Arlington Virginia, USA; 1995.
20. Forster P, Ramaswamy V, Artaxo P, Berntsen T, Betts R, Fahey DW, Haywood J, Lean J, Lowe DC, Myhre G, Nganga J, Prinn R, Raga G, Schulz M, Van Dorland R. Changes in atmospheric constituents and irradiative forcing. Climate change: The physical science basis. Contribution of working group I to the fourth assessment report of the intergovernmental panel on climate change. Solomon S, Qin D, Manning M, Chen Z, Marquis M, Averyt KB, Tignor M, Miller HL. (eds) Cambridge University Press, Cambridge, UK; 2007.
21. SAS. Statistical Analysis System, SAS User's Guide: Statistics, SAS Institute Inc., Cary, USA; 2000.
22. Anikwe MAN, Mbah CN, Ezeaku PI, Onyia VN. Tillage and plastic mulch effects on soil properties and growth and yield of cocoyam (*Colocasia esculenta*) on an ultisol in Southeastern Nigeria. Soil Tillage Res. 2007;93:264-272.
23. Soltani N, Anderson JL, Hamson AR. Growth and analyses of watermelon plants with mulches and row covers. J. Amer. Soc. Hort. Sci. 1995;120:1001-1009.
24. Bashyal LN. Response of cauliflower to nitrogen fixing biofertilizer and graded levels of nitrogen. J. Agric Environ. 2011;12:41-50.
25. Broadley MR, Seginer I, Burns A, Escobar-GutieÂrrez AJ, Burns IG, White PhJ. The nitrogen and nitrate economy of butter head lettuce (*Lactuca sativa* var. *capitata* L.). J. Exp. Bot. 2003;54:2081-2090.
26. Kogbe JOS, Adediran JA. Influence of nitrogen, phosphorus and potassium application on the yield of maize in the savanna zone of Nigeria. African J. Biotech. 2003;2:345-349.
27. Salman SK, Abou-Hadid AF, El-Beltagy IMJ, El-Beltagy AS. Plastic house microclimate as affected by low tunnels

and plastic mulch. Egypt J. Hort. 1992; 2:111-119.
28. Wien HC, Minotti PL. Increasing yield of tomatoes with plastic mulch and apex removal. J. Amer. Soc. Hort. Sci. 1988; 113:342-347.
29. Yuri JE, De Rodrigues MCG, Juarez J. Effect of organic compost on crisp head lettuce production and commercial characteristics. Hort. Bras. 2004;22:117-130.
30. Dalal R, Weijin W, Robertson GP, Parton WJ, Myer CM, Raison RJ. Emission sources of nitrous oxide from Australian agricultural and forest lands and mitigation options. Australian Greenhouse Office; 2003.
31. Rajaniemi M, Mikkola H, Ahokas J. Greenhouse gas emissions from oats, barley, wheat and rye production. Agron. Res. Biosys. Eng. 2011;1:189-195.
32. Drury CF, Yang XM, Reynolds WD, McLaughlin NB. Nitrous oxide and carbon dioxide emissions from monoculture and rotational cropping of corn, soybean and winter wheat. Can. J. Soil Sci. 2008;88: 163–174.
33. Dusenbury MP, Engel RE, Miller PR, Lemke RL, Wallander R. Nitrous oxide emissions from a Northern Great Plains soil as influenced by nitrogen management and cropping systems. J. Environ. Qual. 2008;37:542–550.
34. Halvorson AD, Del Grosso SJ, Reule CA. Nitrogen, tillage, and crop rotation effects on nitrous oxide emissions from irrigated cropping systems. J. Environ. Qual. 2008; 37:1337–1344.

2

Ecological Structure and Fruit Production of Blood Plum (*Haematostaphis barteri Hook. F*) Subpopulations in Benin

Bienvenue Nawan Sourou[1*], Christine Ajokè Ifetayo Nougbodé Ouinsavi[1] and Nestor Sokpon[1]

[1]Faculty of Agronomy, Laboratory of Forestry Studies and Research, University of Parakou, BP 123 Parakou, Benin.

Authors' contributions

This work was carried out in collaboration between all authors. Author BNS designed the study, wrote the protocol, wrote the first French draft of the manuscript and performed statistical analyses of the study. Author CAINO managed the literature searches and the experimental process, translated and wrote the English draft of the manuscript and author NS identified the species of plant, followed up the whole study, read and corrected the manuscript. All authors read and approved the final manuscript.

Editor(s):
(1) Junhong Bai, School of Environment, Beijing Normal University, Beijing, China.
Reviewers:
(1) Raul Leonel Grijalva Contreras, Instituto Nacional de Investigaciones Forestales Agrícolas y Pecuarias, Mexico.
(2) Anonymous, Mediterranea University of Reggio Calabria, Italy.

ABSTRACT

Aims 1: To assess the ecological structure and fruits production of *Haematostaphis barteri* (Blood plum) to provide requisite information for a better management and conservation of the tree subpopulations in Benin.

Study Design: Ecological structure and fruit production were evaluated in randomized design.

Duration of Study: This study was conducted in the department of Atacora and University of Parakou, Benin 2012-2015.

Methodology: The ecological structure of *Haematostaphis barteri* subpopulations was studied on the basis of forest inventory surveys performed in forty six (46) 1-ha plots randomly installed and *Haematostaphis barteri* fruit production was quantified on the basis of a random sampling of the main branches of 126 trees at random from all of these subpopulations.

**Corresponding author: E-mail: bienvenuesourou@yahoo.fr*

Results: Four subpopulations of *Haematostaphis barteri* were identified based on dendrometric variables and among trees distance. Diameter and height classes distribution of the species in each subpopulation adjusted to Weibull distribution showed a bell shaped curve with left dissymmetry, characteristic of young stands (form coefficient between 1 and 3.6). Stand density varies from 12 to 18 stems ha^{-1}. Stand basal area varies from 0.27 to 0.48 m^2 ha^{-1} while mean diameter varies from 16.28 to 19.37 cm. Average number of fruit per panicle varies from 15 to 28 fruits and average fruit number per tree was estimated from 2325 to 7879 fruits. The DBH, TH, average number of panicles per branch and average number of fruit per panicle showed a highly significant difference (P = .000) between subpopulations.
Conclusion: Soil texture, altitude and topography are the factors that best discriminate the different subpopulations and better explain variations among these subpopulations with respect to their structural and production characteristics. Despite the similar production in fruits of *Haematostaphis barteri* subpopulations that of Touncountouna is however the most productive in terms of the number of fruits.

Keywords: Ecological structure; blood plum; fruits production; subpopulation; Benin.

1. INTRODUCTION

Tropical forests are known as the most important ecosystems in the world in terms of species abundance and diversity [1]. They are known as a reservoir of original foods [2]. Despite such importance, these forests cover are subjected to continuous human pressure. According to [3], tropical zone forest destruction rate is about 5.2 million per year between 2000 and 2010. In Benin number of Non timber Forest Products (NTFPs) that are socioeconomically very important in people live, are under treat because of the continuous forest destruction [4,5]. *Haematostaphis barteri*, is one those food specie which provide people with a wide traditional usefulness at a regional scale. However unlike some species such as *Parkia biglobosa, Vitellaria paradoxa, Adansonia digitata, Irvingia gabonensis, Pentadesma butyracea*, that are identified as priority species in Benin [4], *Haematostaphis barteri* has not been giving enough attention. This species is however an endemic species that worth to receive significant attention in terms of species conservation [6]. As most of endemic species, *Haematostaphis barteri* may be subjected to the idea weakness and latent danger [7]. Understanding the plant community's distribution trends and corollary factors of such species are of tremendous for ecosystems conservation and management. It is therefore necessary that appropriate studies and programmes are carried out to enhance the conservation and sustainable use of *Haematostaphis barteri* genetic resources in Benin. Also, sustainable management of renewable natural resources requires analysis and master of the multiple interactions between human, environmental factors and the complex functioning of forest ecosystem.

The current study therefore aims to assess the ecological structure and fruits production of *Haematostaphis barteri* to provide requisite information for a better management and conservation of the tree subpopulations in Benin.

2. MATERIALS AND METHODS

2.1 The Study Area

The current study was carried out in three districts: Natitingou, Toucountouna and Boukombé (Fig. 1) located in the northern-West Benin between the parallels 10° and 10° 35' latitude North in one hand and, between the meridians 1° 06' and 1° 30' longitude East. This study covers the whole distribution range of *Haematostaphis barteri* a species endemic to Atacora mountain, in Benin [6,8]. The study area is under the influence of soudano-guinean climate type that is qualified as Atacorian climate [9]. Two types of winds are dominant: the sea breeze (monsoon) and continental trade winds (Harmattan). There are two well defined seasons: a rainy season and a dry season. The interannual average rainfall is about 1,178 mm water at Natitingou for the period of 1977-2012. The study area is characterized by the Atakora Mountain with general direction SW – NE and which consists of two parallel beads of quartzite and mica with a fairly soft topography. There are two types of soil: raw mineral soils and the leached tropical ferruginous soils concretions and sometimes hydromorphic inside deep. The vegetation is characterized by the presence of woody vegetation (gallery forests, woodlands, savannas and shrublands) of grassland and mixed formations. These courses are dotted with fields and fallows.

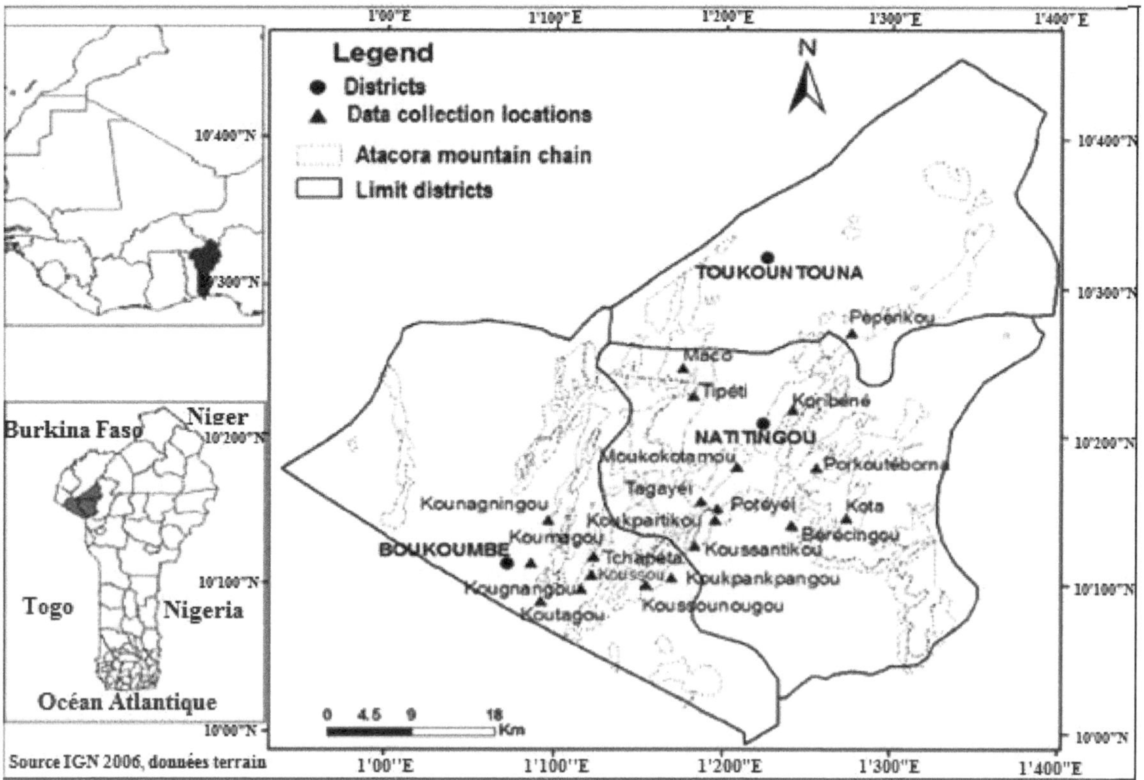

Fig. 1. Map of the study area

2.2 *Haematostaphis barteri* Description

Haematostaphis barteri (Photo 1) is a common species and disseminated, found exclusively in tropical Africa, from Ivory Coast to Sudan, and through Ghana, Togo, Benin, Cameroon and Tchad. It is found in woodlands, especially on rocky hills and crumbling in the Sudanian and Guinean zone [10]. It was reported exclusively in the phytogeographical district chain of Atacora [6,8] *Haematostaphis barteri* is a small tree or shrub of 2.0 to 8.0 m high and with a trunk circumference rarely exceeds 65 cm. The leaves are compound, hairless, alternate and imparipinnate. *H. barteri* is a dioecious species, due to the existence of male plants and female plants. Flowering and fruiting begins in late dry season, usually before the first leaves appear. The fruits are ellipsoid drupe, hairless, purple at maturity, which are carried by long pendulous panicles. They contain a core embedded in a thin pulp, edible, acidic taste.

2.3 Sampling and Data Collection

Square plots of one ha (100 m x 100 m) were established taking into account the various vegetation types, different soil types and different topography type in the study area. A total number of 46 plots were laid out proportionally according to species distribution within the three districts covered by the study area as followed: 18 plots at Boukombé, 18 plots at Natitingou and 10 plots at Toucountouna. Each plot has been georeferenced and the altitude to which it is situated raised thanks to the GPS garmin 60. The topography has also been noted by direct observation supported by the map of relief of the survey area. Young plants consist releases strains (individuals regenerated after cutting a stem) and free strands (1 cm ≤ DBH < 10 cm and height < 1.5 m) were treated regeneration then noted. Given their rarity, the seedlings were noted throughout the area of each inventory plot. Dendrometric parameters including: Diameters at breast height (DBH), total height (TH), crown diameter were measured for all *Haematostaphis barteri* individual tree with DBH ≥ 10 cm.

Fruits production was assessed for *Haematostaphis barteri* subpopulation on the basis of fruits production characteristics: number of main branches bearing fruit, total number of fruits on the two main branches, number of panicles on the ten randomly selected branches bearing panicles, total number of fruits of ten sampled panicles per tree. These parameters were either measured or counted on 126 randomly selected trees within the species distribution rang.

Photo 1. *Haematostaphis barteri* tree bearing fruit

2.4 Data Analysis

2.4.1 Subpopulations of *Haematostaphis barteri* identification

Principal component analysis was performed on the basis of dendrometric parameters (DBH, TH, crown diameter) and among trees distance to partition 704 individual trees collected from the 46 plots using the software Minitab 17.

2.4.2 Characterization of the *Haematostaphis barteri* subpopulations of Atakora Mountain

To determine the dendrometric characteristic of *Haematostaphis barteri,* dendrometric parameters were calculated. These parameters are presented in Table 1. The structural characterization of *Haematostaphis barteri* according to vegetation types and climatic zones was done using the diameter and height class-size distribution. Different histograms of frequency from the diameters and heights were adjusted to Weibull 3-parameter distribution using the software Minitab 17. This distribution was used as it is simple in usage [11]. According to [12], its probability density function is given by the following equation:

$$f(x)=\frac{c}{b}\left(\frac{x-a}{b}\right)^{c-1} exp\left[-\left(\frac{x-a}{b}\right)^{c}\right]$$

a = position parameter; b = scale parameter or size parameter; c = form parameter linked with the observed structure.

A one way Analysis of Variance was performed to test for statistical difference between different subpopulations for variable such as: DBH, TH, Stand density, Stand basal area and Quadratic mean diameter using SPSS for Windows V. 16 Software. Tukey Homogeneity test allows us to separate the site and ecological homogeneous groups. The correlations between growth parameters and fruits production were performed with Minitab 17 Software.

2.4.3 Fruits Production of *Haematostaphis barteri* subpopulations

From the total number of fruit relied on two main branches of each sampled tree, the average number of fruit per branch was determined and the average number of fruit per tree was estimated by multiplying the number of main branches bearing fruit by average number of fruit per main branch. A one way Analysis of Variance was performed using SPSS software version 16.0, to test for statistical difference between subpopulations for some variable such as: average number of panicles per branch, average number of fruit per panicle and per tree. Correlation tests were also performed for the dendrometric data and fruits productions attributes with Minitab 17 Software.

Table 1. Dendrometric parameters and their formula

Parameters	Formula
Stand Density (stem/ha)	N = Total number of stem per ha
Stand Basal Area (m²/ha)	$G = \Sigma \pi D^2 / 4$
Quadratic mean diameter (cm)	$D_g = \sqrt{\frac{4G}{N\pi}}$
Blackman indice [13]	$IB = 6^2 / \mu$: IB = 1 (poisson distribution) ; IB< 1 (regular distribution) and IB>1 (aggregative distribution)

Notes: N = density of subpopulation per hectare, D = diameter of tree, 6^2 = variance of subpopulation tree, μ = mean of subpopulation tree

3. RESULTS

3.1 Principal Component Analysis and Identification of *Haematostaphis barteri* Subpopulations

The partition of *Haematostaphis barteri* species from a matrix of 704 individuals in the first two factorial axes of the Principal Component Analysis (Fig. 2), allowed discriminate four groups of plots (Sub-populations of *Haematostaphis barteri*). The discriminated groups are named according to their location in the presence of the species districts. With an eigenvalue equal to 2.19 and a percentage of variation equal to 36.5%, the axis 1 clearly separated plots executed in Boukoumbé from those executed in Toucountouna and Natitingou. Then we distinguish successively on this axis on the negative side toward the positive side:

- The group of some Toucountouna individuals (subpopulation of Toucountouna) established on gently sloping hillsides at an altitude between 500 m and 550 m. They are distributed over a shallow tropical ferruginous soil with a sandy loam texture.
- The group of trees executed in Natitingou plots (subpopulation of Natitingou 1) whose altitude varies from 415 to 580 m. There are established on the peaks and steep slopes of rocky hills with a tropical ferruginous soil in shallow sandy loam texture.
- Other group of trees executed in Natitingou plots (subpopulation of Natitingou 2). This group is installed on rocky slopes with altitudes ranging between 400 to 600 m, and distributed on the shallow tropical ferruginous soil with sandy loam texture.
- The group of individual trees from Boukombé (subpopulation of Boukombé) established at low altitudes (300-450 m) at the foothills of rocky hills. There are distributed over a hydromorphic tropical ferruginous soil with sandy clay.

So this axis symbolizes a descending gradient to the topography, altitude and soil texture.

3.2 Diameter Classes Distribution of *Haematostaphis barteri* trees

Fig. 3 shows the diameter size classes' distributions of *Haematostaphis barteri* trees for the different subpopulations. The Weibull distribution of *Haematostaphis barteri* trees per diameter classes shows a left dissymmetric bell shaped curve with a c form coefficient comprised between 1 and 3.6 for the four subpopulations. This value of c means that the stands were monospecific and dominated by young trees diameter centered on class [10-20 cm].

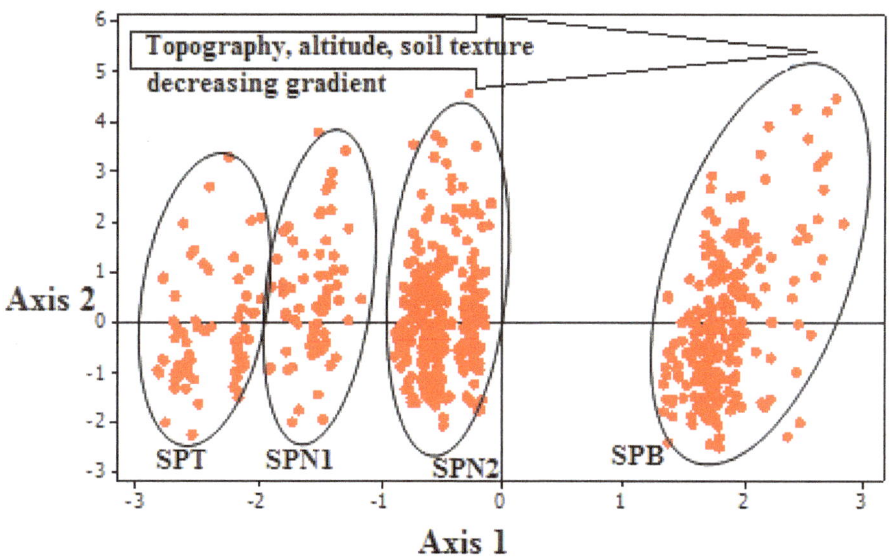

Fig. 2. Projection of *Haematostaphis barteri* trees in the space of first and second principal components

SPT = toucountouna subpopulation, SPN1= Natitingou1 supopulation, SPN2 = natitingou 2 subpopulation, SPB = boukombe subpopulation

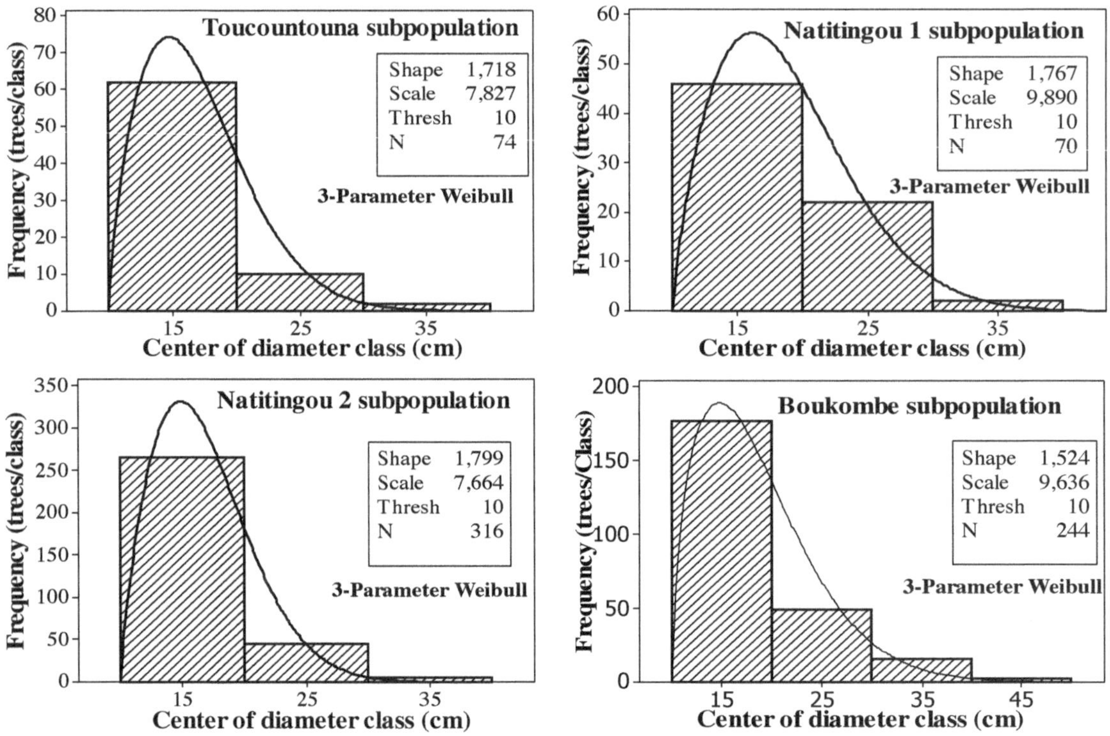

Fig. 3. Diameter class distribution of *Haematostaphis barteri* subpopulations

3.3 Total Height Classes Distribution of *Haematostaphis barteri* trees

Fig. 4 shows the total classes distribution of *Haematostaphis barteri* trees in different subpopulations. This distribution well-adjusted to the Weibull function showed a bell shaped curve with a left dissymmetry ($1 \leq c \leq 3.6$). Like the diameter class distribution, the stands were dominated by young trees. Trees of the total height class [4 – 6 m] were the most abundant.

3.4 Structural Characteristics of *Haematostaphis barteri* Subpopulations

Table 2 shows the structural characteristics of the studied *Haematostaphis barteri* subpopulations. The stand density varied from 12 stems ha^{-1} (Toucountouna subpopulation) to 18 stems ha^{-1} (Natitingou 1 and 2 subpopulation) while the stand basal area ranges from 0.27 m² ha^{-1} (Toucountouna subpopulation) to 0.48 m² ha^{-1} (Natitingou subpopulation). The Quadratic mean diameter ranged from 16.28 cm (Toucountouna subpopulation) to 19.37 cm (Boukombé subpopulation). The diameter at the breast height (DBH) and the total height were respectively 15.55 cm and 4.61 m in Toucountouna subpopulation, 17.82 cm and 4.99 m in Natitingou 1 subpopulation, 15.6 cm and 4.91 m in Natitingou 2 subpopulation, and 17.36 cm and 4.55 m in Boukombé subpopulation (Table 2).

The density, basal area and diameter of the mean tree basal area showed no significant differences between sub-populations. As against the mean diameter and mean total height of *Haematostaphis barteri* trees showed a highly significant difference ($P < .001$) between the subpopulations. The largest trees of *Haematostaphis barteri* (17.82 cm) and the highest (4.99 m) were observed in the subpopulation of Natitingou 1, while the less large trees (15.55 cm) less high (4.61 m) were noted in the sub-population of Toucountouna.

The calculated values of Blackman index were all higher than 1 indicating that *Haematostaphis barteri* trees have aggregative distribution in all of the four subpopulations. This distribution can appear unrealistic for a density of 12 to 18 stems ha^{-1}, but the direct observations made on the land reveal well that to the level of the sites of presence of *Haematostaphis barteri*, several individuals confine themselves on a small surface determining pockets of presence of the species thus.

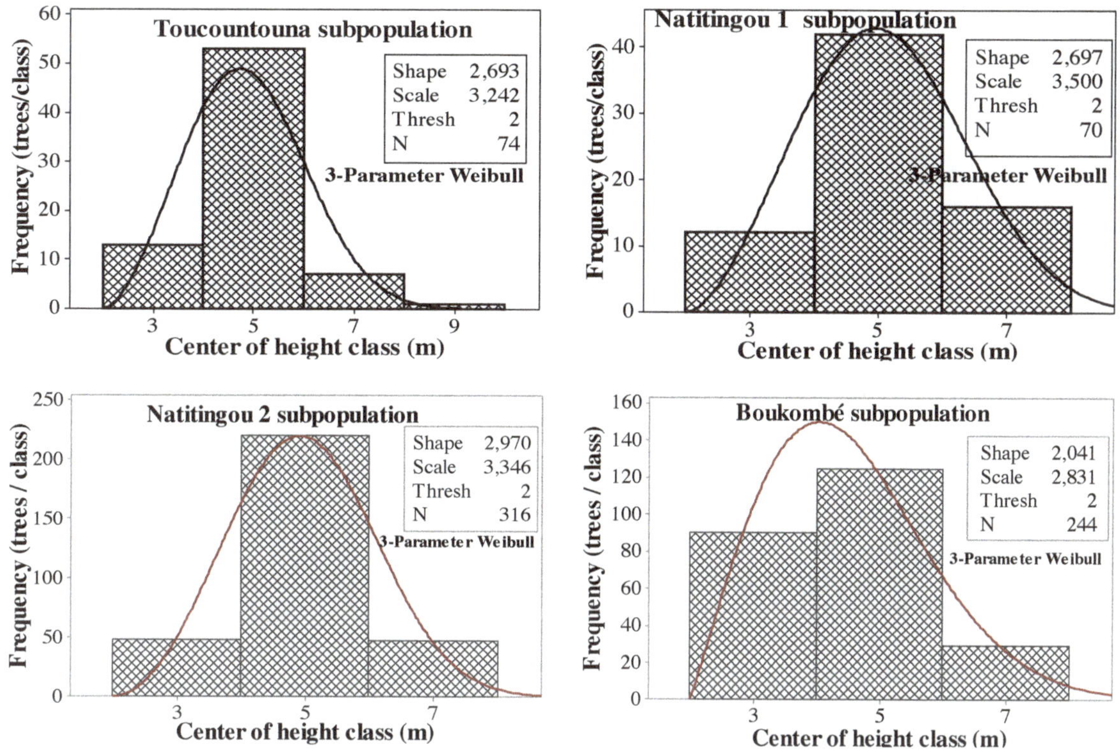

Fig. 4. Total height class distribution of *Haematostaphis barteri* subpopulations

3.5 Fruits Production of *Haematostaphis barteri* Subpopulations

Table 3 shows the fruit production characteristics of *Haematostaphis barteri* trees. The average number of panicle per branch is about 3 panicles of Natitingou 1 subpopulation and 4 panicles for other subpopulations. The average number of fruit per panicle varies from about 15 fruits (subpopulation of Natitingou 1) to about 28 fruits (Natitingou 2 subpopulation). In Boukombé and Toucountouna subpopulations, the average number of fruit per panicle is respectively about 20 and 23 fruits. The average number of fruit per tree varies from 2325 (Natitingou 1 subpopulation) to 7879 fruits (Toucountouna subpopulation). In Natitingou 2 and Boukombé subpopulations, the average number of fruit per tree is respectively 6640 and 7466.

The average number of fruit per tree showed no significant difference between subpopulations. As against the average number of panicle per branch and the average number of fruit per panicle showed a highly significant difference (P = .000) between the *Haematostaphis barteri* subpopulations (Table 3). The branches with the highest number of panicles were found in the Toucountouna subpopulation and those with the lowest number of panicles were observed in Natitingou 1 subpopulation.

3.6 Correlation between Dendrometric Parameters and Fruit Production Variables

The results of correlation analyses between DBH, total height and fruit production variables of *Haematostaphis barteri* are presented in Table 4. The table showed that the DBH is highly correlated with the average number of fruits per tree and the average number of panicles per branch while the total height is significantly correlated with the average number of fruits per tree. However DBH and total height are not correlated with the average number of fruits per panicle. In addition, there is a high correlation between DBH and TH, between the average number of fruits per tree and the average number of panicles per branch, the average number of fruits per panicle. There is also high correlation between the average number of panicles per branch and the average number of fruit per panicle.

Table 2. Structural characteristics of the studied *Haematostaphis barteri* subpopulations

Subpopulations of *Haematostaphis barteri*	Location	DBH (cm)		HT (m)		N (stems ha^{-1})		G (m^2 ha^{-1})		Dg (cm)		IB
		Mean	SE	Mean	SE	Mean	SE	Mean	SE	Mean	SE	
Toucountouna (n = 74)	Homogeneous group made of foot found in the village of pèpèrikou (a urban district of Toucountouna) situated at longitude 1156034 East and at latitude 0331085 North ; (UTM)	15.55^a	0.72	4.61^a	0.13	12^a	3.78	0.27^a	0.09	16.28^a	1.83	5.11
Natitingou 1 (n = 70)	Homogeneous group located in the urban district of Natitingou, in bérécingou village, situated at longitude 1131741 East and latitude 324814 North (UTM), and in Kota village situated at longitude (1132659 East and latitude 0330750 North ; UTM)	17.82^b	0.74	4.99^b	0.13	18^a	4.63	0.48^a	0.11	17.87^a	2.25	8.24
Natitingou 2 (n = 316)	Heterogeneous group made of mixed individuals from subpopulations of Toucountouna and Natitingou, in the village of Maco situated at longitude 1150754 East and latitude 0312859 North, UTM) and those of Tipéti, Tagayéi, Koukpartikou and Koussantikou situated at longitude 1129161 East and latitude 0314320 north ; UTM.	15.6^a	0.35	4.91^b	0.06	18^a	2.18	0.37^a	0.05	17.15^a	1.06	7.19
Boukombé (n = 244)	Homogeneous group located in the urban district of Boukombé, in the village of Koussoucoingou (longitude 1125600 East and latitude 303166 North, UTM), piedmont of the hill range called the « Koumagou » situated at longitude 1127130 East and latitude 0296638 North ; UTM)	17.36^{ab}	0.39	4.55^a	0.07	13^a	2.18	0.38^a	0.05	19.37^a	1.06	3.11
Probability		**0.001*****		**0.000*****		**0,469 (ns)**		**0,585 (ns)**		**0.375 (ns)**		-

: Significant distinction at the level of 5%; **: Significant distinction at the level of 1%; *: Significant distinction at the level of 0, 1%; ns = no significant, the averages of distinguishing features per column among the differents subpopulations of Haematostaphis barteri followed by the same letter are not significanthy different to the level of 5% according to the comparison test of Student Newman Keuls averages. n = nomber of trees; DBH = Diameter; HT = total height; N = Density; G = Stand basal area; Dg= Quadratic mean diameter, IB = the Blackman index, se = Standard error*

Table 3. Fruits production characteristics of *Haematostaphis barteri* subpopulations

Subpopulations of *Haematostaphis barteri*	Average number of panicles per branch	Average number of fruits per panicle	Average number of fruits per tree
Toucountouna subpopulation (n = 30)	4,28±0,14[a]	22,79±0,98[b]	7879±2379[a]
Natitingou1 subpopulation (n = 30)	2,87±0,089[b]	14,64±0,62[d]	2325±1872[a]
Natitingou2 subpopulation (n = 36)	3,92±0.075[a]	27,69±0,52[a]	6640±1265[a]
Boukombé subpopulation (n = 30)	4,00±0,094[a]	19,71±0,65[c]	7466±1566[a]
General Mean	**3,69±0,05**	**21,73±0,35**	**6078±1277**
Probability (P)	**0,000*** **	**0,000*** **	**0,142ns**

*Number with the same letter are not significantly different at the level of 5% according to Turkey comparison tests; ns = no significant; ***: significant distinction at the level of 0, 1%*

Table 4. Correlation between DBH, TH and fruit production characteristics

Parameters	Mean DBH (cm)	Mean TH (m)	Mean number of fruits per tree	Mean number of panicle per branch	Mean number of fruits per panicle
Mean DBH (cm)	1				
Mean TH (m)	0.751***	1			
Mean number of fruits per tree	0.525***	0.486***	1		
Mean Number of panicle per branch	0.308***	0.256	0.560***	1	
Mean Number of fruits per Panicle	0.24	0.205	0.362***	0.314***	1

**** Very significant correlation at the level of 5%*

4. DISCUSSION

4.1 Structural Characteristics of *Haematostaphis barteri* Subpopulations

The average total height and DBH showed a significant difference at the level of 5% (Table 2). Subpopulation of Natitingou 1 distributed at the tops of rocky hills presented larger and higher trees than those of other subpopulations. Similarly, the subpopulation of Natitingou 1 showed the highest values as far as certain structural parameters such as density, basal area are concerned (Table 2). This indicates that the site conditions (peaks and slopes of rocky hills, altitude and sandy loam texture) enhance the development of trees in Natitingou 1 subpopulation. These rocky hills are clearly preferred habitat of *Haematostaphis barteri;* corroborating the previous results on *H. barteri* preferred habitat in West African dry zones [10] and northern Togo [14,15].

The diametric distribution of *Haematostaphis barteri* in the four subpopulations revealed certainly the predominance of young individuals or small diameters but it cannot be interpreted as a good conservation of species. It is probably related to the physiology of the species, which is a shrub whose diameter rarely exceeds 35 cm [15], but also its disruption or vulnerability at certain stages of its development [16].

However, there is generally a relationship between the temperament of the species and its diameter class distribution [17]. According to [18], while the light dependent species has little or no stems in small diameters, a number sometimes are often erratic in mean diameters distributions, sometimes with a maximum frequency, while the shade species all have a high number of small diameter stems. Similarly, [19] indicates that the simple dichotomy acquired between light dependent species and shade-tolerant one (or pioneers or mature) does not reproduce the diversity of species' strategies in tropical forests, primarily because each species has a tolerance range at both side of a physiological optimum and secondly because the temperament of a species increases with age. However [20] showed that the structure of a species may vary at a station scale or across the whole distribution area. For example, in Ghana, [21] showed that nationally, the majority of forest species have sufficient natural regeneration and most of these species showed " inverted J "shape distribution.

Also, finding suitable models in an organism's life history has always been a challenge for modern ecology [22]. According to [23], the 3-parameters probability density function of Weibull is increasingly used to model the diameter class distribution and height class of forest species. The popularity of the probability density function of Weibull is due to its flexibility to consider all models of unimodal diameter class distribution. In addition, it allows quick and easy estimation of tree diameter without integrating the probability density function once the parameters adjusted.

The bell-shaped distribution with left asymmetry obtained with *H barteri* diameter classes and height classes corroborated well with the results gotten in Benin by [17,23-25] respectively on the populations of *Borassus aethiopum*, of *Afzelia African*, *Prosopis africana* and of *Sclerocarya birrea.* Similar results were obtained by [15,26,27] respectively with *Vitellaria paradoxa* and *Haematostaphis barteri* species in Togo. However, this structure cannot be derived only from the temperament of the species but also related to human pressure.

4.2 Fruits Production and Sustainable Management of the *Haematostaphis barteri* Subpopulations on the Atakora Mountain

Even although, the average number of fruit per tree did not vary significantly from one subpopulation to the other, trees in the Toucountouna subpopulation that are significantly less large and high in comparison with other subpopulations trees, were the most productive in terms of number of fruit (7879 fruits per tree). However, less productive were the trees of Natitingou 1 subpopulation (2325 fruits per tree) which are though significantly larger and higher than the others. This suggests that the environment and more specifically the altitude and topography significantly influence the vegetative development of *Haematostaphis barteri* but would have no significant influence on the production of *Haematostaphis barteri* in terms of average number of fruit per tree. The average fruit yield of *Haematostaphis barteri* tree obtained is significantly higher than those obtained by [17,28,29] respectively on the trees of *Borassus aethiopum, Vitellaria paradoxa* and *Parkia biglobosa.* This difference would be a part linked to the physiology of the species and also different methods of quantification used. While fruit quantification method of *B. aethiopum* and *Vitellaria paradoxa* used by [17,28] was based on systematic enumeration of all fruits, that employed for the quantification of *Haematostaphis barteri* fruit production is based on random sampling of main branches [30]. *Haematostaphis barteri* is a species weakly disseminated in the woodlands of the atacora chain mountain. These subpopulations are in a state of vulnerability because of their ageing due to the rarity of natural regeneration and accentuated not only by seed predation by small rodents but also by high pressure. This situation was also emphasized by [15] in Togo. As the principal organs of the species are of great importance [31], long-term management of *Haematostaphis barteri* subpopulations is needed to improve the local population's livelihoods. An effective sensitization of the villagers for the protection of the species in fallow and in situ conservation on the slopes and hilltops of the Atacora could allow future preservation. It is important to protect natural regeneration against fires and wood cutting, but also the seed bearer ones dispersed in the fields. The study of sexual reproduction (mode of reproduction, dormancy, etc) in natural environment and in laboratory and tests vegetative propagation low cost [32,33] are studies needed to be undertaken to consider a possible enrichment of woodlands by *Haematostaphis barteri.*

5. CONCLUSION

This study idenfied four subpopulations of *Haematostaphis barteri* over its natural distribution range in Benin: Toucountouna subpopulation, the subpopulation of Natitingou 1, subpopulation of Natitingou 2 and the Boukombé subpopulation. Soil texture, altitude and topography are the factors that best discriminate the different subpopulations and better explain variations among these subpopulations with respect to their structural characteristics (density, basal area and average diameter). Ecological structure of the Blood prune subpopulations of Benin, adjusted to Weibull distribution showed a bell shaped curve with a left dissymmetry proving the predominance of young trees within these subpopulations. The average yield of fruit per tree did not show a significant difference between the different subpopulations of

Haematostaphis barteri. However Toucountouna subpopulation is the most productive in terms of average number of fruit per tree. For domestication and conservation purposes, more investigations need to be done particularly on fruit morphological characterization, mode of reproduction and socio -economic importance of *Haematostaphis barteri.*

COMPETING INTERESTS

Authors have declared that no competing interests exist.

REFERENCES

1. Phillips OL, Hall P, Gentry AH, Sawver SA, Vásquez MR. Dynamics and species richness of tropical rain forest. Proceedings of the National Academy of the Sciences USA. 1994;9(1):2805-2809. DOI: 10.1073/pnas.91.7.2805
2. Sokpon N, Lejoly J. Les plantes à fruits comestibles d'une forêt semi-caducifoliée de Pobè au sud-est du Bénin, in: Hladik CM, Pagezy H, linares OF, Froment A, Eds. L'alimentation en forêt tropicale: Interactions bioculturelles et perspectives de développement. UNESCO. 1996;1:115-122. French.
3. FAO. Situation des forêts du monde. Rome. 2012;118. French. Available:http://www.fao.org/docrep/016/i3010f/i3010f.pdf (Accessed 30 April 2013)
4. Wong JLG, Thornber K, Baker N. Evaluation des ressources en produits forestiers non ligneux: Expérience et Principes de Biométrie. FAO. 2001;52. French. Available:http://www.fao.org/3/a-y1457f.pdf (Accessed 15 January 2012)
5. Tchatat M. Produits forestiers autres que le bois d'œuvre: Place dans l'aménagement durable des forêts denses humides d'Afrique centrale; Série FORAFRI. 1999; 95. French.
6. Adomou CA. Vegetation pattern and environmental gradients in Benin: Implication for biogeography and conservation. PhD thesis, Wageningen University; 2005.
7. Garnier A. Qu'est-ce qu'une espèce rare? Origines et fonctionnement de la rareté naturelle. Travail Bibliographique, DEA BEFA; 2008. French.
8. Akoègninou A, Van Der Burg WJ, Van Maesen. Flore analytique du benin. Backhuys Publishers. 2006;1034. French.
9. Houssou C. Les bioclimats de l'Atacora (Nord-ouest du Bénin) et leurs implications socio-économiques. Thèse d'état, Dijon, France; 1998. French.
10. Arbonnier M. Arbres arbustes et lianes des zones sèches d'Afrique de l'Ouest. CIRAD-MNHN-UICN. 2002;573. French.
11. Johnson NL, Kotz S. Distributions in Statistics: Continuous univariate distributions -1 and 2. technometrics. 1971; 13(4):919-920. DOI: 10.2307/1266973
12. Rondeux J. La mesure des arbres et des peuplements forestiers. Presses Agronomiques de Gembloux. 1993;521. French.
13. Blackman GE. Statistical and ecological studies in the distribution of species in plant communities: I. dispersion as a factor in the study of changes in plant populations. Ann Bot. 1942;6(2):351-370.
14. Eyana KA. Les anacardiaceae du Togo: Etudes botaniques, ecologiques et proprietés antifongiques. Thèse de doctorat de l'Université de Reims Champagne-Ardenne et de l'Université de Lomé; 2007. French.
15. Agbogan A, Tozo K, Wala K, Batawila K, Dourma M, Akpagana K. Abondance et structure des populations d'un fruitier spontané: *Haematostaphis barteri* Hook F dans deux sites rocheux en région soudanienne au Togo. Int. J. Biol. Chem. 2012;6(6):6042-6048. French. DOI: 10.4314/ijbcs.v6i6.31
16. Ouédraogo A, Thiombiano A, Hahn Hadjali K, Guinko S. Diagnostic de l'état de dégradation des peuplements de quatre espèces ligneuses en zone soudanienne du Burkina Faso. Sécheresse. 2006;17(4): 485-491. French. DOI: 10.1684/sec.2006.0058
17. Ouinsavi C, Gbémavo C, Sokpon N. Ecological structure and fruit production of African fan palm (*Borassus aethiopum*) populations. American Journal of Plant Sciences. 2011;2:733-743. DOI: 10.4236/ajps.2011.26088
18. Rollet B. L'architecture des forêts denses humides sempervirentes de plaines. CTFT, Paris; 1974. French.
19. Lieberman M, Lieberman D, Peralta R, Hartshorn GS. Canopy closure and the distribution of tropical forest tree species at

La Selva. Costa Rica. J. Trop. Ecol. 1995; 11:161-178.
DOI: 10.1017/s0266467400008609
20. Durrieu De Madron L, Forni E. Aménagement forestier dans l'Est du Cameroun; Strucutre du peuplement et périodicité d'exploitation. Bois For. Trop. 1997;254:39-49. French.
21. Hawthorne WD. Ecological profiles of ghanaian forest trees, Tropical Forest. 1995;29.
Oxford Forestry Institute, Dept. of Plant Sciences, South Parks Rd, Oxford OX1 3RB, UK
22. Begon M, Harper JL, Townsend CR. Life-history variation. Blackwell Scientific Publications, Oxford. 1990;473-509.
23. Bonou W, Glèlè Kakaï R, Assogbadjo AE, Fonton HN, Sinsin B. Characterization of *Afzelia africana* Sm. Habitat in the Lama Forest Reserve of Benin. Forest Ecology and Management. 2009;258(7):1084-1092.
DOI: 10.1016/j.foreco.2009.05.032
24. Towanou H, Gbèmavo DSJC, Ouinsavi CAIN, Sokpon N. Structural Characterization of *Prosopis africana* Populations (Guill. Perrott, Rich.) Taub in Benin. International Journal of Forestry Research. 2015;1-9.
DOI: 10.1155/2015/101373
25. Gouwakinnou GN, Kindomihou V, Assogbadjo AE, Sinsin B. Population structure and abundance of *Sclerocarya birrea* (Rich A.) Hochst subsp. birrea in two contrasting land-use systems in Benin. International Journal of Biodiversity and Conservation. 2009;1:194–201.
26. Kperkouma W, Sinsin B, Guelly K, Kokou K, Akpagana K. Typologie et Structure des Parcs Agroforestiers Dans la Prefecture de Donfelgou. Secheresse. 2005;1:3. French.
27. Agbogan A, Tozo K, Kpérkouma W, Bellefontaine R, Dourma M, Akpavi S, Woegan YA, Dimobe K, Akpagana K. Structure des Populations de *Sclerocarya birrea*, *Lannea microcarpa* et *Haematostaphis barteri* au nord du Togo. Journal of Animal & Plant Sciences. 2015; 25(2):3871-3886. French
Available:http://www.m.elewa.org/JAPS. (Accessed July 2015)
28. Lamien N, Mulualem T, Sita G, Per christer O. Variations in dendrometric and fruiting characters of *Vitellaria paradoxa* populations and multivariate models for estimation of fruit yield. Agroforest Systems. 2007;69(1):1-11.
DOI: 10.1007/s10457-006-9013-x
29. Bayala J, Ouedraogo SJ, Teklehaimanot Z. Rejuvenating indigenous trees in agroforestry parkland systems for better fruit production using crown pruning. Agroforest Systems. 2008;72(3):187–194.
DOI: 10.1007/s10457-007-9099-9
30. Shields DA. USDA's method of estimating fruit and tree nut production. Fruit and Tree Nuts Situation and Outlook Report (TFS-261). 1992;58-63.
31. Sourou B. Etude de l'endémisme local du *Haematostaphis barteri* Hook. F (Anacardiaceae) au nord ouest du Bénin. DEA, Université de Parakou (Bénin); 2012. French.
32. Bellefontaine R. Régénération naturelle à faible coût dans le cadre de l'aménagement forestier en zones tropicales sèches en Afrique. Vertigo. 2005;6(2):1-15. French.
DOI: 10.4000/vertigo.4335
33. Dourma M, Guelly KA, Kokou K, Batawila K, Wala K, Bellefontaine R, Akpagana K. Multiplication par drageonnage d'*Isoberlinia doka* et *I. tomentosa* au sein des formations arborées du Nord-Togo. Bois For. Trop. 2006;288(2):49-57. French.

3

Leaching Kinetics of Sulphates in Acidic Soil

Anuj Kumar Garg[1*], A. K. Gupta[1] and Ashu Rani[2]

[1]Department of Chemistry, S.S.V. P.G. College, Hapur, U.P, India.
[2]Department of Pure and Applied Chemistry, University of Kota, Kota, Rajasthan, India.

Authors' contributions

This work was carried out in collaboration between three authors. Author A. K. Garg designed the study, conduct the experimental work, collected the soil samples, wrote the protocol, wrote the first draft of the manuscript and the literature searches. Authors A. K. Gupta and AR managed the analyses of the study and the interpretation of results. All authors read and approved the final manuscript.

Editor(s):
(1) Eliana L. Tassi, Institute of Ecosystem Studies, National Research Council (ISE-CNR), Italy.
Reviewers:
(1) Alok Prasad Das, Siksha 'O' Anusandhan University, Odisha, India.
(2) Yasser Ahmed El-Amier, Mansoura University, Egypt.
(3) Pavlina Simeonova, Bulgarian academy of Sciences, Bulgaria.

ABSTRACT

Aims: Present study is concerned about the leaching kinetics of sulphate anions in soil using glass column method.
Study Design: We have focused our attention on the mechanism of solute transport via leaching.
Place of Study: Soil samples were collected from the barren field located in the industrial area near Hapur district.
Methodology: The leaching study of sulphate is carried out in the laboratory by using the glass column. During the entire course of study the flow rate of deionised water or salts solution from glass column was maintained at 10±0.2 mL/10 min. Three salts of sulphates Na_2SO_4, $MgSO_4$ and $Al_2(SO_4)_3$ were used to study the leaching kinetics of sulphates in soil.
Results: Initial leaching rate profiles are determined for each salt. On increasing $[SO_4^{2-}]_i$, LR_{obs} values are seen to increase for all sulphate salts. The log-log plots of $[SO_4^{2-}]_i$ and LR_{obs} indicated to be the fractional order in $[SO_4^{2-}]_i$. The effect of temperature on sulphate leaching was studied in the range 20-50℃. It is clear that total initially leach able content i.e. $[SO_4^{2-}]_i$ increases with the increase in temperature, probably due to the mobilization and leaching of some insoluble sulphate. LR_{obs}

**Corresponding author: E-mail: anujkgarg02@gmail.com*

values increases with the size of cation. The results can be explained on the basis of covalent character and radii of the hydrated ions. Initial water content or water filled porosity (θ) of the soil column was varied from .33 to .51 $cm^3\ cm^{-3}$. LR_{obs} calculated values were found to increase only slightly on increasing initial water content of the soil column. Leached concentration were found time dependent and varied with time as the pore volume of effluent increased with time. The first order kinetics and Elovich model has been found to have excellent correlation for the desorption/Leaching of the sulphate confirmed by SEE and high r^2 values. However there is little correlation for the zero order and parabolic profiles.

Conclusion: Present study proves that leaching of sulphate salts can be studied by initial rate method, which can also be applied for monitoring the fate of applied sulphate in agricultural fields and its diffusion into soil solution for plants uptake as well as in calculating the downward transportation of sulphate from polluted site to the underground water.

Keywords: Leaching; sulphate; Elovich; first order; dry and wet precipitation.

1. INTRODUCTION

Atmospheric constituents containing sulphates can significantly affect humans and environment. The wet and dry deposition of SO_4^{2-} has received considerable attention throughout the world. Sulphate is a relatively mobile anion. Sulphate desorption can enhance leaching of acid cations (H^+ and Al^{3+}) and base cations such as sodium (Na^+) potassium (K^+), calcium (Ca^{2+}), magnesium (Mg^{2+}). The leaching of base cations from the soil due to the deposition of SO_4^{2-} anions can therefore results in both acidification and loss of fertility of the soil [1-3]. The pH lowering in stream and lake waters is due to the mobility of suphate in the soil which lowers the pH of percolating and run off waters [4,5]. Substantial leaching loss via leaching can occur in soils with low retention capacity observed in New Zealand [6].

The wet and dry depositions of sulphate have changed the soil characteristics in many polluted areas, where acid precipitation and sulphate deposition is a common phenomenon. The tendency of SO_4^{2-} to be removed from soils varies widely and the extent of leaching depends upon the soil properties and fertilization practices. Sulphate is retained to certain extent by most soils [2,3].

To understand the behaviour of sulphate leaching, especially during short period of soil water interaction and at a wide range of water soil ratios, studies on sulphate desorption during continuous flow as well as restricted flow are needed. Literature survey revealed that soil's capacity to adsorb sulfate depends on numerous soil properties, including the concentrations of Al and Fe-hydrous oxides, clay mineralogy, soil pH [6,7] organic matter content and soil temperature [8,9]. Leaching studies of sulphate through soil have significance in environmental as well as in agronomy and agricultural sciences.

The forces involve in SO_4^{2-} adsorption can range from weak, physical, vanderwaals forces and electrostatic outer sphere complexes and chemical interaction. The chemical interactions include the inner sphere complexation which involve ligand exchange mechanism, covalent bonding or there can be precipitation of basic Al sulphate. The ligand exchange mechanism found to be applicable on a time scale of weeks for sulphate adsorption studies [10]. Rajan [11] studied the sorption of sulphate on hydrous alumina and found that the reaction was 90% complete within 10 min and 95% within 60 min. He suggested the formation of inner sphere complexes responsible for sorption. Sulphate ions were found to form bridge between two aluminium ions to form 6 membered rings. Sparks [12] and Jung K, et al. [9] also studied the kinetics of sulphate adsorption and desorption on soil and soil constituents.

Most of the studies consider only the sulphate adsorption kinetics rather than desorption and leaching kinetics. The desorption process is also very important for the release of ions which can increase the nutrients available to plants but also allows them to leach and contaminate the ground water. Advanced techniques like miscible displacement [13], stirred batch procedures [25] pressure jump relaxation [14] etc. have been used by several workers for studying adsorption and desorption of sulphates on soil materials and on rocks and minerals. However, they noted that the lack of suitable techniques to measure the rate of sulphate adsorption and desorption is the major obstacle in understanding the mechanism of sulphate retention and transport in the soil system.

Preliminary investigations on movement of sulphates in Indian soils have been conducted by Malik et al. [15] on soil columns and found that higher amounts of calcium carbonates, phosphates and higher soil pH result in increased sulphate leaching. Although, several studies have been conducted on sulphate adsorption and desorption from soils, but the mechanism of ionic adsorption and desorption is still not well understood. As far as Indian studies are concerned, only very little information is available on the kinetics of sulphate desorption from Indian soils under field and laboratory conditions.

Present work describes the results of investigations carried out on the leaching behavior and kinetics of the sulphate ion in acidic soil of Indian origin. The water soluble salts of sulphates viz. Na_2SO_4, $MgSO_4$ and $Al_2(SO_4)_3$ have been chosen for the present study.

2. MATERIALS AND METHODS

A glass column of 50 cm length and 2.5 cm diameter, For leaching kinetic studies, in which 60 g fine textured soil was filled. The temperature of the column, the flow rate of extractant was found constant in columns of similar sizes. A fixed volume of aqueous salt solutions (slug) with desired anion concentration was added on the top of the soil column in each experiment. Salt solution was allowed to get adsorbed uniformly in the column for 24 h, after which the column was continuously leached with deionised water or with other extractants as per the requirement of the study. The leaching was carried out till the soluble anions were completely removed.

For calculating initial leaching rates, the concentrations of anions were determined in leachates collected periodically at an interval of 10 min. Sulphate ions leached were determined by measuring the degree of turbidity by Systronics (106) UV-VIS spectrophotometer at 440 nm. In order to determine the SO_4^{2-} ions, the buffer solution is added to the suitable volume of sample in 250 mL flask and mix in stirring apparatus. While stirring, a spoonful $BaCl_2$ crystals were added and diluted up to the desired volume and the turbidity was determined [26]. The flow rate was quite slow with only 10±0.2 mL leachate per 10 min. The total leached concentration was taken equal to the total leachable content present at t = 0. After completely removing the soluble ions from column, the soil of columns was transferred into a beaker and suspension was obtained in a known volume of extractant solution or deionised water. This soil suspension was stirred for 2 h. The ion concentration were further determined in the filtrate of the soil suspension to ensure the complete removal of soluble ions from the soil in the column.

The effect of variation of temperature was studied by maintaining the temperature of the column in the range 20℃ to 50℃.The analysis of soil has been carried out by using standard methods [27] and given in Table 1.

Table 1. Physico-chemical characteristics of the column soil

Properties	Values
pH	**6.4**
E C (dsm^{-1})	0.87
Bulk Density (mgm^{-3})	1.44
Colour	Greyish black
Sp. Gravity	2.47
Textural class	Sandy-clay loam
Sand %	48%
Silt %	25%
Clay %	27%
C.E.C.	16.54 cmol kg^{-1}
O.C.%	0.60
Exchangeable cations (mg / 100g)	
Ca^{+2}	20.5
Mg^{+2}	10.4
Na^{+}	4.2
K^{+}	1.9

3. RESULTS AND DISCUSSION

3.1 Data Interpretation

The abbreviations used for the presentation of data are:

$[SO_4^{2-}]_s$ =leachable sulphate present in column soil, 13.6 mg/kg
$[SO_4^{2-}]_{add}$ =SO_4^{2-} concentration introduced in the soil column
$[SO_4^{2-}]_i$ =total leachable content present initially
$[SO_4^{2-}]_t$ =leached concentration at time t
$[SO_4^{2-}]_1$ =$[SO_4^{2-}]_i$ - $[SO_4^{2-}]_t$
=leachable concentration remaining at time t

3.2 Total Leachable Sulphate, $[SO_4^{2-}]_i$

The values of total leachable concentrations of $[SO_4^{2-}]_i$ ions for Na_2SO_4, $MgSO_4$ and $Al_2(SO_4)_3$ are given in Table 2,3,4. It is clear that $[SO_4^{2-}]_i$ increases with the $[SO_4^{2-}]_{add.}$ The soil has itself has leachable SO_4^{2-} present in it i.e. $[SO_4^{2-}]_s$. If we add the salt of sulfate from outside in the soil column, the concentration of leachable SO_4^{2-} increases. In the absence of added SO_4^{2-}, the total leachable contents are equal to the water soluble sulphate presents naturally in soil itself. Similar results have been noted by Rajan [11]. He observed an increase in percentage desorption of SO_4^{2-} from soil when added SO_4^{2-} concentration was increased. He suggested that desorption of sulphate may be related with the surface charges on clay and found that the sulphate quantity left on the surface after desorption is about the same even on increasing the initial SO_4^{2-} concentration.

3.3 Leaching Rate Profiles

Initial rates for leaching, LR_{obs}, (observed leaching rate) represents the rate of change in leachable concentration $[SO_4^{2-}]_1$ with time. LR_{obs} value were obtained from the slope of the plot between $[SO_4^{2-}]_1$ and time as shown in Figs. 1, 2 and 3 .

3.3.1 Leaching of Na_2SO_4

Leaching of SO_4^{2-} ions was studied at different added concentrations of Na_2SO_4 on soil in columns. Amount of $[SO_4^{2-}]_{add}$ was varied from 0-30 mg/kg. The flow rate of percolating water was 10±0.2 ml/10 min. Low width of the column was found advantageous in minimizing the flow rate of percolation of water and maximize the soil water interaction period. The detailed results of Na_2SO_4 leaching for varying $[SO_4^{2-}]_i$ are given in Table 2. Initial rate profiles for Na_2SO_4 leaching at different $[SO_4^{2-}]_i$ are shown in Fig. 1. This figure show that initially leaching is in general relatively fast at almost all the $[SO_4^{2-}]_i$ concentrations.

3.3.2 Leaching of $MgSO_4$

Reaction profiles of SO_4^{2-} leaching in case of $MgSO_4$ are presented in Fig. 2. Detailed results and LR_{obs} Value are given in Table 3.

3.3.3 Leaching of $Al_2(SO_4)_3$

Leaching rate profiles for $Al_2(SO_4)_3$ are given in Fig. 3 and detailed results are given in Table 4.

Table 2. Variation in $[SO_4^{--}]_1$ and $[SO_4^{--}]_t$ with time at different $[SO_4^{--}]_{add}$ and $[SO_4^{--}]_i$ concentrations for Na_2SO_4 leaching at 30℃

$[SO_4^{2}]_{add}$-->	0		10		20		30	
time, s	$[SO_4^{--}]_t$	$[SO_4^{--}]_1$	$[SO_4^{--}]_t$	$[SO_4^{--}]_1$	$[SO_4^{--}]_t$	$[SO_4^{--}]_1$	$[SO_4^{--}]_t$	$[SO_4^{--}]_1$
0	0	13.6	0	23.8	0	32.9	0	30.3
600	3.1	10.5	4.3	19.5	6.0	26.9	4.3	26.0
1200	6.0	7.6	8.4	15.4	12.7	20.2	8.5	21.8
1800	8.0	5.6	10.8	13.0	16.6	16.3	12.4	17.9
2400	10.1	3.5	12.2	11.6	19.5	13.4	15.2	15.1
3000	11.6	2.0	13.8	10.0	22.6	10.3	17.9	12.4
3600	12.6	1.0	15.7	8.1	25.2	7.7	20.9	9.6
$[SO_4^{--}]_i$	13.6		23.8		32.9		30.3	
LR_{obs} $mgkg^{-1}s^{-1}$	0.048		0.099		0.227		0.172	

$[SO_4^{--}]_1$, $[SO_4^{--}]_{add}$, $[SO_4^{--}]_t$ and $[SO_4^{--}]_i$ are in mg/kg

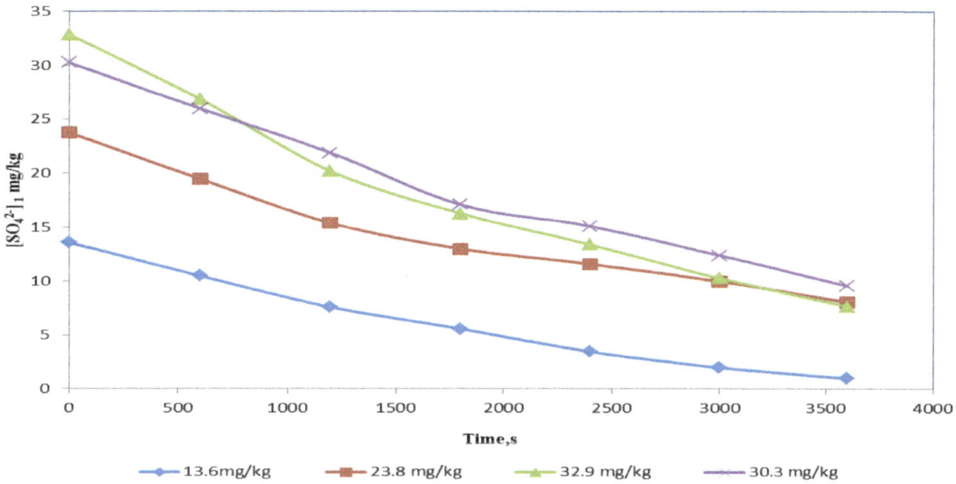

Fig. 1. Initial leaching rate profiles for Na_2SO_4 leaching at different concentration of $[SO_4^{2-}]_i$ at 30℃

Table 3. Variation in $[SO_4^{--}]_1$ and $[SO_4^{--}]_t$ with time at different $[SO_4^{--}]_{add}$ and $[SO_4^{--}]_i$ concentrations for $MgSO_4$ leaching at 30℃

$[SO_4^{2-}]_{add}$-->	0		10		20		30	
time, s	$[SO_4^{--}]_t$	$[SO_4^{--}]_1$	$[SO_4^{--}]_t$	$[SO_4^{--}]_1$	$[SO_4^{--}]_t$	$[SO_4^{--}]_1$	$[SO_4^{--}]_t$	$[SO_4^{--}]_1$
0	0	13.6	0	35.9	0	41.8	0	54.7
600	3.1	10.5	5.8	30.1	4.5	37.3	6.1	48.6
1200	6.0	7.6	11.4	24.5	8.6	33.2	11.8	42.9
1800	8.0	5.6	15.3	20.6	12.3	29.5	18.2	36.5
2400	10.1	3.5	19.1	16.8	16.3	25.5	23.9	30.8
3000	11.6	2.0	22.2	13.7	20.3	21.5	29.8	24.9
3600	12.6	1.0	24.6	11.3	23.6	18.2	36.1	18.6
$[SO_4^{--}]_i$	13.6		35.9		41.8		54.7	
LR_{obs} $mgkg^{-1}s^{-1}$	0.048		0.244		0.275		0.492	

$[SO_4^{--}]_1$, $[SO_4^{--}]_{add}$, $[SO_4^{--}]_t$ and $[SO_4^{--}]_i$ are expressed in mg/kg

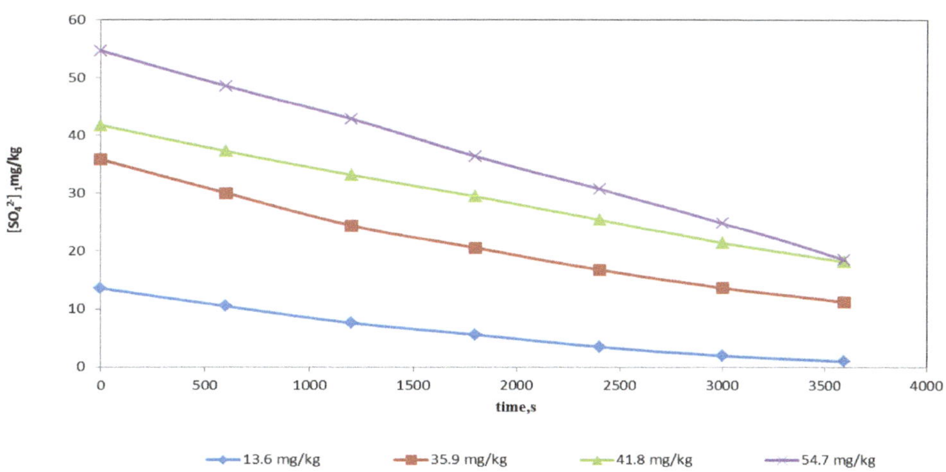

Fig. 2. Initial leaching rate profiles for $MgSO_4$ leaching at different concentration of $[SO_4^{2-}]_i$ at 30℃

Table 4. Variation in $[SO_4^{--}]_1$ and $[SO_4^{--}]_t$ with time at different $[SO_4^{--}]_{add}$ and $[SO_4^{--}]_i$ concentrations for $Al_2(SO_4)_3$ leaching at 30℃

$[SO_4^{2-}]_{ad}$-->	0		10		20		30	
time, s	$[SO_4^{--}]_t$	$[SO_4^{--}]_1$	$[SO_4^{--}]_t$	$[SO_4^{--}]_1$	$[SO_4^{--}]_t$	$[SO_4^{--}]_1$	$[SO_4^{--}]_t$	$[SO_4^{--}]_1$
0	0	13.6	0	30.2	0	37.5	0	43.6
600	3.1	10.5	4.1	26.1	3.1	34.4	4.7	38.9
1200	6.0	7.6	7.1	23.1	6.3	31.2	9.0	34.6
1800	8.0	5.6	10.2	20.0	10.0	27.5	13.2	30.4
2400	10.1	3.5	13.0	17.2	13.5	24.0	17.3	26.3
3000	11.6	2.0	15.6	14.6	16.3	21.2	21.0	22.6
3600	12.6	1.0	17.8	12.4	20.0	17.5	24.5	19.1
$[SO_4^{--}]_i$	13.6		30.2		37.5		43.6	
LR_{obs} $mgkg^{-1}s^{-1}$	0.048		0.147		0.210		0.296	

$[SO_4^{--}]_1$, $[SO_4^{--}]_t$, $[SO_4^{--}]_{add}$ and $[SO_4^{--}]_i$ are expressed in mg/kg

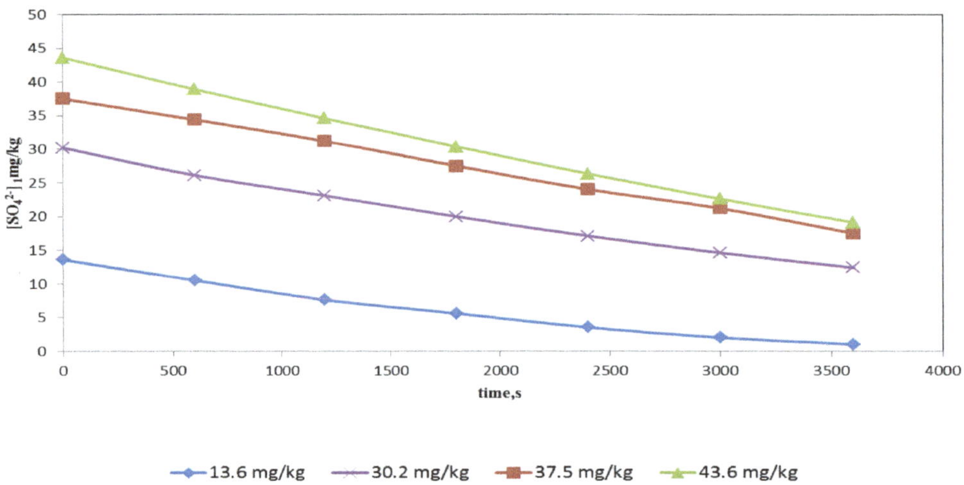

Fig. 3. Initial leaching rate profiles for $Al_2(SO_4)_3$ leaching at different concentration $[SO_4^{2-}]_i$ at 30℃

3.4 Dependence of LR_{obs} on $[SO_4^{2-}]_i$

In general, the leaching profiles for different salts studied are of the same type. On increasing $[SO_4^{2-}]_i$, LR_{obs} values are seen to increase for all sulphate salts. The log-log plots of $[SO_4^{2-}]_i$ and LR_{obs} indicated to be the fractional order in $[SO_4^{2-}]_i$. The leaching rates fitted the rate laws 1 to 3 for different salts.

(i) Na_2SO_4

$$LR_{obs} = k_1 [SO_4^{2-}]_i^n \quad (1)$$

(i) $MgSO_4$

$$LR_{obs} = k_2 [SO_4^{2-}]_i^n \quad (2)$$

(i) $Al_2(SO_4)_3$

$$LR_{obs} = k_3 [SO_4^{2-}]_i^n \quad (3)$$

From the linear plot of LR_{obs} vs $[SO_4^{2-}]_i^n$ as defined by rate laws 1 to 3, the values of k and n for all three salts are given in Table 5.

Table 5. The values of rate constant (k), order (n) and correlation parameters for different salts of sulphates

Parameters	Na_2SO_4	$MgSO_4$	$Al_2(SO_4)_3$
$k\times10^7$	0.36	0.44	1.00
n	1.66	1.63	1.51
r^2	0.971	0.996	0.993
SEE	0.138	0.076	0.079

Units of k is $mg^{1-n}kg^{n-1}s^{-1}$

A comparison of n values shows these to be almost same for Na_2SO_4, $MgSO_4$ but slightly less for $Al_2(SO_4)_3$. The comparison of LR_{obs} values show these to be in order Na_2SO_4>$MgSO_4$ >$Al_2(SO4)_3$. The linear power form equation (1-3) which express the sulphate leaching rates are similar to the equation, proposed by Sharply [16] for desorption of sulphate.

$$S_d = KS_e t^{\alpha} W^{\beta}$$

Where

S_d =amount of sulphate desorbed in time t at a water/ soil ratio W
S_e =Initial amount of extractable SO_4 present in the soil.
α, β and K are constant for given soil

3.5 Effect of Temperature on $[SO_4^{2-}]_i$

The effect of temperature on sulphate leaching was studied in the range 20-50℃. Fig. 4 clearly indicate that total initially leachable content i.e. $[SO_4^{2-}]_i$ concentration increases with the increase in temperature, probably due to the mobilization and leaching of some insoluble sulphates with increase in temperature.

3.6 Effect of Temperature on LR_{obs}

The effect of temperature on LR_{obs} at $[SO_4^{--}]_{add}$ =20 mg/kg is shown in Fig. 5, which show that the temperature has only a small effect on LR_{obs}' although there is only a sizeable effect on $[SO_4^{--}]_i$ at fixed $[SO_4^{--}]_{add}$ concentration.

From the intercept of log LR_{obs} vs log $[S0_4^{--}]_i$ curve, the value of log k were determined at different temperatures. The Values of the rate constant are found out to be different at all temperatures.

Earlier the effect of temperature on sulphate desorption in cecil soil has been studied by Hodge and Johnson [13]. The first order desorption plots show an initial increase with increase in the temperature range 4-20℃ however first order rate constant varied directly with the percentage of sulphate desorbed. Since the temperature changes with the change in climatic condition, soil depth and crop growth. A study of temperature dependence on the rate of leaching is important to understand the effect of climatic factors on leaching.

3.7 Effects of Attached Cation Size on LR_{obs}

The sulphate salts used in this study have the different cations. The role of attached cation on the leaching of sulphate is also very important and has not been studied in the earlier works. In the present study, the effect of attached cation on the sulphate leaching appears to be related to the size of cation. Fig. 6 depicts the plot between LR_{obs} vs radii of the cations. It can be seen clearly from Fig. 6 that LR_{obs} values increases with the size of cation. The results can be explained on the basis of covalent character and radii of the hydrated ions. Smaller the size of cation, larger will be the radius of the hydrated ion and lesser will be its mobility. The ionic size, ionic character, radii of hydrated ions, mobility of ions and LR_{obs} values for different ions are in following order.

Cations size	Na^+>Mg^{2+}> Al^{3+}
Radii of hydrated ion	Al^{3+}> Mg^{2+}> Na^+
Mobility of ion	Na^+>Mg^{2+}> Al^{3+}
Ionic character	Na_2SO_4>$MgSO_4$>$Al_2(SO4)_3$
LR_{obs}	Na_2SO_4>$MgSO_4$>$Al_2(SO4)_3$

3.8 Effect of Water Filled Porosity

Initial water content or water filled porosity (θ) of the soil column was varied from .33 to .51 cm^3cm^{-3}. LR_{obs} calculated values were found to increase only slightly on increasing initial water content of the soil column.

3.9 Application of the Kinetics Models

The data for the leaching of sulphate in acidic Indian soils were used to find the best fit model describing the kinetics of leaching of water soluble sulphates in soil. It appears from the leaching profiles that the rate of leaching in the first few hours is relatively fast becoming quite slow in the later stage. A similar behavior has been noted earlier in the sulphate desorption [17]. While applying the various integrated equations of different kinetics models, the concentration of leachable sulphates was assumed to be the maximum initially, i.e. at t=0 for $[SO_4^{--}]_i$. The concentration terms used in different equations are defined as

$[SO_4^{--}]_i = C_0$ $[SO_4^{--}]_t = C_t$ $[SO_4^{--}]_1 = [SO_4^{--}]_i - [SO_4^{--}]_t = C_0 - C_t$

Zero Order Kinetics Model	$C_0 - C_t = a - bt$
First Order Kinetics Model	$\ln[C_0 - C_t] = a - bt$
Elovich Equation	$C_t = a + b\ln t$
Parabolic diffusion	$C_t = a + b\, t^{1/2}$

These equations were tested by using the statistical analysis on the excel software. The first order kinetics and Elovich model has been found to have excellent correlation for the desorption/Leaching of the sulphate attested by SEE and high r^2 values. However there is little correlation for the zero order and parabolic profiles.

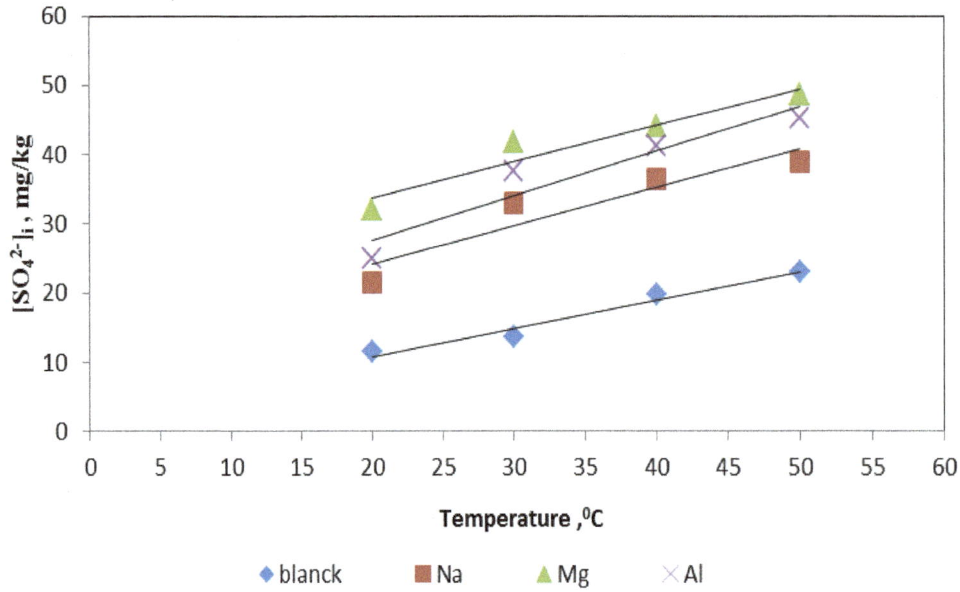

Fig. 4. Variation of $[SO_4^{2-}]_i$ with temperature for leaching of different sulphate salts

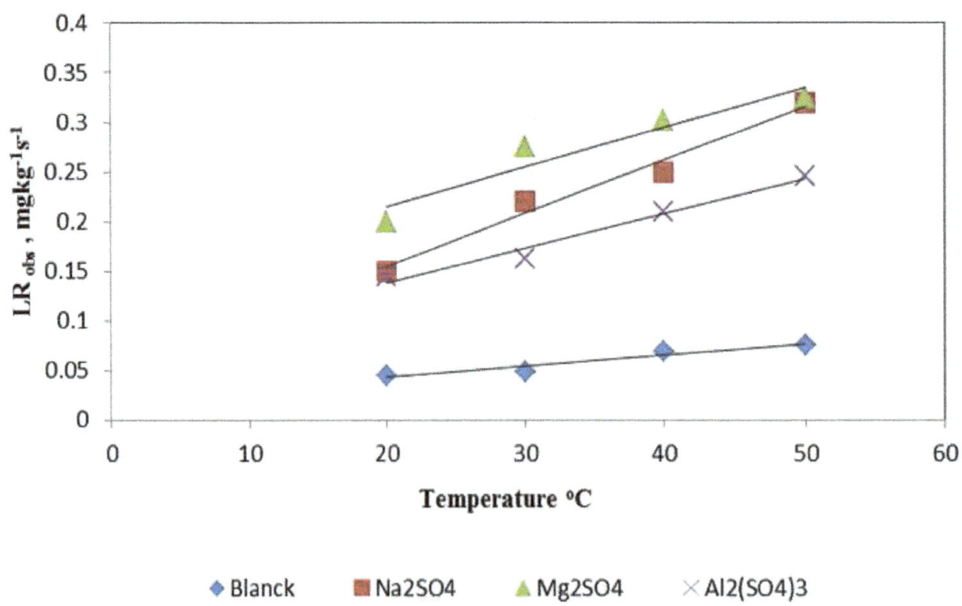

Fig. 5. Variation of LR_{obs} with temperature for the leaching of sulphate salts, $[SO_4^{2-}]_{add}$=20 mg/kg

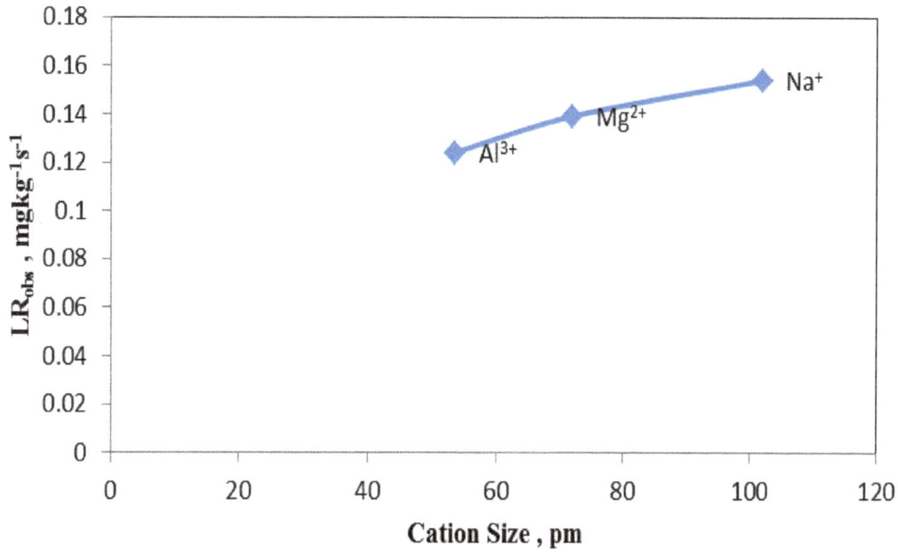

Fig. 6. Change in LR_{obs} with size of cation for different sulphate salts at fixed $[SO_4^{2-}]_i$=20 mg/kg at 30°C

4. CONCLUSION

Investigations have been carried out to study the leaching behavior and kinetics of three different sulphate salts viz. Na_2SO_4, $MgSO_4$ and $Al_2(SO_4)_3$ in Indian soil environment. In this work the initial rate method has been applied for the first time for the study of sulphate leaching using glass column method. The importance of this study is underlined by the preliminary reports of earlier workers that the sulphate desorption is time dependent. Almost all the sulphate desorption studies carried out so far explained that maximum desorption (75% to 100%) has been achieved during the initial period of 30 min [11]. We have also arrived to a similar conclusion in the present study.

The results of the present study show that the sulphate leaching rate is least for Al_2 $(SO_4)_3$ which can be explained as follows-

(i) Al_2 $(SO_4)_3$ in the soil systems is reported to [19] make a complex with hydroxyl ions e.g. Zurbanite $Al(OH)SO_4$, basaluminite $[Al(OH)_{10}SO_4]$ and alunite $[KAl_3(OH)_6(SO_4)_2]$. There is always a possibility of formation of such complexes in the present study too, which may result in the hindrance in leaching, thus resulting in lower leaching rates for Al_2 $(SO_4)_3$.

(ii) In soil, sulphate is reported [20] to form a binuclear bridge complex with aluminium which may also hinder the mobility of $Al_2(SO_4)_3$ in soil.

(iii) So far as, the diffusion of salts in heterogeneous soil is concerned, in order to maintain electrical neutrality during diffusion of an ionic species, inter diffusion of the ions of opposite charge occur in the same direction [21]. It is studied [22] that an anion species while leaching may slowly change the partners with invading cations and become more inhabitant anion. If the salt concentration is high, this change of partners will have insignificant effect on anion diffusion. But, this earlier fact could not explain our results of change in leaching rates of sulphate with change in cation partners. In the present case, increase in LR_{obs} with increase in size of the attached cation can be explained on the basis of the covalent character and radii of the hydrated cation.

As the ionic size increases, the ionic character decreases, but covalent character increases. Moreover, smaller is the size of cation, larger will be the radii of the hydrated ion. Thus the mobility of the ion will decrease with the increase in the size of hydrated ion [22]. Our results support this explanation as maximum leaching rate is observed with Na_2SO_4 while minimum for $Al_2(SO_4)_3$.

The raise in $[SO_4^{2-}]_i$ at constant $[SO_4^{2-}]_{add}$ value with increase in temperature, in the range 20-50°C seems to be due to increase in solubility of salt. It was observed that at fixed $[SO_4^{2-}]_{add}$, LR_{obs}

varied with temperature due to increase in $[SO_4^{2-}]_i$. Temperature variation at fixed $[SO_4^{2-}]_i$ was not possible in our study as soil columns were not saturated with salt solutions prior to leaching.

The present study shows, beyond doubt that first order and Elovich equation are the best fit models for representing sulphate leaching with water in acidic soil. Hodge and Johnson while studying desorption of sulphate with KCl solution found parabolic diffusion to be the best fit, while other studies [18,23,24] support applicability of Elovich and first order equations.

This study suggests that for development of salt transport models in soils, influence of porosity, attached cations and temperature should also be taken into consideration along with other physico-chemical characteristics of the soil [25].

Thus, the present study proves that leaching of sulphate salts can be studied by initial rate method, which can also be applied for monitoring the fate of applied sulphate in agricultural fields and its diffusion into soil solution for plant up take as well as in calculating the downward transportation of sulphate from polluted site to the underground water.

COMPETING INTERESTS

Authors have declared that no competing interests exist.

REFERENCES

1. Swaluw EV, Asman WAH, Jaarsveld HV, Hoogerbrugge R. Wet deposition of ammonium, nitrate and sulfate in the Netherlands over the period 1992-2008. Atmos. Environ. 2011;45:3819-3826.
2. Giannoni SM, Rollenbeck R, Trachte K, Bendix J. Natural or anthropogenic? On the origin of atmospheric sulfate deposition in the Andes of southeastern Ecuador. Atmos. Chem. Phys. 2014;14:11297–11312.
3. Shimadera H, Kondo A, Shrestha KL, Kaga A, Inoue Y. Annual sulfur deposition through fog, wet and dry deposition in the Kinki Region of Japan. Atmos. Environ. 2011;45:6299-6308.
4. Kuribayashi M, Ohara T, Morino Y, Uno I, Kurokawa J. Long-term trends of sulfur deposition in East Asia during 1981-2005. Atmos. Environ. 2012;59:461-475.
5. Sebesta J, Samonil P, Lacina J, Oulehle F, Houska J, Bucek A. Acidification of primeval forests in the Ukraine Carpathians: Vegetation and soil changes over six decades. Forest Ecology and Management. 2011;262:1265–1269.
6. Kopáček J, Hejzlar J, Orcal P, Posch M. leaching from diffuse agricultural and forest sources in a large central European catchment during 1900–2010. Science of the Total Environment. 2014;470,471:543-550.
7. Uzoho BU, Obasi SN, Iwuanyanwu UP. Sulphate sorptivity in relation to land use types of soils of the humid tropical rainforest, southeastern Nigeria. International Journal of Environment and Pollution Research. 2014;2(1):1-8.
8. Alves ME, Lavorenti A. Sulfate adsorption and its relationships with properties of representative soils of the São Paulo State, Brazil. Geoderma. 2004;118:89–99.
9. Jung K, Ok YS, Chang SX. Sulfate adsorption properties of acid-sensitive soils in the Athabasca oil sands region in Alberta, Canada. Chemosphere. 2011;84: 457–463.
10. Scheidegger AM, Sparks DL. A critical assessment of sorption–desorption mechanism at soil mineral/water interface. Soil Science. 1996;161(12):813-831.
11. Rajan SSS. Adsorption and desorption of sulphate and charge relationship in allophanic clays. Soil Sci. Soc. Am. J. 1978;43:65-69.
12. Sparks DL. Kinetics and mechanisms of chemical reactions at the soil mineral /water interface. In Soil Physical Chemistry, Sparks, DL, Ed. CRC Press: Boca Raton, FL. 1999;135-191.
13. Hodge SC, Johnson G. Kinetics of sulphates sorption and desorption by cecil soil using miscible displacement. Soil Sci. Soc. Am. J. 1987;51:323-331.
14. Zhang PC, Sparks DL. Kinetics and mechanism of sulphate adsorption / desorption of Goethite using pressure jump relaxation. Soil Sci. Soc. Am. J. 1990;54: 1266-1273.
15. Malik RS, Karwasra SPS, Khera AP. Movement of sulphate in some Indian soils. J. Ind. Soc. Soil Sci. 1992;40:711-715.
16. Sharply AN. Kinetics of sulphate desorption from soil. Soil Sci. Soc. Am. J. 1990;54:1571-1575.

17. Alves ME, Lavorenti A. Sulfate adsorption and its relationships with properties of representative soils of the São Paulo State, Brazil. Geoderma. 2004;118:89-99.
18. Khanmirzaei A, Akbari H, Rezaei S. Phosphorus desorption kinetics under saturated and field capacity condition in a calcareous soil. Adv. Environ. Bio. 2013; 7(11):3328-3331.
19. Pierzynski GM, Sims JT, Vance GF. Soils and Environmental Chemistry. (3rd Ed.) CRC Press: Boca Raton, FL; 2005.
20. Shustorovich E, Ed. In Metal Surface Reactions Energetic: Theory & Applications to Heterogeneous Catalysis, Chemisorptions and Surface Diffusion, Wiley-VCH, Weinheim; 1991.
21. Jost W. Diffusion in solid, liquid, gases. Academic Press, New York; 1952.
22. Cichota R, Vogeler I, Bolan NS, Clothier BE. Cation influence on sulphate leaching in allophonic soils. Australian Journal of Soil Research. 2007;45:49-54.
23. Brajendra, Shukla LM, Kherawat BS, Lal M. Kinetics of sulphate release in soils of Jharkhand. Internat. J. agric. Sci. 2013; 9(1):139-144.
24. Chao TT, Harward ME, Fang SC. Iron or alumimium coatings in relation to sulphate adsorption characteristics of soils. Soil Sci. Soc. Am. Proc. 1964;28:632-635.
25. Selim HM. Transport and fate of chemicals in soil: Principles and Application. CRC Press; 2014.
26. Benton J Jones Jr. Laboratory guide for conducing soil tests and plant analysis. CRC Press; 2002.
27. American Public Health Association Standard Methods for the Examination of Water and Waste Water including Bottom Sediments and Sludges. APHA, (American Public Health Association, Washington D.C.); 2000.

4

Effect of Nitrogen and Sulphur on Growth and Yield Attributes of Potato (*Solanum tuberosum* L.)

Harinder Singh[1], Madhu Sharma[1], Aakash Goyal[2] and Monika Bansal[1*]

[1]School of Agriculture, Lovely Professional University, Phagwara, Punjab, India.
[2]Biodiversity and Integrated Gene Management Program, International Center for Agriculture Research in the Dry Areas (ICARDA), P.O.Box 6299, Rabat-Institutes Rabat, Morocco.

Authors' contributions

This work was carried out in collaboration between all authors. Author MS designed the study, wrote the protocol and wrote the first draft of the manuscript. Authors MB and AG managed the literature searches, analyses of the study performed and final proof submission and author HS managed the experimental process. All authors read and approved the final manuscript.

Editor(s):
(1) Fatemeh Nejatzadeh, Department of Horticulture, Faculty of Agriculture, Khoy Branch, Islamic Azad University, Iran.
Reviewers:
(1) Denise Aparecida Chiconato, Sao Paulo State University, Brazil.
(2) Raul Leonel Grijalva- Contreras, Instituto Nacional de Investigaciones Forestales Agrícolas y Pecuarias, Mexico.

ABSTRACT

This investigation was carried out to study the effect of four levels of nitrogen (0,90,180,270) and three levels of sulphur (0,25,50) application on growth and yield attributes of potation a split plot design with three replication, at lovely professional university, Phagwara, Punjab, India. The treatments ware replicated thrice in split plot design. The study revealed that application of N 180 $kgha^{-1}$ + S 50 $kgha^{-1}$ significantly enhanced morphological and quality attributes such as plant emergence, number of shoots, periodic plant height, dry matter accumulation, leaf area index, percent reducing sugar and tuber dry matter, there by proving the role of sulphur and nitrogen in high tuber yield in potato 'Kufri-chipsona-3'. Among all treatments,highest total and processable yield was exhibited by treatment T_9 i.e N 180 $kgha^{-1}$+S 50 kg/ha^{-1} expressing the role of S in N uptake and use efficiency. Benefit cost (B: C) ratio was 2.25 which also indicates maximum profitability obtained with this combination.

**Corresponding author: E-mail: monphd@gmail.com*

Keywords: Potato; sulphur; nitrogen; yield; growth; quality; benefit cost ratio.

1. INTRODUCTION

Potato (*Solanum tuberosum*) is a third largest food crop of the world in terms of fresh produce after rice and wheat. India is the third largest producer of potatoes in world after China and Russia with a total production of 45.3 million tonnes from an area of 1992.2 thousand hectares -13). In Punjab, potato is cultivated on an area of 85.21 thousand hectares (5 per cent of total Indian area) with total production of 2.13 million tonnes which contributes 5.07 per cent in the total production of the country [1]. In Punjab, primarily potato belt is confined to Doaba region i.e. Hoshiarpur, Jalandhar, Kapurthala and Nawanshehar districts. It is also grown in Amritsar, Ludhiana, Moga and Patiala districts.

Nitrogen is one of the most essential nutrients required by plant globally. It is an integral component of many compounds such as chlorophyll, nucleotides, alkaloids, enzymes, hormones and vitamins, etc. which are essential for plant growth processes [2]. Nitrogen is valuable nutrient for plants and plays an important role in tuber size development but overdose of nitrogen lowers the tuber dry matter [3]. Proper level of nitrogen has a positive impact on quality and yield of potatoes. Appropriate use of nitrogen expanded the leaf area index and increases photo assimilates [4]. Excessive application of nitrogen decrease starch content and also spoil the taste while cooking. Industry requires 40-80 mm size of potatoes, high in dry matter and low reducing sugar for quality processing. Sulphur is one of sixteen essential nutrient elements and fourth major nutrient after NPK, required by plants for proper growth and yield as it is known to take part in many reactions in all living cells [5]. Sulphur deficient plants have poor utilization of nitrogen, phosphorus and potash at all age [6].

Intensive cropping and use of high-grade fertilizers have caused the depletion of sulphur in soils. Decrease in tuber dry matter yield particularly cystine and leucine were observed with sulphur deficiency [7]. Sulphur has a direct effect on soil as it may reduce pH which improves the availability of microelements such as Fe, Zn, Mn and Cu as well as crop yield and its related characteristics [8]. The need of application of sulphur along with its beneficial effects on yield and quality has been reported by earlier workers[9]. Sulphur also has influence on potato flower by involvement in the volatile S^- compound [10]. Therefore the present study was envisaged to determine the effect of nitrogen and sulphur on the yield and quality of potato 'Kufri chipsona-3' in split plot design, and to determine best treatment in terms of benefit-cost ratio.

2. MATERIALS AND METHODS

2.1 Experimental Site and Location

Field experiment was carried out during 2013-2014 at the Agriculture Research Field, Lovely Professional University(31°15' N 75°41 E) Phagwara, (Punjab) under irrigated conditions. The experiment was conducted in Split plot Design having 12 treatments with three replications. The experiment was laid out on Sandy loam soil. The planting of crop was done on October 11, 2013. Seed tubers were planted by dibbling on ridges at the spacing of 60 X 20 cm. Chemical analysis of the soil showed a neutral pH (7.1), 0.46% organiccarbon,156 kg nitrogen 28.9 $kgha^{-1}$phosphorus, and 356 $kgha^{-1}$ exchangeable potassium. Recommended dose of N, P and K (180:60:120 kg N, P_2O_5 and K_2O ha^{-1}) were applied. Full dose of P and K were applied along with 50 per cent of N at the time of planting. The remaining 50 per cent N was applied at time of earthling up.

2.2 Analysis of Variables

2.2.1 Growth characteristics

The emergence of potato seedlings from each plot was recorded on alternative day, starting from the day when sprouting start emerging above ground, and days taken to complete 50% and 100% emergence were counted for each treatment. The number of shoots emerged from the tubers were counted in a one row length from each plot and the plants data was recorded on five selected plants. Height of five randomly selected and tagged plants from each plot were measured at 30, 60 and 90 DAS (Days after sowing) as shown in (Fig. 1.1, 1.2 and 1.3). Dry matter was recorded from the randomly selected plants and then mean values were worked out. Fresh detached leaves were taken to measure leaf area index at 60 DAS with the leaf area meter.

$$\text{Leaf Area Index} = \frac{\text{Leaf Area } (cm^2)}{\text{Ground Area } (cm^2)}$$

2.2.2 Yield attributes

The number of tubers per plant ware calculated from same five randomly selected plants meant for plant height and number of stems per hill in each plot. Total tuber yield was calculated separately from each net plot for computation of yield in Kg ha^{-1}. Tuber diameter between 40 mm to 80 mm is considered as processable tuber. These are sorted out after harvest with help of Vernier Caliper. Then there average weight is calculated as per different treatments.

The tubers below 40 mm and above 80 mm are considered as non processable.

Non processable yield = Total yield – Processable yield.

Table 1. Different grades of potato

Serial number	Diameter	Grade
1	Above 80 mm	Large
2	Between 80 mm-40 mm	Medium
3	Below 40 mm	Small

2.2.3 Quality characters

Tuber dry matter accumulation was analysed after interval of 60 DAS, and 90 DAS and was determined by oven drying 50 g finely chopped and mixed tuber piecesat 65°C till constant weight.The reducing sugars were estimated by Nelson-Somogi method [11]. The reducing sugars when heated with alkaline copper tartrate reduce the copper from the cupric to cuprous state and thus cuprous oxide isformed when cuprous oxide is treated witharsenomolybdic acid, the reduction of molybdic acid to molybdenum blue take place. The blue colour developed is compared with a st of standards in colorimeter at 620 nm. Digging of potatoes was done manually on January 23, 2014.

2.2.4 Economics

The gross monetary returns in rupees per hectare wereworked out on the basis of potato yield. The prevailing market price of potato in the phagwara district of Punjab, India was considered.

2.2.4.1 Gross returns ($ ha^{-1})

The gross returns were calculated by considering the prices of potato at the time of harvest.

Gross returns = Yield according to grades X Market price according to grade.

2.2.4.2 Net returns ($ ha^{-1})

The net return was calculated by deducting the cost of cultivation from the gross returns.

Net returns ($.) = Gross income -Total cost of cultivation.

The benefit cost ratio was calculated as follows;

$$\text{B: C ratio} = \frac{\text{Gross returns (\$ ha}^{-1})}{\text{Cost of cultivation (\$ ha}^{-1})}$$

2.3 Data Analysis

Statistical analysis of the data recorded was done as per split plot design (Cochran and Cox, 1963).

3. RESULTS AND DISCUSSION

3.1 Growth Analysis

A good and uniform emergence is required for the successful raising of any crop, which ultimately determines the crop yield. The data in Table 1 reveals that different treatments have insignificant effect on emergence. It means that there was almost similar level of plant population in all the treatment. This may be due to the fact that growing seedlings get their food from the tuber. [12] also reported that N application had insignificant effect on plant emergence under their conditions. Number of shoots per plant or per unit area is an index of growth and adaptability of the plant to the soil and climatic conditionand has a direct bearing on development of potato yield.The data revealed that treatment T_9 (Nitrogen 180 kg N and Sulphur 50 kg ha^{-1}) gave significantly maximum number of shoots. Number of shoots in T_3, T_6 T_7 T_{12} was statiscally at par to each other but significantly lower than T_{11}. T_{10} is significantly superior over T_4, T_5 and T_8. This might be due to the fact that fertilization application encouraged more number of independent stems. The results are in consonance with those of [13]. Who reported that the highest number of stems per hill (4.43) obtained when the highest rate of nitrogen (254 kg ha^{-1}) was applied. There was a periodic increase in plant height and it differs significantly among all the treatments. After 40 days maximum plant height (40.7) was recorded in treatment T_9, which was statically at par with T_{12}. T_6 was also at par with T_8 and T_{10} whereas, T_1 (33) recorded minimum plant height. Minimum plant height (50) was recorded in the control. After 80 days T_9 gave significantly higher plant height (65.5) which was superior to all the treatments. The probable reason for increasing plant height might be due to more uptake of N during growth period resulting in increase in cell size, elongation and enhancement of cell division which ultimately increase the plant growth. [14] found that the application of 100% N of recommended dose significantly increased the plant height with their genetic material under their conditions. The results are further supported by [15] who stated that application of N up to 240 kg ha^{-1} significantly increase the plant height. After 40 days dry matter accumulation (1.24) was recorded significantly higher in treatment T_9, followed by T_{12} which was statiscally at par with T_{11} and T_{10}. Treatment T_1 (0.68) recorded minimum dry matter accumulation. After 60 days highest dry matter accumulation (6.60) was recorded in treatment T_9, which is at par with T_{12}. Minimum dry matter accumulation (4.34) was recorded in control plot treatment. After 80 days T_3 gave significantly highest dry matter accumulation (8.02). Treatment T_1 gave lowest dry matter (6.05). The results are further supported by [16] who stated that application of N up to 200 kg/ha significantly increase the dry matter content with their genetic material under their conditions.The data of leaf area index was recorded in cm 2 at 60 days after sowing. The data is presented in Table 2 After 60 days highest leaf area index (5.12) was recorded in treatment T_9, which was at par with T_{12}. Whereas, minimum dry matter accumulation (1.17) occurred in control plot treatment. Researcher reported that application of N fertilizer give significantly higher leaf area index with their genetic material under their conditions. Dry matter accumulation is good growth index to express the photosynthetic efficiency of the plant. The dry matter accumulation was measured at 40, 60 and 80 DAS and the mean was worked out from five plants, which were selectedat randomly in each treatment. It was observed that after 60 days, maximum tuber dry matter was observed in T_9 (41.72 g $plant^{-1}$) and differ significantly from all other treatments. The minimum tuber dry matter was recorded in T_1 (17.85). Maximum dry matter accumulation in tubers after 90 days was observed in T_9 (80.15 g $plant^{-1}$) followed by T_{12}, T_{11} and T_8 whereas, T_5 and T_6 were at par to each other but significantly higher than T_2 and T_1. As above given discussion show that there is an increasing trend in dry matter accumulation up to 90 days. Due to high LAI, the crop under this treatment might be able to intercept relatively higher solar energy resulting in increased dry matter production in all the plant parts. [17] and [18] reported that application of N fertilizer and sulphur can significantly influence the dry matter accumulation of potato tubers with their genetic material under their conditions.

3.2 Yield Studies

The data on number of tuber at harvest was recorded. The number of tubers produced ware

Table 2. Effect of Nitrogen and sulphur on morphological characters of potato

Treatment	Days taken to 50% emergence	Days taken to 100 % emergence	No. of shoots 80 das	Dry matter (gram $plant^{-1}$)			Leaf area index	Plant height (cm)		
				40 days	60 days	80 days		40 days	60 days	80 days
No + S0	5.75^{a}	9.85^{a}	2.7^{c}	7.73^{f}	68.56^{g}	100.60^{i}	1.17^{e}	33.00^{h}	50.00^{a}	58.70^{f}
No + S25	5.58^{a}	9.58^{a}	2.7^{c}	9.26^{e}	69.53^{f}	102.73^{g}	1.45^{e}	34.70^{g}	50.30^{a}	58.90^{f}
No + S50	4.50^{a}	8.58^{a}	3.2^{b}	9.86^{e}	70.63^{e}	104.70^{e}	2.34^{d}	36.50^{ef}	52.60^{a}	59.80^{e}
N90 + S0	5.18^{a}	9.25^{a}	3.0^{b}	11.20^{d}	70.33^{e}	100.40^{i}	3.71^{c}	36.80^{e}	52.60^{a}	58.40^{f}
N90 + S25	4.90^{a}	9.0 ^{a}	3.0^{b}	11.70^{d}	71.46^{d}	101.70^{h}	3.81^{c}	36.00^{f}	53.60^{a}	60.20^{e}
N90 + S50	4.58^{a}	8.90^{a}	3.2^{b}	12.00^{d}	72.60^{c}	103.73^{f}	3.95^{c}	38.40^{cd}	55.80^{a}	61.80^{d}
N180 +S0	5.50^{a}	9.33^{a}	3.2^{b}	12.43^{cd}	71.46^{d}	100.76^{i}	4.04^{bc}	38.00^{d}	52.00^{a}	60.40^{e}
N180 + S25	5.0 ^{a}	8.90^{a}	3.0^{b}	12.70^{cd}	73.06^{c}	106.40^{d}	4.15^{bc}	38.50^{cd}	56.20^{a}	63.50^{c}
N180 + S50	4.93^{a}	8.83^{a}	4.4^{a}	12.96^{c}	74.10^{b}	110.53^{a}	5.12^{a}	40.70^{a}	59.80^{a}	65.50^{a}
N270 + S0	4.85^{a}	8.60^{a}	3.4^{b}	13.56^{bc}	71.50^{d}	103.70^{f}	4.06^{bc}	38.50^{cd}	56.90^{a}	63.50^{c}
N270 + S25	5.25^{a}	9.10^{a}	4.0^{a}	14.06^{b}	74.70^{b}	107.53^{c}	4.45^{b}	38.9^{c}	58.4^{a}	65.1^{a}
N270 + S50	5.02^{a}	8.70^{a}	3.3^{b}	15.00^{a}	75.63^{a}	109.14^{b}	5.04^{a}	39.8^{b}	59.2^{a}	64.4^{b}
C.D 5% main	NS	NS	0.55	0.86	0.65	0.78	0.46	0.5	NS	0.67
Sub	NS	NS	0.64	0.38	0.4	0.47	0.5	0.36	NS	0.28
Main X Sub	NS	NS	NS	NS	0.8	0.95	NS	NS	NS	0.56

counted from five randomly selected plants and the mean was worked out. Data in Table 3 reveals that higher yield of small sized tubers was recorded with T_9 (51.92) which was significantly higher than all other treatments. T_8, T_{10}, T_{11} and T_{12} were observed at par but significantly higher than T_3, T_2 and T_1. T_3 and T_4 were also found at par but significantly higher than T_6. [19] also found significant difference among different genotypes for grade wise tuber yield with their germplasm. The data on the effect of different fertilizer treatments on total tuber yield of potato are therefore, presented in Table 3 and are depicted in Fig. 1.4. Highest total yield was recorded in T_9, which was at par with T_8, T_{12} and T_{11}, whereas minimum total yield was reported in control plot treatment. These results are in line with [20] who reported that increase in tuber yield with increasing sulphur levels may be attributed to its role in better partitioning of the photosynthates in the shoot and tubers. Similarly, [21] have also reported significant effect on grade wise tuber yield and increase in bulking rate with sulphur application with their genetic material under their conditions. The increase in yield with the application of recommended doses of NPK by fertilizers and FYM (Farm yard manure) could be attributed to corresponding increase in leaf area, which is responsible for synthesizing photosynthates and increase in number and weight of tubers as reported by [22]. Processing industry requires 40-80 mm size of potatoes. Those size in between this range are consider fit for purpose of processing. Results in Table 3 revealed that T_9 have highest processable yield whereas, T_8 and T_{12} were statically at par to each other. All the treatments were found significantly higher than T_1 (control).

3.3 Quality Parameters

Dry matter accumulation by tubers is an important growth characteristic in potato which ultimately affects yield of the crop. Data on dry matter accumulation by tubers were recorded at 60 and 90 DAS. Maximum dry matter accumulation in tubers after 90 days was observed in T_9 (80.15 gplant^{-1}) followed by T_{12}, T_{11} and T_8 whereas, T_5 and T_6 were at par to each other but significantly higher than T_2 and T_1. The data clearly reveal that application of nutrients at different doses of nitrogen and sulphur significantly influenced the dry weight. As above given discussion show that there were an increasing trend in dry matter accumulation up to 90 days. Due to high LAI (Leaf Area Index), the crop under this treatment might be able to intercept relatively higher solar energy resulting in increased dry matter production in all the plant parts. Per cent reducing sugar was found maximum and significantly higher in T_9 treatment as compared to all other treatments. Reducing sugar content in treatment T_9 (0.37) was statistically at par to treatment T_8. Data Table 3 further revealed that treatment T_4 (0.31) along with T_3 (0.33) and T_1 (0.33) had the minimum reducing sugar. These results were in consonance to the findings of [23] with their genetic material. Raghav et al. [24] observed that there was significant effect on net income by using organic and inorganic manures with their genetic material under consideration.

Table 3. Effect of nitrogen and sulphur on yield and quality attributes of potato

Treatment	No. of tubers per plant	Percent reducing sugar	Processable yield (qha^{-1})	Grade wise yield			Non-processable yield(q ha^{-1})	Dry matter (gram plant^{-1})	
				Large	Medium	Small		60 days	80 days
No + S0	5.3^{a}	0.31^{d}	96.46^{h}	50.67^{c}	68.45^{f}	45.04^{b}	67.70^{e}	17.85^{e}	40.16^{e}
No + S25	6.6^{a}	0.32^{c}	107.80^{f}	55.90^{b}	85.78^{e}	42.22^{b}	75.10^{d}	18.19^{e}	41.11^{e}
No + S50	6.7^{a}	0.31^{d}	112.10^{e}	51.22^{c}	98.23^{d}	44.95^{b}	82.30^{c}	19.12^{e}	42.07^{e}
N90 + S0	5.6^{a}	0.31^{d}	104.86^{g}	58.33^{ab}	71.53^{f}	44.81^{b}	69.56^{e}	28.02^{d}	67.28^{d}
N90 + S25	6.0^{a}	0.33$^{b c}$	135.06^{d}	44.64^{d}	117.3^{c}	41.84^{b}	68.76^{e}	32.93^{c}	70.09^{c}
N90 + S50	6.3^{a}	0.33$^{b c}$	142.60^{c}	45.22^{d}	129.34^{a}b	40.58^{b}	72.73$^{d e}$	36.44^{b}	70.41^{c}
N180 + S0	5.6^{a}	0.32^{c}	108.66^{f}	54.97^{b}	124.87^{b}	45.04^{b}	116.06^{a}	35.15^{bc}	71.26^{c}
N180 + S25	6.6^{a}	0.34^{b}	151.53^{b}	57.68^{ab}	128.58^{ab}	50.56^{a}	85.00^{c}	37.33^{b}	76.88^{b}
N180 + S50	7.6^{a}	0.37^{a}	160.60^{a}	61.02^{a}	133.06^{a}	51.92^{a}	85.40^{c}	41.72^{a}	80.15^{a}
N270 + S0	6.6^{a}	0.31^{d}	110.73^{ef}	53.04^{bc}	125.45^{b}	50.55^{a}	118.63^{a}	34.08^{c}	71.85^{c}
N270 + S25	7.0^{a}	0.33$^{b c}$	140.80^{c}	55.01^{b}	126.32^{b}	52.23^{a}	92.73^{b}	36.63^{b}	72.4^{c}
N270 + S50	6.6^{a}	0.32^{c}	149.56^{b}	57.7^{ab}	129.58^{ab}	50.62^{a}	88.36^{b}	41.15^{a}	76.12^{b}
C.D 5% Main	NS	0.0116	3.085	3.72	4.63	4.94	4.96	1.96	2.64
Sub	0.70	0.0104	1.487	1.24	2.26	1.46	2.26	0.71	1.12
Main X Sub	NS	0.0207	2.974	2.65	2.34	2.93	4.52	1.43	2.25

Significance at 5% of level of significance

Table 4. Economic analysis (in \$ ha^{-1}) of nitrogen and sulphur for potato crop

Treatment	Land rent (\$ ha^{-1})	Seed bed preparation (\$ ha^{-1})	Labour (\$ ha^{-1})	Harvesting labour	Seed (\$ha^{-1})	Manure and fertilizer (\$ ha^{-1})	Total input cost (\$ ha^{-1})	Total income (\$ ha^{-1})	Net income (\$ ha^{-1})	B:C ratio
T_1	390.71	62.51	35.95	128.93	250.05	0.00	868.15	1394.15	525.99	1.60
T_2	390.71	62.51	35.95	128.93	250.05	51.25	919.40	1581.77	662.37	1.72
T_3	390.71	62.51	35.95	128.93	250.05	71.57	939.72	1677.49	737.77	1.78
T_4	390.71	62.51	35.95	128.93	250.05	34.35	902.52	1487.25	584.74	1.64
T_5	390.71	62.51	35.95	128.93	250.05	54.68	59,048.8	1777.88	855.04	1.92
T_6	390.71	62.51	35.95	128.93	250.05	75.00	922.83	1886.90	943.75	2.00
T_7	390.71	62.51	35.95	128.93	250.05	37.80	905.95	1960.16	1054.21	2.16
T_8	390.71	62.51	35.95	128.93	250.05	58.11	926.26	2057.06	1130.80	2.22
T_9	390.71	62.51	35.95	128.93	250.05	78.43	946.58	2137.34	1190.76	2.25
T_{10}	390.71	62.51	35.95	128.93	250.05	41.23	909.38	1987.76	1078.37	2.18
T_{11}	390.71	62.51	35.95	128.93	250.05	61.54	929.70	2024.66	1094.97	2.17
T_{12}	390.71	62.51	35.95	128.93	250.05	81.86	950.01	2067.03	1117.02	2.17

1. Selling cost ofone kg of potato, Small size = 0.07 \$/kg, Medium size = 0.09 \$ kg^{-1}, Large size = 0.09 \$ kg^{-1},
2. COST of fertilizers nitrogen = 0.04 \$ kg^{-1}, Phosphorus = 0.17 \$ Kg^{-1}, Potash = 0.17 \$ kg^{-1} Sulphur = 52 \$ Kg^{-1}

Fig. 1.1. Crop growth at 30 days

Fig. 1.2. Crop growth at 60 days

Fig. 1.3. Plant height at 60 days

Fig. 1.4. Variation in tubers

3.4 Economic Analysis

Economic analysis of using different levels of nitrogen and sulphur for potato crop was determined by total input cost, total income and net income in $/ha in Phagwara district, Punjab, India. From data Table 4 total cost was determined by adding all six classes (Land rent, seed bed preparation, and labour, harvesting labour, seed, manure and fertilizer). Total income was calculated by multiplying total yield with marketable price of potato crop. Net income was also found out by reducing total cost from total income. Data in Table 4 revealed that treatment T_9 gave the maximum income followed by T_8, T_{12}, T_{11}, T_{10}, T_7 and T_6. The benefit cost ratio was found maximum in treatment T_9 followed by T_8, T_{10}, T_{12}, T_{11}, T_7 and T_6 whereas control (T_1) recorded minimum benefit cost ratio. Data showed that all treatments gave benefit cost ratio above 1, but application of 180 $kgha^{-1}$ of N and 50 kg/ha^{-1} of S recorded highest net income and benefit cost ratio. [24] observed that there was significantly effect on net income by using of organic and inorganic manures with their genetic material under their conditions.

4. CONCLUSIONS

Analysis of variance revealed significant difference for total and processable yield and other component traits. T_9 encompassing of N180+S50 exhibited highest number of shoots, plant height, dry matter accumulation, leaf area index, number of tubers per plant, grade wise yield, total yield, processable yield, non processable yield, percent reducing sugar, tuber dry matter. Treatment T_8 (N180 + S25) revealed the same trend in yield and its component traits but lesser than T_9 (N180+S50). There was an enhancement of morphological and quality attributes such as plant emergence, number of shoots, periodic plant height, dry matter accumulation, leaf area index, and percent reducing sugar and tuber dry matter proved the role of S in N use efficiency. In term of economics, crop grown under treatment T_9 (N180 +S50) gave higher total income and net income as compared with other treatments. Benefit cost ratio was significantly higher in T_9 than other treatments expressing importance of nitrogen and sulphur in enhancing yield and economic returns.

COMPETING INTERESTS

Authors have declared that no competing interests exist.

REFERENCES

1. Anonymous. Area production report of vegetables (potato), National Horticulture Board; 2012-13. Available:www.nationalhorticultureboard.com/statistics/area production statistics
2. Brady NC, Weil RR. The nature and properties of soils. Revised 14th ed. Pearson Prentice Hall. New Jersey; 2008.
3. Zelalem A, Tekalign T, Nigussie D. Response of potato (*Solanum tuberosum* L.) to different rates of nitrogen and phosphorus fertilization on vertisols at Debre Berhan, in the central highlands of Ethiopia. African J. of Plant Sci. 2009; 3(2):16-24.
4. Vaezzadeh M, Naderidarbaghshahi M. The effect of various nitrogen fertilizer amounts on yield and nitrate accumulation in tubers of two potato cultivars in cold regions of Isfahan (Iran). Int J Agri Crop Sci. 2012; 4(22):1688–91.
5. Sud KC, Sharma RC. Sulphur needs of potato under rainfed conditions in shimla hills. In: Potato Global Research and Development, Paul Khurana SM, Shekhawat GS, Pandey SK, Singh BS, (Eds.). Indian Potato Association, Shimla. 2002;2:889-899.
6. Nasreen S, Haq SMI, Hossain MA. Sulphur effects on growth responses and yield of onion. Asian J. Plant Sci. 2003;2: 897-902.
7. Eppendorfer WF, Eggum BO. Effect of Sulphur nitrogen, phosphorus, potassium and water stress on dietary fibre fractions, starch, amino acids and on biological value of potato protein. Plant Foods Human Nutr. 1994;45:299–313.
8. Tantawy E, El-Beik EM, El-Beik AK. Relationship between growth, yield and storability of onion (*Allium cepa* L.) with fertilization of nitrogen, sulphur and copper under calcareous soil conditions. Res J Agric. Biological. Sci. 2009;5:361-371.
9. Singh JP, Marwaha RS, Srivastava OP. Processing and nutritive qualities of potato tubers as affected by fertilizer nutrients and sulphur application. J. Indian Potato Assoc. 1995;22:32-37.
10. Forney CF, Jordan MA, Campbell-Palmer L, Fillmore S, McRae K, Best K. Sulfur fertilization affects onion quality and flavor chemistry during storage. Acta Hort. 2010; 877:163-168.

11. Pearson D. The chemical analysis of foods. 7th edn., Churchill Livingstone, Edinburgh, Scotland; 1976
12. Kumar P, Panday SK, Singh SV, Kumar D, Singh BP, Singh S, Rawal S, Meena RL. Influence of N and K rate of yield and quality of chipping variety Kufri Chipsona-3. Potato J. 2012;39(2):191-196.
13. Zamil MF, Rahman MM, Rabbani MG, Khatun T. Combined effect of nitrogen and plant spacing on the growth and yield of potato with economic performance. Bangladesh Res. Public. J. 2010;3(3): 1062-1070.
14. Singh SK, Gupta VK. Influence of farmyard manure, nitrogen and bio-fertilizer on growth, tuber yield of potato under rain-fed condition in east Khasi hill district of Meghalaya. Agriculture Science Digest. 2005;25(4):281-83.
15. Rajanna KM, Shivasankar KT, Krishnappa KS. Effect of different levels of nitrogen, phosphorus and potassium on growth, yield and quality of potato. South Indian Hort. 1987;35(5):347–355.
16. Sandhu KS, Chinna GS, Marwaha RS, Kumar Parveen, Pandey SK. Effect of Nitrogen fertilization on yield and chipping quality of processing varieties grown in coller north Indian plains. Potato J. 2010; 37(3-4):143-150.
17. Sharma DK, Kushwah SS, Nema PK, Rathore SS. Effect of sulphur on yield and quality of potato. Int. J. Agric. Res. 2011; 6(2):143-148.
18. Lei S, Liulian G, Xianlong P, Yuanying L, Xuezhan L, Xiufeng Y. Effect of nitrogen fertilizer application time on dry matter accumulation and yield of Chinese potato variety KX 13. Potato Res. 2012; 55(3-4):303-13.
19. Bhardwaj V, Pandey SK, Manivel P, Singh SV, Kumar D. Stability of indigenous and exotic potato processing cultivars in Himachal Pradesh hills. Proceedings of the Global Potato Conference, Dec. 9-12, New Delhi. 2008;22-22.
20. Sud KC, Sharma RC. Sulphur needs of potato under rainfed conditions in shimla hills. In: Potato Global Research and Development, Paul Khurana SM, Shekhawat GS, Pandey SK, Singh BS, (Eds.). Indian Potato Association, Shimla. 2002;2:889-899.
21. Lalitha BS, Nagaraj KH, Anand TN. Effect of source propagation, level of potassium and sulphur on potato (*Solanum tuberosum* L.). Mys. J. Agric. Sci. 2002;36: 148-153.
22. Shah MA, Ismail M. The effect of different level of NPK on Potato yield. 1983;10: 136-139.
23. Reshu C, Manoj R. Yield and quality of potato cultivar as affected by organic and inorganic source of nutrition. Progress Hort. 2007;39(1):58-62.
24. Raghav M, Kumar T, Kamal S. Effect of organic sources of nutrients on growth, yield and quality of potato. Progress. Hort. 2007;39(1):95-100.

5

Participatory Selection of Cowpea Varieties in Kilifi County of Kenya

J. B. Ndiso[1*], G. N. Chemining'wa[2], F. M. Olubayo[2] and H. M. Saha[1]

[1]Department of Crop Sciences, Pwani University, P.O.Box 195 – 80108, Kilifi, Kenya.
[2]Department of Plant Science and Crop Protection, University of Nairobi, P.O.Box 30197 00100 GPO, Nairobi, Kenya.

Authors' contributions

This work was carried out in collaboration between all authors. All authors read and approved the final manuscript.

Editor(s):
(1) Scd Victoria Anatolyivna Tsygankova, Department for Chemistry of Bioactive Nitrogen-Containing Heterocyclic Compounds, Institute of Bio-organic Chemistry and Petrochemistry, Plant Physiology, NAS, Ukraine.
Reviewers:
(1) Edward Missanjo, Malawi College of Forestry andWildlife, Malawi.
(2) Wissem Hamdi, High Institute of Agronomy, Tunisia.
(3) Anonymous, University of Tsukuba, Japan.

ABSTRACT

Aim: To identify farmer preferred cowpea varieties in coastal lowland Kenya.
Study Design: The experiment was laid out in a randomized complete blocks design with three replications.
Place and Study Duration: Tezo location, Bahari division, Kilifi County, Kenya during July – October 2011/2012 cropping seasons.
Methodology: Thirty nine farmers (30 female and 9 male) from three farmer groups in Tezo location, Kilifi County participated in the establishment and evaluation of 11 cowpea varieties using their own selection criteria. The varieties were evaluated at flowering, podding, maturity and post harvest stages. Farmers' cowpea selection criteria before flowering and at podding were high grain yield, drought tolerance, early maturity, ease of harvesting and leafiness.
Results: Kutambaa, KVU 27-1 and Nyeupe were rated top varieties at these stages. Farmers' selection criteria at maturity and after harvest included grain yield, color, taste and cooking duration. KVU 419, Kaima koko and Nyeupe were rated top varieties at these stages. Grain yield varied from 3.3 t ha^{-1} (KVU 419) to 0.48 t ha^{-1} (Kaima koko).

**Corresponding author: E-mail: j.ndiso@pu.ac.ke*

Conclusion: The results of this study have demonstrated the need for plant breeders to integrate grain color, taste and cooking duration traits in cowpea improvement programmes. Integration of KVU 419, Nyeupe, KVU 27-1 and Kutambaa cowpea varieties into the maize-based system is likely to improve cowpea productivity in Kilifi County, Kenya.

Keywords: Grain yield; maize-based; Nyeupe; drought; leafiness; color.

1. INTRODUCTION

Plant breeding and agricultural promotion programs continue to be patterned after those in western industrialized countries, emphasizing the use of modern innovations that practice the development of high yielding varieties that perform well in environments that are stabilized through the use of irrigation, fertilizers, pesticides and other inputs [1]. Perhaps the most significant realization at the beginning of the 21st century is the fact that the areas in the developing world, characterized by traditional/subsistence agriculture, remain poorly served by the top-down transfer-of-technology approach, due to its bias in favor of modern scientific knowledge and its neglect of local participation and traditional knowledge [2]. Single genotypes have been widely promoted, to be grown in pure stands regardless of the system in which the crop is currently being grown or the availability of risk reducing inputs [1]. This could be the reason why many promising crop varieties coming out of national agricultural research institutes and universities are often not taken up by farmers. For example, despite the availability of improved maize varieties farmers still grow coastal maize landraces in coastal lowland Kenya [3]. According to [4], the disconnect that has been there between the crop varieties which have been released by plant breeders for a particular region and what the farmers in that region grow raised two major challenges. First new varieties can be disappointing to farmers in case undesirable traits go undetected during the breeding process. Second, breeders discard many crosses and varieties during the selection process because of traits considered undesirable. However, these traits may actually be of interest to farmers, thus indicating the communication gap between breeders and farmers [5].

The top-down transfer-of-technology approach as a strategy in promoting agricultural programs has not produced the desired results of sustainable and increased crop yields. It has also failed to make use of the significant agricultural biodiversity available and its potential to address food security concerns and issues in the region [1]. The most effective way to ensure adoption of improved cultivars is involving farmers in the process of development. Participatory varietal selection (PVS) is a simple way for breeders and agronomists to learn which varieties perform well on-station and on-farm and to obtain feedback from the potential end users in the early phases of the breeding cycle. It is a means for social scientists to identify the varieties that most men and women farmers prefer, including the reasons for their preference and constraints to adoption [6]. It has shown success in identifying an increased number of preferred varieties by farmers in a shorter time than the conventional system, in accelerating their dissemination and increasing cultivar diversity [7].

[8] reported that complex farming systems, adapted to the local conditions, have helped small scale farmers to sustainably manage harsh environments and to meet their subsistence needs, without depending on mechanization, chemical fertilizers, pesticides or other technologies of modern agricultural science. According to [1] a small farmer deals with a variable environment and has multiple production objectives that will affect his or her choice of crops and selection of genotypes. Next to yield, which in formal breeding programmes is by far the most important objective, yield stability, adaptation to production techniques and conditions, and various consumption purposes are selected for. This range of objectives often results in the use of a large number of varieties by individual farmers and the use of genetically heterogeneous varieties. Farmers need adaptation to the local and variable water and soil conditions in combination with a variety of characteristics related to labour and food availability, intercropping and weed competition [9]. Consumption objectives include culinary and cultural preferences regarding taste, color, consistence, size, cooking time, processing quality and suitability for preparation of traditional dishes or drinks. Consumption criteria also include secondary uses, such as leaves of sweet potato, cowpea and cassava as vegetable or animal feed [9]. Cowpea is the most important grain legume in coastal Kenya, mainly grown as an intercrop with maize [10,11]. There are many

cowpea varieties with potential for high yields tolerance to drought and resistance to insects in the region but have not been evaluated by farmers. The current study was carried out to identify cowpea varieties most preferred by coastal farmers through participatory variety selection.

2. MATERIALS AND METHODS

2.1 Experimental Site

The participatory varietal selection study was carried out on-farm at Majaoni in Tezo location, Kilifi County which is 68 km north of Mombasa. It lies between latitudes 3° S and 4° S and longitudes 39° E and 40° E. The region receives an average annual rainfall of 600–1100 mm that comes in two seasons [12]. The long rains are received in March/April through August while the short rains are received in October, November and December. The long rains season is the most important cropping season and 75% of the annual rainfall is usually received during this time [13]. Mean monthly minimum and maximum temperatures are about 22°C and 30°C, respectively, and the mean relative humidity is 80% [14]. According to [12] the soils in coastal lowland Kenya are mostly ferralsols. These soils have low organic matter content, are deficient in essential plant nutrients (especially nitrogen), are prone to leaching, and have a pH ranging between 5 and 7 [15].

2.2 Experimental Design, Treatments and Crop Husbandry

Needs of farmers were assessed using the participatory rural appraisal technique to set goals and identify farmer's preferences and perceptions on ideotype cowpea cultivar [16]. Farmers who participated and evaluated the cowpea varieties were representative of the area. The participant farmers were selected based on their indigenous knowledge [1] and from the most active farmer groups growing maize and cowpea in the region. The farmer groups were identified through the County Director of Agriculture's office in Kilifi. Participants in the participatory varietal selection were drawn from Lamukani, Dziuzeni and Sife moyo farmer groups, all in Bahari Division, Kilifi County. The farmer groups were briefed by the researcher on what was involved in varietal selection exercise. Only farmers from each group who accepted to take part in the study were involved in participatory variety selection. Traits preferred by farmers were identified and prioritized for both men and women. Cowpea varieties were established and evaluated in one of the farmer group's field in Tezo location based on yield performance and farmers' selection criteria, according to [1]. The test varieties included the most popular local varieties and improved varieties which are recommended for the region. Cowpea varieties were laid out in a randomized complete block design (RCBD) and replicated three times.

The cowpea varieties evaluated were (plate 1 - 11): (i). KVU 419 (improved variety from KALRO Katumani); (ii). Khaki (local variety); (iii). K80 (improved variety for the coastal region); (iv). Macho (local variety); (v). Kaima-koko (local variety); (vi). Nyeupe (local variety); (vii). KVU 27-1 (improved variety from Katumani); (viii). Nyekundu (local variety); (ix). M 66 (improved variety from Katumani); (x). Kutambaa (local variety); and (xi). Mwandato (local variety). Each variety was grown in a 5 m x 5 m plot size. The plant spacing was 60 cm by 30 cm with two seeds per hill. Land was ploughed using a tractor and the farmers participated in the leveling and planting. There were two weedings: Three and seven weeks after planting, respectively. Insect pests were controlled by spraying with a pesticide cypermethrin 100/1 EC (pestox ® 100 EC) two weeks after planting then fortnightly up to podding stage. Pestox insecticide, whose active ingredient is cypermethrin, is a synthetic pyrethroid that belongs to a group of insecticides used widely as industrial and agricultural pesticides [17].

Thirteen (13) farmers (10 female and 3 male) from each farmer group participated in the exercise. The farmers participated in land preparation, crop establishment, weeding, pest control and harvesting. Evaluation was done before flowering and at podding, maturity and post-harvest stages. The female and male farmers were assisted to develop selection criteria independently and then jointly. At every evaluation time all farmers were given voting cards. Males were given red cards while the females were given green cards. All the farmers were given a voting card to pick the variety which was best for a certain trait, before choosing the second best and so on. After harvest, all the varieties were cooked in pots before the organoleptic test. During cooking, the water boiled in the pot first before the cowpeas were placed in. The time taken from when the cowpeas were put in the boiling water to when they were ready for eating was recorded. Farmers were later allowed to taste all the varieties and rank them in order of decreasing palatability. Cooked cowpea samples were scored by semi-trained sensory panel using a modified version of quantitative descriptive analysis (QDA) since standards were not provided [18]. The sensory panel, which consisted of ten (10) panelists, was conducted at one of the farmer's homes near the farmer group field where the evaluation was conducted [19]. The languages used during the sensory testing were Kiswahili (national language) and Giriama (local language). The panelists had been screened for familiarity with the cowpea dish and ability to determine differences between cowpea dishes from different cowpea varieties. Sensory attributes evaluated were color and taste.

2.3 Data Collection

Data were collected before flowering, at podding, maturity and post harvest stages. Before flowering and at podding the farmers evaluated the varieties on the basis of grain yield (pod load), drought tolerance (based on leaf phenoology such as leaf senescence) [20], time to maturity, and ease of harvesting (plants with large pods were considered easy to harvest). At maturity and post harvest stages the farmers evaluated the varieties based on grain color, grain yield (in terms of kg per plot), taste (organoleptic test after cooking) and cooking time (time taken to cook in boiling water). Each criterion was scored on a scale of 1 to 5 (1 = very poor, 2 = poor, 3 = average, 4 = good and 5 = very good) for each variety [4].

Plate 1. KVU 419 cowpea variety

Plate 2. Nyeupe cowpea variety

Plate 3. Macho cowpea variety

Plate 4. Kaima koko cowpea variety

Plate 5. Khaki cowpea variety

Plate 6. K80 cowpea variety

Plate 7. Kuhambala cowpea variety

Plate 8. M66 cowpea variety

Plate 9. Mwandato cowpea variety

Plate 10. KVU 27-1 cowpea variety

Plate 11. Nyekundu cowpea variety

2.4 Data Analysis

Collected data were analyzed by the general linear model (GLM) procedure for analysis of variance using SAS statistical package (SAS Institute, 1993). Where the F values were significant, means were compared using the least significant difference (LSD) test, at p = 0.05.

3. RESULTS

3.1 Farmers' Rating of Cowpea Varieties before Flowering and at Podding Stages

The male farmers' selection criteria in order of decreasing preference were early maturity, high grain yield, drought tolerance, ease of harvesting and leafiness (Table 1). The female farmers' selection criteria in order of decreasing preference were high grain yield, drought tolerance, pest and disease resistance, ease of harvesting and leafiness. Both men and women ranked leafiness fifth. Early maturity was a top priority for the men but it did not feature in the women's criteria, while pest and disease resistance featured in the women criteria as a priority but not in the men selection criteria. Farmers' collective selection criteria knocked out the female farmers' criteria of pest and disease resistance and reorganized the male farmers' criteria by considering the female farmers' priority criteria of high grain yield and drought tolerance. Most of the traits in the farmers' selection criteria, such as high grain yield, drought tolerance and early maturity, were similar with the traits considered by plant breeders. The farmers' selection criteria differed from those of plant breeders in that the former included traits such as ease of harvesting and leafiness (Table 1).

In Table 2, male farmers rated M66 to be earliest maturing cowpea variety followed by KVU 27-1 and Kutambaa. The female farmers rated KVU 27-1 and Kutambaa the best for that trait followed by Nyekundu. Kutambaa was rated the highest yielder by both male and female farmers, but male farmers also ranked Kaima koko as the highest yielder.

The female farmers rated KVU 27-1 as drought tolerant while male farmers rated Kutambaa as drought tolerant. Khaki variety was rated the best in terms of ease of harvesting by male farmers while female farmers ranked Nyeupe as the best in this attribute. The mean scores in Table 2 indicate that the best varieties were Kutambaa, KVU 27-1 and Nyeupe with mean scores of 1.9, 2.3 and 3.0, respectively.

3.2 Farmers' Rating of Cowpea Varieties at Maturity and Post Harvest Stages

The variety selection criteria used by farmers at maturity and post harvest stages were grain color, taste, grain yield and cooking time (Table 3).

Unlike most plant breeding programmes, farmers considered taste and cooking time as key criteria in cowpea variety selection. Male and female farmers had the same scores for each of the varieties evaluated. Farmers identified Nyeupe as having the best color, KVU 419 as the highest grain yielder, Macho as tastiest and Khaki as the fastest cooking (Table 3). Across the selection criteria, the most preferred cowpea varieties were KVU 419, Macho and Nyeupe with mean scores of 2.5, 2.8 and 3.3, respectively.

Table 1. Male and female farmers' criteria for selecting cowpea varieties and the most preferred traits

	Criteria		
Ranking	**Men**	**Women**	**Men and women**
1	Early maturity	High grain yield	High grain yield
2	High grain yield	Drought tolerance	Drought tolerance
3	Drought tolerance	Pest and disease resistance	Early maturity
4	Ease of harvesting	Ease of harvesting	Ease of harvesting
5	Leafiness	Leafiness	Leafiness

Note: pest and disease resistance was based on pest and disease damage scores on leaves, pods and grains

Table 2. Farmers' rating of cowpea varieties before flowering and at podding stages

Cowpea variety	Early maturity		High yield		Drought tolerance		Ease of harvesting		Mean score
	Men	Women	Men	Women	Men	Women	Men	Women	
KVU - 419	4	4	2	5	3	5	3	3	3.6
Khaki	3	5	2	5	2	5	1	5	3.5
K80	4	5	2	5	4	3	3	5	3.9
Macho	5	5	2	5	5	5	5	5	4.6
Kaima- koko	5	4	1	3	3	3	3	5	3.4
Nyeupe	4	5	4	2	2	3	3	1	3.0
KVU 27-1	2	1	2	4	3	1	2	3	2.3
Nyekundu	5	2	5	2	5	4	5	3	3.9
M 66	1	4	4	5	3	3	5	5	3.8
Kutambaa	2	1	1	1	1	3	3	3	1.9
Mwandato	5	3	5	2	5	2	5	2	3.6

**Overall Scores: (1 – 5) Scales; 1 = Very good, 2 = Good, 3 = Average, 4 = Below average and 5 = Poor*

Table 3. Farmers' rating of cowpea varieties at post harvest stage

Cowpea variety	Colour (score)	Grain yield (score)	Taste (score)	Cooking duration (minutes)	Cooking (score)	Mean score	Actual grain yield (t/ha)
KVU - 419	2	1	5	39	2	2.5	3.30
Khaki	5	5	5	29	1	4.0	1.40
K80	5	5	5	40	3	4.5	0.90
Macho	3	2	1	63	5	2.8	2.66
Kaima- koko	5	5	3	55	5	4.5	0.48
Nyeupe	1	5	2	52	5	3.3	1.04
KVU 27-1	4	4	5	55	5	4.5	1.70
Nyekundu	5	5	2	57	5	4.3	1.68
M 66	5	3	5	47	4	4.3	1.80
Kutambaa	5	5	4	55	5	4.8	1.20
Mwandato	5	5	5	82	5	5.0	1.06

**Scores scales: (1 – 5); 1 = Very good, 2 = Good, 3 = Average, 4 = Below average and 5 = Poor; cooking duration was the time the cowpea variety took to be ready for eating while cooking score was the score based on the time taken to be ready for eating*

Table 4. Farmers' overall rating of the cowpea varieties

Cowpea variety	Mean scores		Mean scores
	Flowering and podding stages	Maturity and post harvest stages	
KVU - 419	3.6	2.5	3.1
Khaki	3.5	4.0	3.8
K80	3.9	4.5	4.2
Macho	4.6	2.8	3.7
Kaima- koko	3.4	4.5	4.0
Nyeupe	3.0	3.3	3.2
KVU 27-1	2.3	4.5	3.4
Nyekundu	3.9	4.3	4.1
M 66	3.8	4.3	4.1
Kutambaa	1.9	4.8	3.4
Mwandato	3.6	5.0	4.3

**Scores scales: (1 – 5): 1 = Very good, 2 = Good, 3 = Average, 4 = Below average and 5 = Poor;*
Note: ranking dependent on the stage of evaluation

3.3 Overall Ranking of the Cowpea Varieties

Farmers' overall evaluation before flowering, at podding, maturity and after harvest ranked KVU 419, Nyeupe, KVU 27-1 and Kutambaa as the most preferred varieties with mean scores of 3.1, 3.2, 3.4 and 3.4 respectively (Table 4). Improved varieties K80, M66 and local varieties Nyekundu, and Kaima koko were ranked below average.

4. DISCUSSION

From the study the farmers' most preferred trait was grain yield. Similar observations were made by [21] and [22] who reported that high grain yield was ranked top because all the farmers who participated in the evaluation used cowpea mostly for grain production. The different patterns in the gender division of labour and final forms of consumption of cowpea influence men's and women's varietal selection criteria [2]. Women were also concerned about storage quality and this is reflected in the fact that they considered resistance to diseases and insect pests as very important criteria. This is because women have more knowledge of cooking, food processing, preservation and storage [1]. Although most of the variety selection criteria for men and women were similar, the slight variation is an indication that male and female farmers have particular preferences for certain traits. They have different preferences because they are related to the food chain in different ways, and often at different times and places. According to [2] men have more selection criteria for production performance and storage while women have more criteria related to processing and food preparation, where final forms of consumption are also considered. [23] argued that men and women play different roles and responsibilities within the household, in farming, and in society, yet the operational implications are often obscured, by gender bias and ignorance on the part of plant breeders.

Farmers differed from researchers by considering grain color, ease of harvesting, taste and cooking time as key variety selection criteria. According to [24] farmer's selection criteria are based on ecological needs, livelihood uses, and gender, social and economic needs. This shows the significance of involving farmers in varietal selection to complement plant breeders' selection processes; thereby enhancing adoption and diffusion of any recommended variety [21]. At podding stage the farmers selected the high yielding varieties based on pod load. The number of pods per plant has been reported to be the main yield component with direct effects on cowpea grain yield [25].

The low broad sense heritability for pod load implies that the trait is influenced by environmental effects [26]. The varieties which both male and female farmers selected at podding stage were not the ones they selected at maturity and post harvest stages. The overall rating identified four cowpea varieties namely KVU 419, Nyeupe, KVU 27-1 and Kutambaa based on the farmers' most preferred traits of early maturity, pod load, drought tolerance, ease of harvesting, grain color, grain yield, taste and cooking time. From the findings of this study farmers may be inclined to integrate KVU 419, Nyeupe, KVU 27-1 and Kutambaa cowpea varieties in their maize-cowpea intercropping system because they participated in the selection process. The recommended cowpea varieties in the region K80 and M66 were rated below average and this could partly explain why many farmers have not fully adopted these varieties. This study reinforces the suggestions of [16] that farmer's participation in varietal selection improves crop development and adoption [6].

5. CONCLUSION

The farmers' most preferred cowpea traits were early maturity, pod load, drought tolerance, ease of harvesting, grain color, grain yield, taste and cooking time. The four most preferred cowpea varieties were KVU 419, Nyeupe, KVU 27-1 and Kutambaa. The farmers' most preferred cowpea traits differed with those of plant breeders in that farmers' considered traits such as colour, taste, cooking duration and ease of harvesting. Therefore, plant breeders could consider these traits in their cowpea breeding programmes.

ACKNOWLEDGEMENTS

Pwani University Research and Innovation Committee and National Commission for Science, Technology and Innovation are appreciated for providing funds for research.

COMPETING INTERESTS

Authors have declared that no competing interests exist.

REFERENCES

1. Nkongolo KK, Chinthu KKL, Malusi M, Vokhiwa Z. Participatory variety selection and characterization of sorghum (*Sorghum bicolor* (L.) Moench) elite accessions from Malawian gene pool using farmer and breeder knowledge. African Journal of Agricultural Research. 2008;3(4):273-283.
2. Miguel AA. Agroecology: The science of natural resource management for poor farmers in marginal environments. Agriculture, Ecosystem and Environment. 2002;93:1–24.
3. Ndiso JB, Kibe AM, Mugo S, Pathaka RS. Influence of drought stress on growth, yield and yield components of selected maize genotypes in coastal lowland Kenya. International Journal of Agricultural Sciences. 2012;2(6):178–185.
4. Girma A, Teshale A, Hussein H, Tewodrose M, Abdel Rehman MA. Participatory selection of drought tolerant maize varieties using mother and baby methodology: A case in the semi arid zones of the central Rift Valley of Ethiopia. World Journal of Agricultural Sciences. 2005;1(1):22–27.
5. Kamara A, Defoer T, De Groote H. Selection of new varieties through participatory research, use of corn in South Mali. Tropicultura. 1996;14:100–105.
6. Paris TR, Manzanilla D, Tatlonghari G, Labios R, Cueno A, Villanueva D. Guide to participatory varietal selection for submergence- tolerant rice. International Rice Research Institute. 2011;1–111.
7. Weltzien E, Smith ME, Meitzien LS, Sperling L. Technical and institutional issues in participatory plant breeding – Farm the perspective of farmer plant breeding. A global analysis of issues, results and current experience. PPB Monograph No. 1. Cali: PRGA; 2003.
8. Denevan WM. Prehistoric agricultural methods as models for sustainability. Advanced Plant Pathology. 1995;11: 21–43.
9. Almekinders CJM, Louwaars NP, De Bruijn GH. Local seed systems and their importance for an improved seed supply in developing countries. Euphytica. 1994;78: 207-216.
10. Njunie MN, Saha HM, Muli BM, Lewa KK. Soil and water management to enhance food security in coastal Kenya: End of project technical report (2002–2006). Technical report No. 5. KARI Mtwapa; 2007.
11. Ministry of Agriculture (MOA); 2010. Kenya. Annual Food and Crop Situation Report; 2009.
12. Sombroek WG, Braun HMH, Van Der Pouw BJ. Exploratory soil survey report no. E1, Kenya Soil Survey, Nairobi; 1982.
13. Saha HM. Improving resource use under maize – green legume systems in coastal lowland Kenya. Ph. D Thesis. Jomo Kenyatta University of Agriculture and Technology; 2007.
14. Jaetzold R, Schmidt H, Hornetz B, Shisanya C. Farm Management handbook of Kenya. Vol. II/ Part C, Sub-Part C2, coast province. Ministry of Agriculture, Kenya, in corporation with the German Agency of International Cooperation (GIZ); 2012.

15. Mureithi JG, Tayler RS, Thorpe W. Productivity of alley farming with leucaena (*Leucaena leucocephala* Lam. de Wit) and Napier grass (*Pennisetum purpureum* K. Schum) in coastal lowland Kenya. Agroforestry Systems. 1995;31:59-78.
16. Adu Dapaah HK, Asibuo JY, Danquah OA, Asumadu H, Haleegoah J, Asafo Adjei B. Farmer participation in groundnut varietal selection. African Crop Science Conference Proceedings. 2007;8:1435–1439.
17. Chibuike SU, Parker EJ. Comparative effect of "pestox" powdered insecticide on both serum aspartate aminotransferase and alanine aminotransferase of male and female albino rats. Research Journal of Pharmacology and Pharmacodynamics. 2010;2(4):309–313.
18. Tomlins KI, Manful JT, Larwer P, Hammond L. Urban consumer acceptability and sensory evaluation of locally produced and imported parboiled and raw rice in West Africa. Food Quality Preference. 2005;16:79–89.
19. Tomlins K, Ndunguru G, Stambul K, Joshua N, Ngendello T, Rwiza E, Amour R, Ramadhani B, Kapande A, Westby A. Sensory Evaluation and consumer acceptability of pale-fleshed and orange-fleshed sweet potato by school children and mothers with preschool children. Journal of Science, Food and Agriculture. 2007;87(13):2436-2446.
20. Bolanos J, Edmeades GO. The importance of the anthesis-silking interval in breeding for drought tolerance in tropical maize. Field Crops Research. 1996;48:65-80.
21. Adu Dapaah HK, Asibuo JY, Danquah OA, Asumadu H, Haleegoah J, Asafo Adjei B. Farmer participation in groundnut rosette resistant varietal selection in Ghana. In: Fisher, et al. (Eds). New direction for a diverse planet: Proceedings for the 4th International Crop Science Congress, Brisbane, Australia; 2004.
22. Truong TNC, Phan VL, Thelma P. Farmers' participation in rice variety selection. Omonrice. 2007;15:159–163.
23. Franworth CR, Jiggins J. Participatory plant breeding and gender analysis. CGIAR system wide program on participatory research and gender analysis (PRGA Program), Cali, Colombia. 2003;116. (PPB Monograph No. 4).
24. Halleegoah J, Adu Dapaah HK, Danquah OA. Farmers preferences for bambara groundnut ideo type in two districts in the forest savannah transition zone of Ghana. Ghana Journal of Horticulture. 2005;4: 82–92.
25. Almeida WS, Fernandes FRB, Teofilo EM, Bertini CHCM. Correlation and path analysis in components of grain yield of cowpea genotypes. Revista Ciencia Agronomica. 2014;45(2):726–736.
26. Singh P, Narayanna SG. Biometrical techniques in plant breeding. Kalyani Publishers. New Delhi, India. 2000;74–84.

6

Effect of Spacing on Grain Yield and Yield Attributes of Three Rice (*Oryza sativa L.*) Varieties Grown in Rain-fed Lowland Ecosystem in Ghana

Buri Mohammed Moro[1*], Issaka Roland Nuhu[1] and Essien Ato Martin[1]

[1]CSIR–Soil Research Institute (SRI), Academy Post Office, Private Mail Bag, Kwadaso, Kumasi, Ashanti Region, Ghana.

Authors' contributions

This work was carried out in collaboration between all authors. Author BMM designed the study, wrote the protocol and wrote the first draft of the manuscript. Authors BMM and IRN managed the literature searches and analyses of the study while author EAM managed the experimental process. All authors read and approved the final manuscript.

Editor(s):
(1) Eliana L. Tassi, Institute of Ecosystem Studies, National Research Council (ISEâ€ CNR), Italy.
Reviewers:
(1) Crepin b. Pene, Ivorian Academy of Sciences (ASCAD), Ivory Coast.
(2) Tunira Bhadauria, Kanpur University, Uttar Pradesh, India.

ABSTRACT

Row transplanting is one of the good agronomic practices being promoted for lowland rice production in Ghana. The recently introduced 'Sawah' system promotes effective land preparation, good water management and row transplanting as well as the use of hand tools for weed control in lowland rice production. To determine the appropriate crop density, a trial was conducted to evaluate the effect of different spacing on three rice varieties (Sikamo, Jasmine 85, Marshall). The experimental design was a split plot with four replications. Spacing which served as main treatments included: S_1 (20 cm x 20 cm); S_2 (20 cm x 25 cm); S_3 (25 cm x 15 cm); S_4 (15 cm x 15 cm) and S_5 (30 cm x 10 cm). Rice was transplanted at 2 seedlings per stand. Results showed that spacing significantly affected tiller production, number of panicles per m^2, total biomass and paddy yield. Both the number of tillers per stand and number of panicles per m^2 were significantly reduced under closer spacing than wider spacing in the order: $S_1 = S_2 > S_3 > S_4 = S_5$. Highest grain yield was obtained at 20 cm x 25 cm (11.4 t ha^{-1}) and 20 cm x 20 cm (10.9 t ha^{-1}) for all the three varieties while lowest grains yield (3.0" t ha^{-1}) was recorded at 30 cm x 10 cm spacing. For

*Corresponding author: E-mail: moro_buri@yahoo.com

increased yields and easy adoption of simple hand tools for weed control, wider spacing of 25 cm x 20 cm and 20 cm x 20 cm is recommended for rain-fed lowland rice production in Ghana.

Keywords: Ghana; grain yield; plant density; rain-fed lowland; row transplanting; yield attributes; sawah technology.

1. INTRODUCTION

Even though rice has been listed as one of the commodities for increased food security and import substitution for Ghana under the National Rice Development Strategy of Ministry of Food and Agriculture, the country still produces less than 40% of her domestic requirements [1]. The country, however, possesses a huge and vastly untapped potential for lowland rice production with an estimated irrigable land area of over seven hundred thousand ha, out of which less than 20,000 ha is currently under rice cultivation [1]. Rice is cultivated in virtually all the agro-ecological zones across the country.

Rice production in Ghana is categorized according to agro-ecology namely: Irrigated, rain-fed lowland and rain-fed upland. The lowland rain-fed system covers 78% of the arable area; the irrigated system covers 6% while the upland system covers 16%. Maximum levels of production are expected to be obtained through increased cultivation of lowlands, particularly inland valleys using appropriate and affordable technologies. Even though rice production seems to be picking up over a couple of years now, there still remains a considerable gap between production and demand levels. Currently the national rice supply demand gap is bridged by importation. The annual per capita rice consumption in Ghana was 17.5 kg (1999-2001) rising to 22.6 kg (2002-2004) and it is expected to further rise to 63.0 kg by 2015 with demand increasing to 1, 680,000 tons per year by the same period [1].

The forest agro-ecological zone is one of the major rice growing areas and is more endowed with higher rainfall and guaranteed water availability throughout the growing season. The zone has the potential for achieving national self-sufficient in rice production. However, rice producers in Ghana are mostly peasant farmers who are poorly resourced. Production is constrained by such factors as lack of modern technology and/or machinery. Yields are still relatively low. Promoting the local rice industry will therefore enhance output and income of small holder farmers' thus promoting economic growth in these communities. While the national average paddy yield is about 2.0 t ha^{-1}, rice production in some parts of the country has recorded paddy yields of over 5.0 t ha^{-1} with the introduction of the 'Sawah' system [2-5]. The "Sawah" system (bunded, puddled and levelled fields) which employs simple but effective water harvesting and improved land preparation methods leads to high use of nutrients. An evaluation of rice production where "Sawah" was introduced showed significant increases in rice grain yields and higher returns on investment for farmers [4]. There was over a fourfold increase in grain yields from less than 1.0 t ha^{-1} grain under the traditional system of rice cultivation to over 4.0 t ha^{-1} under the "Sawah" system at the end of the first year of its introduction.

Efforts to ensure increases in rice yield through proper water, crop and soil management have been going on for some time now in the country and results have been positive. However, lack of knowledge and/or necessary information on plant population density management practices is still a major constraint. Rice growth is known to be affected both qualitatively and quantitatively by plant population densities. However, many contradictions exist regarding the correct spacing of rice. Many authors [6-8] have indicated that closer spacing of 15 cm x 10 cm, 15 cm x 15 cm and 15 cm x 20 cm were superior to wider spacing of 30 cm x 10 cm, 20 cm x 20 cm and 15 cm x 25 cm by producing more effective tillers per unit area, higher plant height, higher leaf area index and total dry matter accumulation. However, other authors [9,10] on the other hand have reported that wider spacing (25 cm x 25 cm, and 30 cm x 30 cm) produced significantly higher rice tillers, panicles per m^2, longer and weighty panicles, and higher grain yield than closer spacing (15 cm x 15 cm). In Ghana, most rice farmers' practice random spacing and also do not transplant in rows. In the savannah agro-ecological zone, where water limitations are high, direct seeding either through broadcast or dibbling are the preferred methods, while within the forest agro-ecological zone where water is easily available, random transplanting and dibbling are the commonest method. With the introduction of "Sawah" system and other good agricultural practices, row transplanting is being encouraged and promoted. For maximum yields, farmers need to plant at optimum densities. With the introduction of improved varieties and promotion of row transplanting which is being rapidly adopted by farmers coupled with the desire to promote/encourage mechanization (introduction of hand tools for weed control), there is therefore the major challenge of providing options on spacing (planting density options) so that farmers can plant at optimum plant densities if maximum yields are to be obtained.

2. MATERIALS AND METHODS

2.1 Location

The experiment was conducted in a rain-fed valley in Kumasi in the Forest agro-ecological zone of Ghana. The valley lies on latitudes 6° 40' 59" and longitude 1° 37' 0". It is a long valley that runs several kilometres but measures less than 300 m across.

2.2 Experimental Design and Land Preparation

The design was a randomized complete block arranged in a split plot arrangement with four replications. The site was initially slashed and vegetative cover removed. The area was then ploughed using a power tiller. The ploughed site was then divided into four main blocks using bunds (about 100 cm high) to represent four replications. Each block was then divided into five (5) main plots by using bunds to represent the different spacing of S_1 (20 cm x 20 cm), S_2 (20 cm x 25 cm), S_3 (30 cm x 15 cm), S_4 (25 cm x 15 cm) and S_5 (15 cm x 15 cm). Each main plot was again sub-divided to three micro-plots measuring 2.0 m x 2.0 m for the three rice varieties (Sikamo, Jasmine 85, Marshall). Each sub-plot was then manually puddled and levelled.

2.3 Transplanting

Three weeks old rice seedlings of the three varieties were manually transplanted to each micro-plot. Seedlings were transplanted at two per stand (hill) for all the different spacing adopted.

2.4 Mineral Fertilizer Application

Mineral fertilizer was applied at 90 kg N ha^{-1} as Urea, 60 kg P_2O_5 ha^{-1} as Triple Super Phosphates and 60 kg K_2O ha^{-1} as Muriate of Potash. All P and K and 50% N were applied immediately after transplanting. The remaining 50% N was applied at panicle initiation stage.

2.5 Weed Control

Weeds were manually controlled through occasional hand picking.

2.6 Data Collection

2.6.1 Growth characteristics

Number of tillers was counted after maximum tiller formation stage and mean number of tillers determined while plant height was measured at harvest.

2.6.2 Yield characteristics

At maturity, an area of 1.0 m^2 excluding border rows was measured out in each sub-plot, number of panicles counted and harvested (grain and stover). Grain and stover yield were measured and yield per ha estimated. Panicles were also collected from non-border rows and mean individual weight per panicle determined. The weight of 1000 grains was measured using an electronic balance. Grain harvest index (GHI) was calculated as ratio of grain yield to total yield (grain + stover).

2.7 Statistical Analysis

The statistical software STATISTIX 8 was used to analyse the data and LSD at 5% probability level was used as the mean separator.

3. RESULTS AND DISCUSSION

The forest agro-ecological zone has a bimodal rainfall pattern as shown in Fig. 1. Soils of the site have low to moderate levels of soil nutrients. Organic Carbon level was 21 g kg^{-1}, total Nitrogen level was 2.3 g kg^{-1} while available Phosphorus was 5 mg kg^{-1}. Exchangeable Potassium was 0.24 cmol (+) kg^{-1} with effective Cation Exchange Capacity (eCEC) being 11.5 cmol (+) kg^{-1}. Soil clay content was also very low (80 g kg^{-1}). With such low contents of both clay and organic matter, water and nutrient retention capacities are most likely low. Under such conditions, not only is it essential to adopt water and nutrient conservation methods but also the establishment of optimum plant populations if higher grain yields are to be achieved.

3.1 Effect of Spacing on Growth Factors

3.1.1 Tiller number and plant vigour

The number of tillers produced per stand under the different spacing adopted is presented in Fig. 2. Tiller number was significantly affected by spacing. Highest mean tiller number per hill (12.1) was recorded in the 20 cm x 25 cm spacing for all the varieties which was significantly higher than all the spacing but similar to the 20 cm x 20 cm spacing. The 20 cm x 20 cm spacing (10.6) and 25 cm x 15 cm spacing (10.1) had similar number of tillers per hill but the 20 cm x 20 cm spacing produced significantly more tillers than the 30 cm x 10 cm spacing. All the different spacing produced significantly more tillers than the 15 cm x 15 cm spacing. As the distance between plant stands decreases competition for space, light and soil nutrients sets in, resulting in lower tiller production. The number of tillers formed is a major factor that affects grain yield in rice hence grain yield decreases with decreasing number of tillers per hill. In addition, plants under wider spacing (25 cm x 20 cm and 20 cm x 20 cm) exhibited more vigorous growth than plants under closer spacing (30 cm x 10 cm and 15 cm x 15 cm). Growth attributes were significantly affected by spacing. Wider spacing resulted in the production of more tillers per stand than closer spacing. There was a significant increase in the number of tillers per stand with increased spacing. Larger plants with more vigorous growth were produced under wider spacing than closer spacing. Between varieties, however, spacing showed no significant effect as trend was similar.

3.1.2 Plant height

The effect of spacing on plant height is presented in Fig. 3. Mean plant height across the different spacing ranged from a lowest of 97.1 cm (15 x 15 cm) to the highest of 100.1 cm (20 x 20 cm). Between varieties, Sikamo recorded a fairly uniform plant height for all the spacings ranging from 98.5 cm to 102.6 cm. Plant height under Marshall ranged from 95.6 cm (30 cm x 15 cm) to 100.1 cm (20 cm x 25 cm) while Jasmine recorded plant height ranging from 93.1 cm (15 cm x 15 cm) to 102.3 cm (25 cm x 15 cm). Spacing seemed to have a little effect on plant height with no significant differences. While Ogbodo et al. [11] observed that plant height and number of tillers were significantly higher when crops were transplanted at 30 cm x 30 cm than at 10 cm x 10 cm and 20 cm x 20 cm, Nwokwu [12], on the contrary, reported that plant spacing at 40 cm x 40 cm only produced higher number of tillers, number of leaves, shorter days to 50% heading and 1000 grain weight whereas plant spacing at 30 cm x 30 cm recorded longest panicles, more number of panicles and paddy yield but with no significant differences in plant height for the two different spacing. The authors attributed the observation to the presence of adequate plant nutrients in the soil for normal growth. In this study, plant height was less affected by spacing. As indicated earlier, adequate mineral fertilizer was applied to all treatments and therefore adequate nutrients were provided for optimum crop nutrition and growth. Hence such factors that affect plant height were probably not limiting.

3.2 Effect of Spacing on Yield Factors

3.2.1 Number of panicles per m^2

The effect of spacing on the number of panicles produced per m^2 is presented in Table 1. The 20 cm x 25 cm (605) and 20 x 20 cm (563) spacing produced significantly higher panicles per m^2 than all the spacing (15 cm x 15 cm; 30 cm x 10 cm; 20 cm x 15 cm). The lowest number of panicles per m^2 was recorded under 30 cm x 10 cm spacing (318) which was similar to panicles per m^2 under 15 cm x 15 cm spacing. The three varieties, however, recorded similar number of panicles per m^2. For spacing by variety interaction, all the varieties interacted with 20cm x 20 cm and 20 cm x 25 cm spacing to give significantly higher number of panicles per m^2 than the other spacing. Mirza et al. [13] observed that closer spacing reduced the number of effective tillers and increased tiller mortality, hence lower number of panicles. In this study, the more vigorous plants and greater tiller numbers produced under wider spacing would have a better photosynthetic ability with a wider feeding area, more accessibility to soil nutrients and increased availability of nutrient. While under closer spacing, plants were denied more access to solar radiation due to closeness of plant canopies, wider spacing provided a more conducive environment where plants are exposed to more solar radiation which promotes more photosynthetic production and better growth due to increased nutrient availability and uptake.

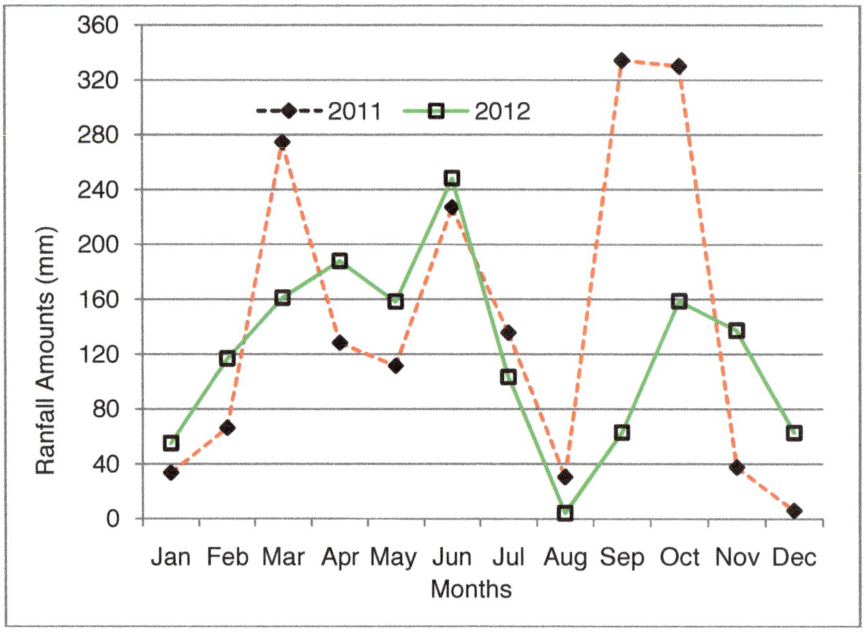

Fig. 1. Rainfall amounts and distribution during the experimental period

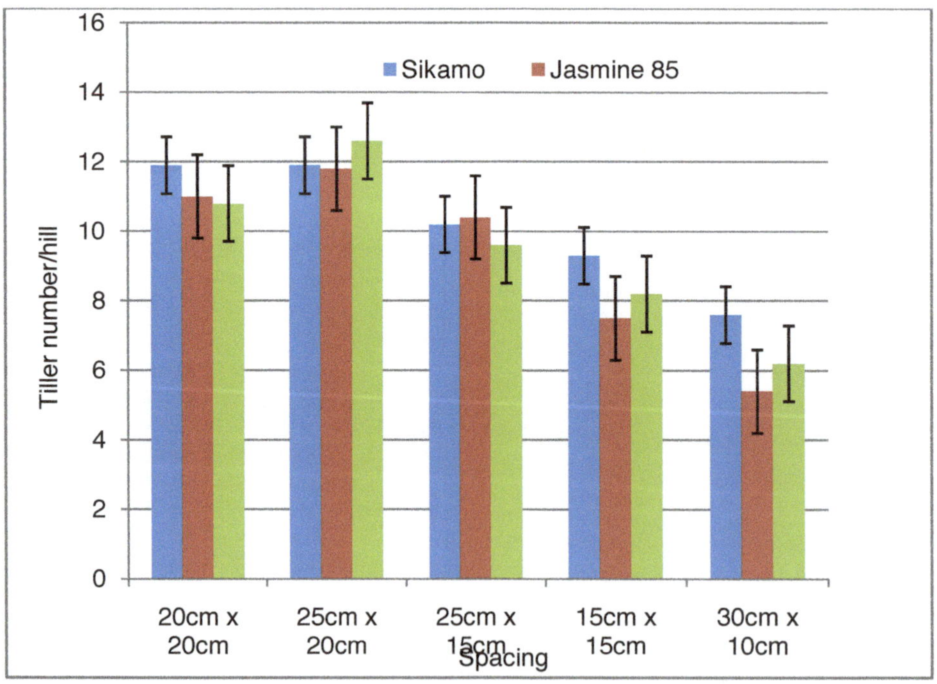

Fig. 2. Effect of spacing on number of tillers per hill of three lowland rice varieties in a rain-fed ecosystem in Ghana

3.2.2 Weight per panicle

The effect of spacing on individual panicle weight is presented in Table 2. Highest mean panicle weight was 5.33 g at a spacing of 20 cm x 25 cm which was similar to mean panicle weight under 20 cm x 20 cm spacing (5.19 g) which were significantly heavier than the panicle weight of the three lower spacing. Mean panicle weight was significantly affected by spacing. The effect of spacing in individual panicle weight was in the order: S2 = S1 > S3 = S4 = S5. Between varieties there was no significant difference in panicle weight. All the three varieties interacted with 25 cm x 20 cm and 20 cm x 20 cm spacing to produce panicles which were significantly heavier than panicles produced under closer spacing of 30 cm x 10 cm sand 15 cm x 15 cm. The general observation was that wider spacing produced heavier panicles than closer spacing.

3.2.3 Weight of 1000 grains

Table 3 shows the effect of spacing on the weight of 1000 grains. There were no significant differences between spacing and varieties. The interaction between spacing and variety gave similar results (no significant difference) between 1000 grain weight.

3.2.4 Grain yield

The effect of spacing on paddy yield is presented in Fig. 4. Higher grain yields were produced under wider spacing (20 cm x 25 cm; 20 cm x 20 cm) than closer (20 cm x 15 cm; 15 cm x 15 cm, 30 cm x 10 cm) spacing. The three varieties interacted with wider spacing (25 cm x 20 cm and 20 cm x 20 cm) to give similar paddy yields of over 10.0 t ha^{-1} which was significantly higher than grain yields produced for any interactions at closer spacing. Grain yield is the ultimate product that is of value and economic importance to the producer or consumer. The three varieties interacted with 25 cm x 20 cm and 20 cm x 20 cm spacing to give significantly heavier and more panicles per m^2. Among others factors, these two combinations of heavier and more panicles per m^2 under wider spacing (25 cm x 20 cm, 20 cm x 20 cm) largely explains the significantly higher grain yields observed for these combinations. Grain yield showed a similar pattern to tiller production and number of panicles per m^2. Grain yield increased significantly with decreasing plant density (Fig. 4). According Ogbodo et al. [11] rice transplanted at 30 cm x 30 cm spacing yielded 2.36 and 1.32 t ha^{-1} significantly higher grains than those transplanted at 10 cm x 10 cm and 20 cm x 20 cm respectively. In this study, a similar observation was made where wider spacing of 25 cm x 20 cm and 20 cm x 20 cm, produced 8.06 and 7.56 t ha^{-1} more grain over the closer spacing of 30 cm x 10 cm and 15 cm x 15 cm. There was less than a 1.0 t $ha-^{1}$ yield difference between 15 cm x 15c m and 30 cm x 10cm spacing. This is in conformity with earlier observations of Bishnu et al. [10] and Mondal and Putch [14] who reported that lower seeding density resulted in the formation of more productive tillers, superior performance for all morpho-physiological and yield components, resulting in higher grain yields over higher

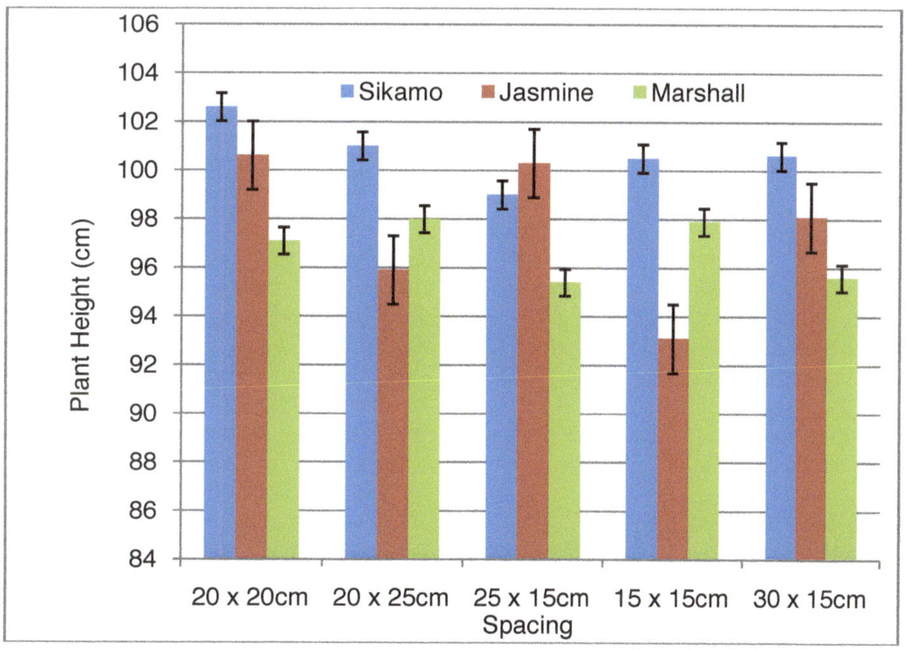

Fig. 3. Effect of spacing on plant height of three lowland rice varieties in a rain-fed ecosystem in Ghana

Table 1. Effect of spacing on number of panicles per m² of three lowland rice varieties in a rain-fed ecosystem in Ghana

Spacing	Rice variety			
	Sikamo	Jasmine 85	Marshall	Mean
20 cm x 20 cm (S1)	597	552	542	563
20 cm x 25 cm (S2)	595	592	630	605
25 cm x 15 cm (S3)	510	522	480	504
15 cm x 15 cm (S4)	467	375	427	423
30 cm x 10 cm (S5)	378	268	308	318
Mean	509	462	477	

LSD ($p \leq 0.05$): Spacing = 96.2; Variety = 73.1; Spacing x Variety = 163.8

Table 2. Effect of spacing on individual panicle weight (g) of three lowland varieties in a rain-fed ecosystem in Ghana

Spacing	Rice variety			
	Sikamo	Jasmine 85	Marshall	Mean
20 cm x 20 cm (S1)	4.93	5.25	5.41	5.19
20 cm x 25 cm (S2)	5.59	5.07	5.34	5.33
25 cm x 15 cm (S3)	4.52	3.93	4.52	4.32
15 cm x 15 cm (S4)	4.46	4.44	4.02	4.31
30 cm x 10 cm (S5)	4.06	3.99	3.94	4.00
Mean	4.71	4.54	4.65	

LSD ($p \leq 0.05$): Spacing = 0.592; Variety = 0.449; Spacing x Variety = 1.011

Table 3. Effect of spacing on 1000 grain weight (g) of three improved lowland varieties in a rain-fed ecosystem in Ghana

Spacing	Rice variety			
	Sikamo	Jasmine 85	Marshall	Mean
20 cm x 20 cm (S1)	26.18	26.50	24.96	25.88
20 cm x 25 cm (S2)	26.92	26.08	25.00	26.17
25 cm x 15 cm (S3)	26.89	24.32	26.12	25.78
15 cm x 15 cm (S4)	25.83	27.51	24.72	26.02
30 cm x 10 cm (S5)	26.77	25.04	25.04	25.62
Mean	26.52	25.89	25.27	

LSD ($p \leq 0.05$): Spacing = 1.877; Variety = 1.313; Spacing x Variety = 3.042

seeding density. As outlined earlier, more vigorous plants and greater number of tillers were produced under wider spacing. This resulted in the production of more dry matter (more photosynthetic material) than the less vigorous plants under closer spacing, thus accounting for the significantly higher grain yields that were recorded under the wider spaced crops over the closer spaced crops.

3.2.5 Dry matter yield (straw + grain)

This is a measure of the total above ground dry matter produced (Fig. 5). Generally, the three varieties interacted with 20 cm x 25 cm spacing to give the highest dry matter which was significantly higher than all the other spacing and varietal combinations. This was followed by dry matter produced due to interaction between any of the varieties and 20 cm x 20 cm spacing. Interaction between the varieties and the three closer spacing gave similar dry matter which was significantly lower than dry matter produced under the two wider spacing. Between varieties, Sikamo produced the highest biomass with Marshall and Jasmine 85 following in that order respectively. Mirza et al. [13] studying the effect of tiller dynamics and dry matter production in transplanted rice as affected by spacing and number of seedlings per hill, observed that wider spacing coupled with higher number of seedlings per hill accumulated maximum amount of dry matter, emphasizing that productivity of tillers as well as dry matter yield was lower with closer spacing and transplanting single seedlings per hill.

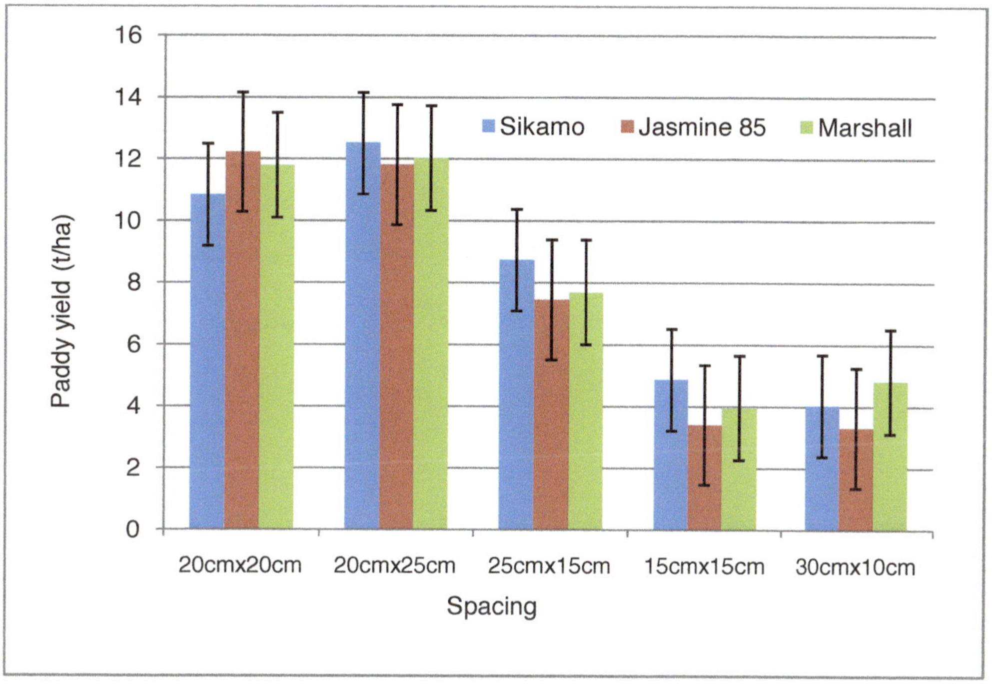

Fig. 4. Effect of spacing on grain yields of three lowland rice varieties in a rain-fed ecosystem in Ghana

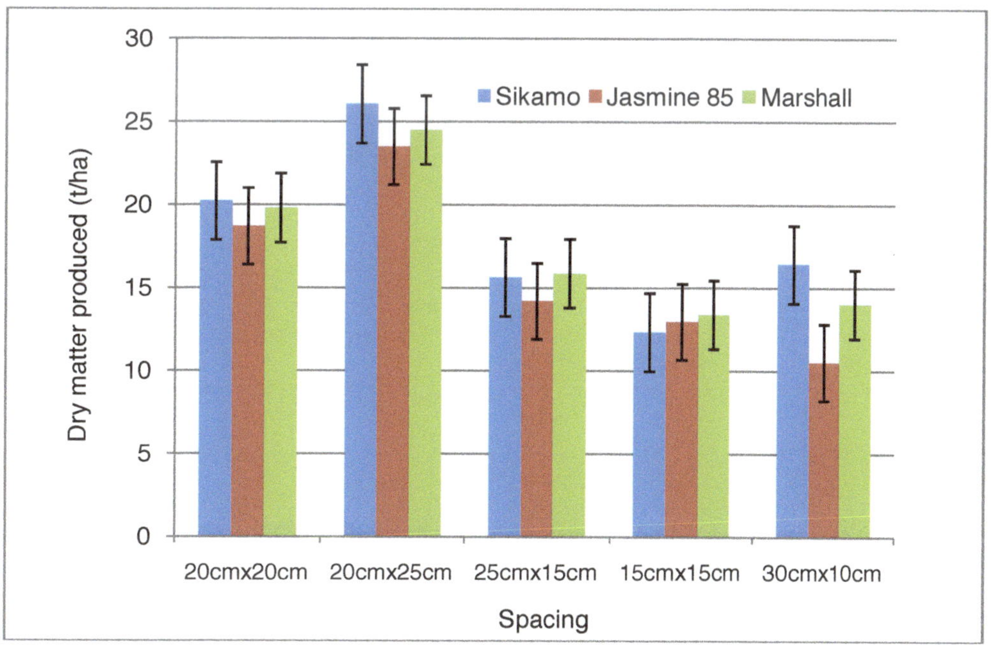

Fig. 5. Effect of spacing on total dry matter production (t/ha) of three lowland rice varieties in a rain-fed ecosystem in Ghana

3.3 Soil and Plant Analysis Development (SPAD) Values

The effect of spacing on SPAD (photosynthetic activity) values during the growth period is presented in Fig. 6. During the first week after transplanting (WAT), SPAD values were relatively similar across the different spacing adopted ranging from 39 (30 x 10 cm) to 41 (20 cm x 25 cm). Increase in SPAD values became rapid

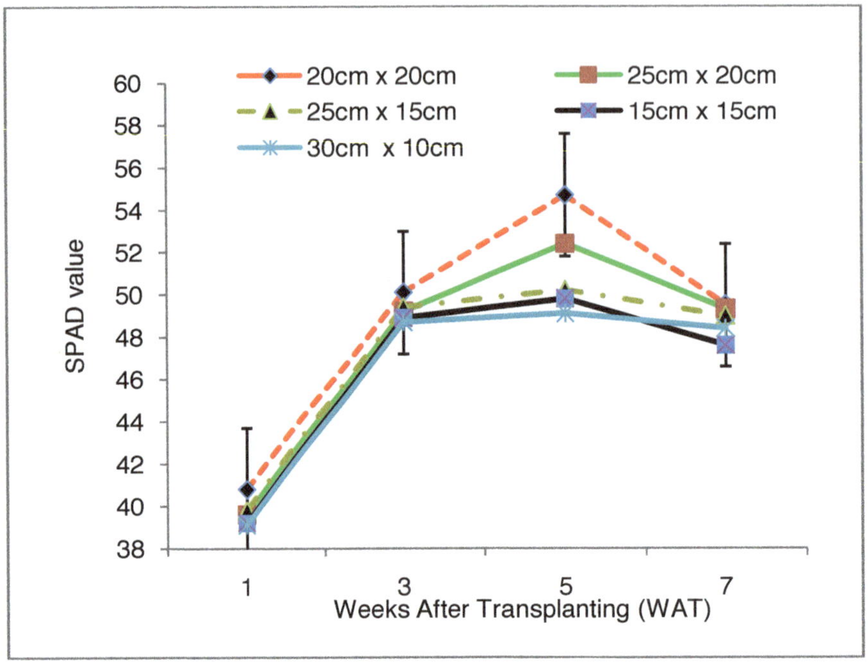

Fig. 6. Effect of spacing on SPAD values of three lowland rice varieties in a rain-fed ecosystem In Ghana

for all spacing up to the third week after transplanting. After the third week after transplanting, wider spacing showed higher SPAD values for the two wider spacing (25 cm x 20 cm, and 20 cm x 20 cm) up to the fifth week and begun to fall sharply till the seventh week after transplanting. Spacing at 25 cm x 20 cm recorded the highest SPAD values throughout the growth period, increasing from 41 during week 1 to 51 in week 3 and to a peak during week 5. This trend was similar to the 20 cm x 20 cm spacing. However, 3 WAT SPAD values for closer spacing showed minimal increases in value in the order: (25 cm x 15 cm) > (15 cm x 15 cm) > (30 cm x 10 cm). By inference, the more vigorous plants and greater number of tillers recorded under wider spacing resulted in more leaves exposed to sunlight, higher and more effective mobilization of photosynthetic matter and better grain filling compared to closer spacing.

4. CONCLUSION

Transplanting at wider spacing produced significantly higher panicles per m^2, heavier individual panicle weights and higher yields than closer spacing. The spacing of 20 cm x 20 cm and 20 cm x 25 cm which corresponds to 20 and 25 stands per m^2 produced significantly higher grain yields and therefore were recommended for the rain-fed lowlands (inland valleys and flood plains). The three varieties, however, showed a similar response pattern to spacing.

ACKNOWLEDGEMENT

The authors are most grateful to all those who contributed in diverse ways towards this study particularly workers of the Soil Chemistry, Fertility and Plant Nutrition Division of the CSIR-Soil Research Institute, Ghana.

COMPETING INTERESTS

Authors have declared that no competing interests exist.

REFERENCES

1. MoFA. National Rice Development Strategy of the Ministry of Food and Agriculture. Ghana; 2010.
2. Buri MM, Issaka RN, Wakatsuki T. Determining optimum rates of mineral fertilizers for economic rice grain yields under the “Sawah” system in Ghana. West African J. App. Ecol. 2008;12;19–31.
3. Buri MM, Issaka RN, Fujii H, Wakatsuki T. Comparison of soil nutrient status of some

rice growing environments in the major agro-ecological zones of Ghana. J. Food, Agric. Environ. 2009;8(1):384-388.

4. Buri MM, Issaka RN, Wakatsuki T, Kawano N. Intensifying and sustaining rice production in inland valley ecosystems in Ghana. Proc. of First International Conference on Rice for Food, Market and Development. (Eds: Onwualu AP, Obinne CPO, Jolaoso MA, Idu EE, Olugbemi BO, Onjewu SS, Sheriff-Musa. M). 2011;87-91. ISBN 978-978-50745-7-4
5. Issaka RN, Buri MM, Wakatsuki T. Effect of soil and water management practices on the growth and yield of rice in the forest agro-ecology of Ghana. J. Food, Agric. Environ. 2008;7(1):214–218.
6. Chandrakar BL, Khan RA. Optimum Spacing for Early, Medium and Late Duration Tall Indica Rice Cultivars. Oryza, 1981;18:108-110.
7. Jashim UM, Sultan A, Harun-ur-Rashid, Mahmudul HM, Asaduzzaman M. Effect of spacing on the yield and yield attributes of transplanted aman rice cultivars in medium Lowland Ecosystem of Bangladesh. J. of Agric. Res. 2011;49(4).
8. Faisul-ur-Rasool, Raihana H, Bhat MI. Agronomic evaluation of rice (*Oryza sativa.* L) for plant spacing and seedlings per hill under temperate conditions. African J. of Agric. Res. 2013;8(37):4650–4653.
9. Weewaroth K, Kadkoa S, Sindhusake C, Chantarated J. Effect of weeding density on gull midge infestation and rice yields in Norther Thailand. IRRN. 1979;4(5):19.
10. Bishnu BA, Biswarup M, Stephan H. Impact of rice nursery nutrient management, seeding density and seedling age on yield and yield attributes. America J. of Plant Sci. 2013;4:146-155.
11. Ogbodo EN, Ekpe II, Utobo EB, Ogah EO. Effect of plant spacing and nitrogen rates on the growth and yield of rice at Abakaliki, Ebonyi Sate, Southeast Nigeria. Research J. of Agric. and Bio. Sci. 2010;6(5):653.
12. Nwokwu GN. Performance of lowland rice (*Oryza sativa*) as affected by transplanting age and plant spacing in Abakaliki, Nigeria. Biology, Agriculture and Healthcare. 2015;5;(9):165-171.
13. Mirza H, Kamrunt N, Roy TS, Rahman ML, Hossain MZ, Ahmed JU. Tiller dynamics and dry matter production of transplanted rice as affected by plant spacing and number of seedlings per hill. Academic J. of Plant Sci. 2009;2(3):162-168.
14. Mondal MMA, Putch AB. Optimizing plant spacing for modern rice varieties. International J. of Agric. & Bio. 2013;15: 175–178.

7

Responses of *Mucuna flagellipes* to Phosphorus Fertilizer Rates in an Ultisol

O. A. Agba[1], E. E. Ikenganyia[2*] and J. E. Asiegbu[2]

[1]Department of Agronomy, Cross River State University of Science and Technology, Obubra, Cross River State, Nigeria.
[2]Department of Crop Science, University of Nigeria, Nsukka, Enugu State, Nigeria.

Authors' contributions

This work was carried out in collaboration between all authors. All authors read and approved the final manuscript.

ABSTRACT

Two field experiments were conducted to investigate the responses of *Mucuna flagellipes* to single super phosphate fertilizer rates. The experiment consisted of five rates of single super phosphate fertilizer namely 0 kg P ha^{-1}, 30 kg P ha^{-1}, 40 kg P ha^{-1}, 50 kg P ha^{-1} and 70 kg P ha^{-1} laid out in a randomized complete block design and replicated four times in 2010 and 2011 cropping season at the Teaching and Research Farm of the Department of Agronomy, Cross River State University of Science and Technology, Obubra, Cross River State, South East Nigeria. The obtained results showed that number of leaves per plant, plant height, leaf area index, number of nodules per plant and nodule dry weight per plant were most significantly ($p < 0.05$) pronounced at 164 days after planting at the highest application rate of 70 kg P ha^{-1} than the other rates in 2010 and 2011 planting season. Pod yield per plant, seed yield per plant and seed yield per hectare were significantly ($p < 0.05$) the highest at 50 kg P ha^{-1}. Generally, there was a consistent significant

**Corresponding author: E-mail: eejike43@yahoo.com*

difference ($p < 0.05$) among the single super phosphate fertilizer rates on the growth and yield of *Mucuna flagellipes*. Application of 50 kg P ha^{-1} and 70 kg P ha^{-1} of single super phosphate fertilizer should be used for the cultivation of *Mucuna flagellipes.*

Keywords: Mucuna flagellipes; single super phosphate fertilizer; ultisol; legume; nodulation.

1. INTRODUCTION

Mucuna flagellipes (vogel ex Hook) is of *Fabaceae* family and *papilioniodeae* sub family [1]. It is a climbing perennial legume with compound trifoliate leaves and is indigenous to Nigeria [2]. The crop is commonly known by the following names: *Ukpo', 'Ibaa' or Okobo* by the Igbos of eastern Nigeria, the *Yorubas* call it '*agbarin*' while the *Hausas* call it *'Karangiwa"* and the *Efiks* call it *"ibaba"* [3]. Unfortunately, the crop is lesser known, under-exploited and has not been fully integrated into the regular farming system. According to [4], the seed is milled into powdery flour and utilized in soup preparation for human consumption. *Mucuna flagellipes* has high nutritional value (55.10% carbohydrate, 21.2% protein, 6.5% dry matter, 2.4% ash and high in minerals) but low in fibre content. The leaves are used to formulate local hair dye [5]. The dye produced from the leaves is used for dyeing pillars of houses and for other home decorations.

Fabaceae is a family with large group of plants that form a cheap source of protein for man and livestock [6]. The group is estimated to contain between 16,000 and 19,000 species in which about 750 genera which add fix nitrogen in symbiosis with *Rhizobia* [7].

Literature shows that the application of phosphorus fertilizers significantly increased the growth and yield of legumes [8-10]. [11,12] working at the humid forest/derived savannah of Nsukka agro-ecology in Nigeria have shown that the soils (ultisols) are predominantly sandy loam, highly leached and acidic. Phosphorus under such a situation constitutes a problem in crop production due to fixation leading to low availability.

There is paucity of literature on the use of phosphorus fertilizer for the regular cultivation of *Mucuna flagellipes.* Therefore, this work aimed to investigate the effects of single super phosphate fertilizer rates on the growth and yield of *Mucuna flagellipes.*

2. MATERIALS AND METHODS

2.1 Description of the Experimental Site

Field trials were conducted at the Teaching and Research Farm of the Department of Agronomy, Cross River State University of Science and Technology, Obubra, Cross River State in 2010 and 2011 cropping season, respectively. Obubra is located at latitude 05° 59' N and longitude 8° 15' E. [13]. Rainfall distribution pattern in this region is bimodal with peaks in July and September and a short dry spell around mid August. The mean rainfall amount is 2000 mm per annum and the mean annual temperature is 27° Celsius with a mean monthly relative humidity of 79%. The experimental site was under two years grass fallow which was cleared and the land ploughed and harrowed in April, 2011.

2.2 Experimental Design and Field Operations

The experiment consisted of five rates of single super phosphate fertilizer namely 0 kg P ha^{-1}, 30 kg P ha^{-1}, 40 kg P ha^{-1}, 50 kg P ha^{-1} and 70 kg P ha^{-1} laid out in a randomized complete block design and replicated four times in 2010 and 2011 cropping season respectively. A total of 20 beds were made, each measured 4 m by 4 m (16 m^2). Each of the beds was separated from one another by a spacing of 1 m and 1 m pathway between replications. *Mucuna flagellipes* seeds were pre-soaked in water at room temperature for 24 hours. Those seed that floated on the water were removed while those seeds that sank in water were selected for sowing. Planting was done on 21st April in 2010 and 2011 cropping season respectively at a depth of 0 - 10 cm and at the rate of two seeds per hole at a plant spacing of 1.0 m by 0.6 m (inter and intra row spacing respectively). Seedlings were thinned to one plant per stand after two weeks of field emergence. Single super phosphate fertilizer rates (0 kg P ha^{-1}, 30 kg P ha^{-1}, 40 kg P ha^{-1}, 50 kg P ha^{-1} and 70 kg P ha^{-1}) were randomly applied to the experimental plots by side banding (5 cm distance from the seedling) at 60 days after planting. Vertical staking of the seedlings was done at three weeks after planting using bamboo stakes. Weeding was done manually by the use of traditional hoe. Basal dressing with 100 kg K ha^{-1} of potassium was applied to all the experimental plots.

2.3 Soil Sample Collection and Analysis

Soil samples were collected with steel auger from the top soil to a depth of 0 to 20 cm before planting. Three representative soil samples were randomly collected per plot and bulked to form a composite soil sample for each plot. A total of 20 composite soil samples were collected. Samples were air dried, ground and passed through a sieve with 2mm standard mesh size. The soil pH was determined with a pH meter using 1:2.5 soil to water ratio and 1: 2.5 soil to 0.1 N KCl (potassium chloride) suspension according to [14]. Organic carbon was determined using the Walkley and Black wet digestion method [15]. Soil organic matter content was obtained by multiplying the value of organic carbon by 1.724 (Van Bemmeler factor). Total nitrogen was determined by micro-kjeldahl procedure [14]. Available phosphorus was extracted with Bray II extractant as described by [16] and determined colorimeterically using ascorbic acid method [17]. Exchangeable potassium was extracted using 1 Nammonium acetate (NH_4OAC) solution and determined by the flame emission spectroscopy as outlined by [18]. Aluminum and Hydrogen content (exchangeable acidity) were determined by titrimetric method after extraction with 1.0 N KCl [19]. The cation exchange capacity was determined by NH_4OAC displacement method [20]. Calcium and magnesium were determined by the complexiometeric titration method as described by [21]. Particle size distribution analysis was done by the hydrometer method [22] and the corresponding textural class determined from the United States Department of Agriculture Soil Textural Triangle. Base saturation was determined by the method outline by [14].

2.4 Growth and Yield Data Collection

Morphological and physiological growth parameters were collected at 45, 60, 75, 90 and 164 days after planting. Number of leaves per plant and number of branches per plant were determined by taking a visual count of the green leaves and the branches, plant height was determined by measuring the length of the plant from the soil level to the tip of the topmost leaf using a measuring tape, Leaf area per plant was obtained by destructive sampling of the leaves per plant taken to the laboratory for leaf area determination using leaf area meter (model Mk – 2). Leaf area index was determined as total leaf area per plant divided by the feeding area available for the plant (inter row spacing multiplied by intra row spacing of each plant) according to [23]. The destructively sampled plants were separated into fractions (leaves, stems, nodules and roots) and put in a paper envelope and oven dried at 80° Celsius to a constant weight for three days for the dry matter determination of, nodules and root. Crop growth rates (nodule growth rate and root growth rate) were determined by the equation stated below.

Crop growth rate (CGR)

$$CGR = \frac{W_2 - W_1}{SA\,(t_2 - t_1)} \ g/m^2/\ day.$$

Where:

CGR = crop growth rate
W_1 and W_2 = dry weight at beginning and end of the interval of growth period
t_1 and t_2 = sampling time 1 and 2.
SA = the area occupied by the plant at sampling.

Yield data were collected after harvest (full maturity). Each plant was harvested separately. Number of pods per plant was obtained by visual counting of pods per plant, number of seeds per plant was determined by visual counting of the seeds in the pod per plant and number of pods without seeds was determined by visual counting pods without seeds. Pod yield per plant, seed yield per plant and seed yield per hectare were recorded at harvest.

Using electronic weighing balance. Days to first and 50% anthesis and pod formation (days to first pod set and 50% pod set)was calculated by counting the number days starting from the day of planting to anthesis and pod formation respectively. Number of inflorescences per plant, number of flowers per plant and number of flowers aborted per inflorescence were computed by visual counting of the inflorescences, flower and flower abortion.

2.5 Statistical Analysis

Data collected were subjected to analysis of variance (ANOVA) as outlined by [24]. Significant means were separated using Fishers least significant difference (F-LSD) at 5% probability level. Statistical analysis was executed using GENSTAT Release 7.2DE Discovery Edition 3 [25] statistical software.

3. RESULTS

The data shown in Table 1 indicated that the soil of the study area before planting was acidic (pH 4.5 and 3.4 in water and potassium chloride respectively in 2010 and 4.3 and 3.2 in water and potassium chloride, respectively in 2011). The soil textural class was sandy loam in 2010 and 2011 planting season, which contained 45% (2010) and 46% (2011) coarse sand, 26% (2010) and 28% (2011) fine sand,26% (2010) and 24% (2011) clay and 2% (2010) and 2% (2011) silt. The organic carbon content was found to be 0.73% (2010) and 0.72% (2011), organic matter content was 1.31% (2010) and 1.24% (2011) and total nitrogen contents were 0.0058% (2010) and 0.0064% (2011). The exchangeable base [sodium 0.8 meg/100 g soil (2010) and 0.8 meg/100 g soil (2011), potassium 0.18 meg/100 g soil (2010) and 0.16 meg/100 g soil (2011) calcium 1.2 meg/100 g soil (2010) and 1.0 meg/100 g soil (2011) and magnesium 0.7 meg/100 g soil (2010) and 0.9 meg/100 g soil (2011).] The cation exchange capacity of the soil was 6.8 meg/100 g soil (2010) and 7.0 meg/100 g soil (2011) for the base saturation 45% (2010) and 43% (2011). The hydrogen 2.6 meg/100 g soil (2010) and 2.4 meg/100 g soil (2011) and aluminum content was found to be 3.4 meg/100 g soil (2010) and 3.6 meg/100 g soil (2011) and available phosphorus (Bray 11) was found to be 8.6 parts per million (2010) and 8.8 parts per million (2011).

Table 2 revealed that the effects of single super phosphate fertilizer treatment on the number of leaves per plant (410.4 in 2010 and 223.1 in 2011), leaf area index (9.16 in 2010 and 5.89 in 2011), number of branches per plant (41.1 in 2010 and 39.4 in 2011) and plant height (554.8 cm in 2010 and 532.4 cm in 2011) of *Mucuna flagellipes* were most significantly ($p < 0.05$) pronounced at 164 days after planting (DAP) than the other periods of observation (45, 60, 75, 90 DAP) at the highest application rate of 70 kg P ha^{-1}. The highest application rate of 70 kg P ha^{-1} at 164 DAP significantly ($p < 0.05$) gave the highest number of nodules per plant (34.2 in 2010 and 32.4 in 2011), nodule dry weight per plant (8.81 g $plant^{-1}$ in 2010 and 6.91 g $plant^{-1}$ in 2011), nodule growth rate (0.931 $g/m^2/day$ in 2010 and 0.913 $g/m^2/day$ in 2011) and root growth rate (1.24 $g/m^2/day$ in 2010 and 1.19 $g/m^2/day$ in 2011), than in the plot which received 0 kg P ha^{-1}single super phosphate fertilizer in 2010 and 2011 cropping season (Table 3).

In Table 4, the number of days to first anthesis (118.7 in 2010 and 125.0 in 2011) and number of days to 50 percent anthesis (135.2 in 2010 and 146.2 in 2011) were significantly ($p < 0.05$) reduced at 50 kg P ha^{-1}as compared with 0 kg P ha^{-1}. The effects of single super phosphate fertilizer rates on number of days to first pod formation (149.9 in 2010 and 160.4.0 in 2011) and number of days to 50 percent pod formation (163.4 in 2010 and 172.6 in 2011) were significantly ($p < 0.05$) reduced at 50 kg P ha^{-1}as compared with 0 kg P ha^{-1}. Application of 50 kg P ha^{-1} of single super phosphate fertilizer rate significantly ($p < 0.05$) enhanced number of inflorescence per plant in 2010 (45.1) and in 2011 cropping season 40 kg P ha^{-1} of single super phosphate fertilizer rate significantly ($p < 0.05$) increased the number of inflorescence per plant (47.3) as compared with 0 kg P ha^{-1}. More so, in Table 4 50 kg P ha^{-1} of single super phosphate fertilizer rate significantly ($p < 0.05$) reduced the number of flowers aborted per plant in *Mucuna flagellipes*in 2010 (3.1) and 2011 (3.0) cropping season respectively as compared with 0 kg P ha^{-1}.

The results in Table 5 indicated, that the number of pods without seeds per plant (4.5 in 2010 and 4.2 in 2011) was significantly ($p < 0.05$) reduced at 50 kg P $ha^{-1}$of single super phosphate fertilizer rate in comparison with 0 kg P ha^{-1}. Furthermore, the number of pods per plant was significantly ($p < 0.05$) the highest (34.6) in plot amended with 40 kg P ha^{-1} as compared with 0 kg P ha^{-1} in 2010 but in 2011, 70 kg P ha^{-1} of single super phosphate fertilizer rate significantly ($p < 0.05$) gave the highest (31.4) number of pods per plant than the other rates. The number of seeds per plant was significantly ($p < 0.05$) the highest (4.6) in plot treated with 50 kg P ha^{-1} as compared with 0 kg P ha^{-1} in 2010 both in 2011, 70 kg P ha^{-1} of single super phosphate fertilizer rate significantly ($p < 0.05$) gave more number of seeds per plant than the control treatment. Pod yield per plant (372.2 g $plant^{-1}$ in 2010 and 368.0 g $plant^{-1}$ in 2011), seed yield per plant (283.5 g $plant^{-1}$ in 2010 and 287.3 g $plant^{-1}$ in 2011), seed yield per hectare of *Mucuna flagellipes* (3.20t ha^{-1} in 2010 and 3.25 t ha^{-1} in 2011), were significantly ($p < 0.05$) the highest at 50 kg P ha^{-1}as compared with 0 kg P ha^{-1}.

4. DISCUSSION

The results of this investigation revealed that *Mucuna flagellipes* benefited from the application of single super phosphate fertilizer in terms of increase in the number of leaves per plant, number of branches per plant, plant height and leaf area index. This finding is in tandem with the observations of [8], who reported an increase in the number of leaves per plant and branches per plant in cowpea as a result of increasing phosphorus fertilizer application rate. *Mucuna flagellipes* growth rate, dry matter accumulation and leaf area increased with successive increase in application rates of single super phosphate fertilizer. This finding is due to greater accumulation of photosynthates in *Mucuna flagellipes* sown in single super phosphate fertilizer treated soil than untreated soil [26].

Fertilization with single super phosphate fertilizer at the rate of 70 kg P/ha in 2010 and 2011, respectively produced the highest number of nodules per plant 34.2 (2010) and 32.4 (2011) and dry weight per plant 8.81 g $plant^{-1}$ (2010) and 6.91 g $plant^{-1}$ (2011) at 164 days after planting, respectively than in the untreated soil. [27], obtained higher number of nodules per plant and dry weights of nodules per plant in groundnut grown on sandy loam soil amended with 38 kg P ha^{-1} of single super phosphate fertilizer than unamended soil.

Amending the soil with single super phosphate fertilizer improved flower production in *Mucuna flagellipes*. Earlier attainment of 50% anthesis and pod initiation, higher percentages of pod set and low flower and pod abortion were observed in single super phosphate fertilizer treated soil than in the untreated soil. Such responses tallied with the results of [28], who reported increased flower production, retention of pods and reduction in the incidences of flower and pod abortion in cowpea with phosphorus fertilizer treatment.

Table 1. Initial soil characteristics before planting in 2010 and 2011 cropping season

Parameters	2010	2011
Particle size distribution (%)		
Coarse sand	45	46
Fine sand	26	28
Clay	26	24
Silt	2	2
Textural class	Sandy loam	Sandy loam
pH (water)	4.5	4.3
pH (KCl)	3.4	3.2
Organic carbon (%)	0.73	0.72
Organic matter (%)	1.31	1.24
Total nitrogen (%)	0.0058	0.0064
Available phosphorus (ppm)	8.6	8.8
Exchangeable bases (meg/100 g soil)		
Calcium	1.2	1.0
Magnesium	0.7	0.9
Potassium	0.18	0.16
Sodium	0.8	0.8
Exchangeable acidity (meg/100 g soil)		
Hydrogen	2.6	2.4
Aluminum	3.4	3.6
Cation exchangeable capacity (meg/100 g soil)	6.8	7.0
Base saturation (%)	45	43

Table 2. Effects of single super phosphate fertilizer rates on vegetative growth of *Mucuna flagellipes* in 2010 and 2011 cropping season

Rates	Number of leaves per plant					Leaf area index					Number of branches per plant					Plant height (cm)				
(kg P ha^{-1})	45	60	75	90	164	45	60	75	90	164	45	60	75	90	164	45	60	75	90	164
	DAP					DAP					DAP					DAP				
2010 cropping season																				
0	9.8	16.8	17.8	27.9	173.3	0.10	0.15	0.29	0.39	3.03	1.6	1.7	2.8	6.6	17.0	122.2	194.3	249.3	284.2	431.2
30	12.1	17.6	20.7	29.2	274.6	0.20	0.22	0.32	0.48	5.92	1.4	1.5	3.6	13.1	29.9	128.3	200.0	250.3	301.5	461.1
40	13.2	17.7	20.2	31.4	309.8	0.22	0.32	0.35	0.49	6.35	1.6	1.7	4.0	15.1	33.4	124.1	203.7	266.9	312.5	481.3
50	13.8	18.0	21.3	35.2	330.3	0.24	0.37	0.39	0.54	7.46	1.6	1.9	4.2	16.7	36.6	128.2	224.0	282.9	331.4	493.2
70	15.2	19.5	25.2	41.3	410.4	0.27	0.39	0.43	0.68	9.16	3.1	2.1	5.2	18.3	41.1	131.6	243.9	317.5	406.2	554.8
$LSD_{(0.05)}$ for 2 rates	1.8	NS	2.1	2.3	11.2	NS	0.02	0.10	0.04	0.2	NS	NS	NS	1.1	1.6	NS	NS	2.1	3.1	2.2
2011 cropping season																				
0	6.3	10.1	12.2	18.4	101.1	0.03	0.11	0.17	0.26	2.10	1.0	1.3	2.0	4.1	11.1	101.5	162.2	206.5	251.3	322.1
30	9.2	13.2	15.5	23.1	152.3	0.11	0.18	0.20	0.34	3.23	1.3	1.4	2.5	7.3	21.3	114.8	183.7	221.7	286.4	413.2
40	11.1	14.3	17.2	27.3	171.5	0.19	0.23	0.27	0.37	4.16	1.4	1.6	3.1	9.2	30.2	120.2	195.1	243.4	301.7	444.8
50	12.3	16.1	18.1	30.2	192.4	0.21	0.29	0.31	0.48	5.57	1.5	1.7	3.5	13.5	33.0	123.6	203.5	264.6	315.5	473.7
70	14.2	17.3	18.9	43.6	223.1	0.32	0.33	0.37	0.51	5.89	1.8	1.9	3.9	16.3	39.4	131.3	241.2	281.6	341.5	532.4
$LSD_{(0.05)}$ for 2 rates	1.1	1.2	1.0	1.5	4.1	NS	0.02	0.11	0.02	0.1	NS	NS	NS	1.1	1.2	NS	1.0	1.3	2.2	1.4

NS – Non significant at 0.05 probability level, DAP – days after planting

Table 3. Effects of single super phosphate fertilizer rates on nodulation of *Mucuna flagellipes* in 2010 and 2011 cropping season

Rates (kg P ha^{-1})	Number nodules per plant			Nodule dry weight (g $plant^{-1}$)			Nodule growth rate ($g/m^2/day$)		Root growth rate ($g/m^2/day$)	
	75	90	164	75	90	164	45-90	90-164	45-90	90-164
	DAP			DAP			DAP		DAP	
2010 cropping season										
0	2.7	3.2	11.1	0.02	0.52	2.02	0.004	0.404	0.63	0.40
30	2.7	3.2	27.3	0.02	0.61	3.61	0.006	0.698	0.65	0.69
40	3.3	4.3	34.3	0.17	0.65	7.75	0.007	0.882	0.82	0.88
50	3.0	6.5	32.3	0.31	0.69	8.39	0.008	0.905	0.91	0.98
70	3.4	7.1	34.2	0.37	0.75	8.81	0.009	0.931	0.97	1.24
$LSD_{(0.05)}$ for 2 rates	NS	0.02	0.4	NS	0.02	0.4	0.03	0.02	0.02	0.02
2011 cropping season										
0	1.2	2.2	8.4	0.01	0.24	1.43	0.803	0.213	0.35	0.49
30	1.8	2.7	15.2	0.02	0.41	2.26	0.004	0.432	0.517	0.48
40	2.4	3.6	21.5	0.12	0.53	5.11	0.005	0.671	0.73	0.10
50	2.8	4.9	29.7	0.23	0.59	6.78	0.006	0.814	0.84	0.88
70	3.2	5.3	32.4	0.31	0.64	6.91	0.009	0.913	0.93	1.19
$LSD_{(0.05)}$ for 2 rates	NS	0.01	0.2	0.01	0.03	0.2	0.03	0.01	0.03	0.01

NS – Non significant at 0.05 probability level, DAP – days after planting

Table 4. Effects of single super phosphate fertilizer rates on flower and pod formation of *Mucuna flagellipes* in 2010 and 2011 cropping season

Rates (kg P ha^{-1})	Days to first anthesis	Days to 50% anthesis	Days to first pod set	Days to 50% pod set	Number of inflorescence per plant	Number of flower aborted per inflorescence
2010 cropping season						
0	140.7	156.8	163.0	178.0	32.3	5.3
30	124.3	146.1	156.7	170.2	37.4	4.2
40	123.3	141.9	153.2	168.6	40.3	4.1
50	118.7	137.3	149.9	163.4	45.1	3.1
70	120.2	135.2	150.3	170.1	38.1	3.5
$LSD_{(0.05)}$ for 2 rates	0.18	1.11	1.33	1.48	0.7	0.1
2011 cropping season						
0	147.7	170.5	171.2	185.5	32.2	5.4
30	135.3	157.4	165.8	180.3	38.1	4.3
40	131.5	150.4	163.2	176.4	47.3	3.4
50	125.0	146.2	160.4	172.6	41.2	3.0
70	129.3	152.4	166.1	174.2	39.3	3.4
$LSD_{(0.05)}$ for 2 rates	0.94	1.35	1.43	1.65	0.8	0.1

NS – Non significant at 0.05 probability level

Application of single super phosphate fertilizer gave higher seed yield and yield parameters in comparison with the unamended soil. This increase in yield and plant size parameters could be related to the greater production of nodules and consequently the higher nitrogen fixation in *Mucuna flagellipes* with single super phosphate fertilizer. This could also be attributed to the greater light interception and high photosynthetic capacity of the plant. [29] had concluded that nitrogen fixation by legumes gave high yields predominantly through its effects on enhancing the leaf area index and net assimilation rate.

Table 5. Effects of single super phosphate fertilizer rates on the yield of *Mucuna flagellipes* in 2010 and 2011 cropping season

Rates (kg P ha^{-1})	Number of pod per plant	Number of pods without seed per plant	Number of seeds per plant	Pod yield per plant (g plant^{-1})	Seed yield per plant (g plant^{-1})	Seed yield per hectare (t ha^{-1})
2010 cropping season						
0	16.3	6.7	3.3	153.2	106.3	0.98
30	25.7	5.3	2.7	257.9	201.3	2.17
40	34.6	5.2	3.2	281.2	236.9	2.71
50	29.4	4.5	4.6	372.2	283.5	3.20
70	27.1	4.6	3.1	243.6	221.5	2.1
LSD $_{(0.05)}$ for 2 rates	3.16	0.51	NS	6.8	11.1	0.05
2011 cropping season						
0	15.2	7.0	3.0	143.3	88.2	0.75
30	22.5	5.3	3.0	197.2	174.7	2.00
40	26.3	4.6	3.1	210.7	231.1	2.50
50	28.0	4.2	3.0	368.0	287.3	3.25
70	31.4	5.1	3.2	227.4	201.4	2.0
LSD $_{(0.05)}$ for 2 rates	2.1	0.30	NS	6.4	10.2	0.03

NS – Non significant at 0.05 probability level

5. CONCLUSION

The results of this study revealed that there was a consistent significant difference ($p < 0.05$) among the single super phosphate fertilizer rates (0 kg P ha^{-1}, 30 kg P ha^{-1}, 40 kg P ha^{-1}, 50 kg P ha^{-1} and 70 kg P ha^{-1}) on the growth and yield of *Mucuna flagellipes* grown in an utisol in south east Nigeria in 2010 and 2011 cropping season, respectively. For optimum crop growth and yield in the cultivation of *Mucuna flagellipes*, the application rate of 50 kg P ha^{-1} and 70 kg P ha^{-1} of single super phosphate fertilizer should be used.

COMPETING INTERESTS

Authors have declared that no competing interests exist.

REFERENCES

1. Polhill RM, Raven PH (eds). Advances in Legume Systematic. Part. I. Royal Botanic Garden. 1981;256.
2. Anonymous. Tropical legumes: Resources for the future. Reports of an Adhoc panel of the Advisory committee on technological innovation. National Academy of Science Washington. D. C. 1979;214.
3. Enwere JN. Foods of plant origin. Afro-orbis publisher Ltd. 1998;245.
4. Ene-Obong HN, Carnovale E. Nigeria soup and condiments. Traditional processing and potential as dietary fibre sources. Food Chemistry. 1992;43:29-44.
5. Okoro C. The darkening material in *Mucuna flagellipes* (*Papilionoceae*), it prevention and prospect of its use as hair coloring. B Pharm. Thesis. University of Nigeria, Nsukka. 1989;63.
6. Henzell EF. The role of biological nitrogen fixation research in solving problems in tropical agriculture. Plant and Soil. 1988; 108:15-21.
7. Allen ON, Allen EK. The leguminosae and nitrogen fixation. University of Wisconsin Press, Madison, Wisconsin. USA. 1989;35.
8. Kumar RM, Pillai. Effect of phosphorus on symbiotic fixation of nitrogen by leguminous crops. Junior Indian Soc. Soil Sci. 1989;16:111-117.
9. Singh GS, Adams M, Kaplan L. Variability of phosphorus intake and translocation in phosphorus *Phaseolus vulgaris* (L.). J. Amer. Soc. Hort. Soc. 1996;102:74-79.
10. Brown AT. Effect of different phosphorus levels on the yield of four varieties of cowpea. Horticultural Society of Nigeria. 2000;15:65-66.
11. Enwezor WO. Soils testing for phosphorus in some Nigeria soils. 3 forms of phosphorus in soil of south eastern Nigeria

and their relationship to plant available phosphorus. Soil Sci. 1977;124(1):27-33.
12. Asiegbu JE. Responses of onion to lime and fertilizer N in a tropical ultisol. Tropical Agric. Trinnidad 1989; 66(2):161-166.
13. Cross River State Agricultural Development Project. Report on wet lands of Cross River State, Nigeria; 1992.
14. Page JR, Miller RH, Keeney DR, Baker DE, Roscoe Ellis JR, Rhoades JD. Methods soil analysis 2. Chemical and Microbiology Properties (2nd Edn.) Madison, Wisconsin, U.S.A. 1982;1159.
15. Bremmer JM, Mulvaaney CS. Total nitrogen. In: Page AL. (eds.). Methods of soil analysis, Part 2. Chemical and Microbial Properties, Second edition Agronomy Series no. 9 Madison, WI, USA, ASA, SSSA; 1982.
16. Bray RH, Kurtz LT. Determination of total, organic and available forms of phosphorus in soils. Soil Science. 1945;91-96.
17. Murphy J, Riley JP. A modified single solution method for determination of phosphate in natural waters. Anal. Chem. Acta. 1962;27:31-36.
18. Anderson JM, Ingram JSI (eds). Tropical soil biology and fertility: A handbook of methods (2nd edition) CAB International. 1993;221.
19. McLean EO. Soil pH and lime requirements. In: Page AL. (eds.). Methods of soil analysis, Part 2. Chemical and Microbial Properties, Second edition Agronomy Series no. 9 Madison, WI, USA, ASA, SSSA; 1982.
20. Rhoades JD. Cation exchange capacity. In; Page AL, Miller RH, Keeney DR (eds.). Methods of soil analysis, Part 2: Chemical methods. Agronomy Monograph no. 9, American Society of Agronomy Madison, Wisconsin, USA; 1982.
21. Chapman HD. Total exchangeable bases. In. Black CA (ed.). Methods of soil analysis, PartT 2. ASA, Madison, USA. 1982;9:902-904.
22. Gee GW, Bauder D. Particle size analysis. In: Dane JH, Topp GC. (eds.). Methods of soil analysis. Part 4, Physical Methods. Soil Sci. Soc. Am. 2002;5:255-293.
23. Redford PJ, Brown WA. Growth analysis formula: Their use and abuse. Crop Sci. 1984;7:171-175.
24. GENSTAT. GENSTAT Release 7.2 DE, Discovery Edition 3, Lawes Agricultural Trust, Rothamsted Experimental station; 2007.
25. Gomes KA, Gomes AA. Statistical producers for Agricultural Research. 2nd edition. John Wiley and Sons. Inc. New York, USA; 1984.
26. Snapp SS, Lynch JP, Toro A. Phosphorus distribution and remobilization in bean plants as influenced by phosphorus nutrition. Crop Sci. 1996;36:929-935.
27. Singh GS, Lamber SPS. Application of phosphorus to legume (cowpea) for higher yields. Farm. J. Calcutta. 1997;14:26-28.
28. Chien SH, Menon RG. Factors affecting agronomic of phosphatic rock and cowpea growth and yield. Fertilizer Research. 1998;41(3):227-234.
29. Cromer RN, Kreedemann PE, Sands PJ, Stewart L. Leaf growth and photosynthetic response to nitrogen and phosphorus in cowpea. Crop Sci. 1993;36:1271-1296.

8

Evaluation of Nitrogen Effects on Yield and Quality of Watermelon {*Citrullus lanatus* (Thunb.) Matsumara & Nakai} Grown in the Coastal Regions of Kenya

Martin Maluki[1], Joshua Ogweno[1] and Robert Morwani Gesimba[1*]

[1]Department of Crops and CHS, Horticulture and Soils, Egerton University, P.O.Box 536 – 20115, Egerton, Kenya.

Authors' contributions

This work was carried out in collaboration between all authors. Author MM designed the study, wrote the protocol and wrote the first draft of the manuscript. Author JO managed the literature searches. Author RMG managed the experimental process. All authors read and approved the final manuscript.

Editor(s):
(1) Fatemeh Nejatzadeh, Department of Horticulture, Faculty of Agriculture, Khoy Branch, Islamic Azad University, Iran.
Reviewers:
(1) Ndereyimana Assinapol, Agricultural University, Coimbatore, India.
(2) Walid Mosa, Alexandria University, Alexandria, Egypt.

ABSTRACT

Field experiment was conducted in KARI, Matuga, Kwale District in the coastal region of Kenya for two seasons to test the response of 'sugar baby' watermelon [*Citrullus lanatus* [Thumb]. Mansf.] to four levels of nitrogen (0,40,80 and 120 kgN/ha). The treatments were laid out in a randomized complete block design with three replications. Nitrogen had positive significant effect on days to flowering, sex expression ratio, number of fruits/plant, fruit weights, firmness, rind thickness, total soluble solids. Application of 80 and 120 kgN/ha increased the fruit sugar content by 23% and 28% compared with zero application. The total fruit yield was highest at 120 kgN/ha and lowest at zero application. For improved growth yield and quality of water melons in the coastal region of Kenya 120 kgN/ha was recommended.

Keywords: Watermelon; 'sugar baby'; nitrogen; growth; yield, quality.

**Corresponding author: E-mail: gesimb@yahoo.com*

1. INTRODUCTION

Watermelon (*Citrullus lanatus*) is one of the most widely cultivated crops in the world [1]. Its global consumption is greater than that of any of the cucurbit family member [2]. It accounts for 6.8% of the world area devoted to vegetable production [3,4]. China is reportedly the leading country in production of watermelon followed by Turkey, United States, Iran and Republic of Korea [1]. There are over 1,200 varieties of watermelon worldwide [5] and a wide variety of watermelons have been cultivated in Africa. Several of these varieties have been recommended for Kenyan range of climate. These include 'Sugarbaby', 'Crimson Sweet', 'Charleston Gray', 'Chilean Black', 'Congo', 'Fairfax' and 'Tom Watson' [6]. However, among these cultivars, only the first three are available in Kenyan markets with 'Sugarbaby' being the most popular because of its sugary taste, smaller size, earlier maturity and higher yields [7]. Watermelons are a good source of water in the diets of human beings, but can also be used for making jams and other preservatives including jell preparation. They are also useful in flavoring drinks and smoothies.

The demand for watermelon in Kenya is higher than production resulting in the fruit being very expensive. With local demand unsatisfied, its potential for export cannot be realized. To meet the local demand and create surplus for export, production of watermelon in Kenya needs to be increased [7]. One of the major challenges currently facing watermelon farmers in Kenya is low yields and poor quality due to heavy or low fertilization especially of nitrogen [7]. Nitrogen fertilization when applied at correct rates increases vegetative growth and hence high yields and quality [8]. The current recommended fertilizers are calcium ammonium nitrate (CAN) at the rate of 80 Kg N/ha and triple super phosphate (TSP) at the rate of 100 Kg P_2O_5/ha and 0 Kg/ha of K_2O [9]. Many farmers have, however, complained of heavy vegetative growth, low fruit yield and poor quality upon use of fertilizer rates. On the other hand, Potassium, has been tested on most Kenyan soils especially in the coastal areas, and found to be enough, thus no need for additional K fertilization [10,11]. Warncke [8] reported that effective and economic crop fertilizer management begins with an understanding its nutrient requirements and the nutrient status of the soil. Crop nutrient requirements vary and in many occasions depend on management practices and variety [12]. Because farmers have complained of low yield and quality of water melons, this research was carried out with the objective of establishing the effect of nitrogen nutrition at different levels on growth, yield and quality of watermelons in the Kenyan coastal region.

2. MATERIALS AND METHODS

2.1 Site and Materials

Field experiments were conducted at KARI, Matuga, Kwale District in the coastal regions for two seasons April to August 2012, February to June 2013. Watermelon variety 'sugar baby' was tested under four levels of nitrogen (0, 40, 80 and 120 kgN/ha). The treatments were laid out in a randomized complete block design replicated three times.

2.2 Soil Testing

Before land preparation, pre-plant soil samples were obtained from the upper soil surface layer (0 - 20 cm), after-planting soil sample were collected on each individual plot treatment bulked together to make a representative sample as per treatment using a 5 cm diameter soil auger. The soil samples were air dried for analysis to establish the initial soil N and P. About 5 gram of the air-dry soil was taken in a glass beaker and 10 ml of distilled water was added. The contents were mixed thoroughly with a glass rod and allowed to stand for 30 minutes. The soil pH was measured using EQUIP-TRONICS Digital pH meter model EQ-610. The soil samples were digested on a labcon digester at 300℃ in a mixture of hydrogen peroxide, sulphuric acid, selenium and salicylic acid [13]. The digests were analyzed for total N and P. The total N content in the digests were determined by Kjeldahl method, 10 ml of the digest solution was taken in the distillation flask, 20 ml of 40% NaOH was added and the NH_3 evolved was collected in a flask containing 4% H_3BO_3. Finally, the distillate was titrated against 0.1N H_2SO. Total phosphorous was determined using the ascorbic acid blue color procedure and the absorbance measured at 880 nm wavelength UV-spectrophotometer [13].

2.3 Plant Tissue Analysis

At 6 weeks after planting (WAP), leaf samples were collected by cutting using a sharp sterilized knife from the three central watermelon plants, oven-dried at 80°C for 48 hours and grounded.

The mineral constituents of the plant were determined by digesting the samples on a labcon digester at 300°C in a mixture of hydrogen peroxide, sulphuric acid, selenium and salicyclic acid [13]. The digests were analysed for total N and P. The total N content in the digests were determined by Kjeltec method using FOSS instrument as described in the ASN3201 as total Kjeldahl nitrogen (TKN). Total phosphorous was determined using the ascorbic acid blue colour procedure and the absorbance measured at 880 nm wavelength UV-spectrophotometer [13].

2.4 Culural Practices

The field was ploughed and harrowed once before planting on the flats in April 2012 and February 2013. Two seeds were planted per hill with a spacing of 1.5×0.6m. The seedlings were later thinned to one per hill two weeks after sowing to give 21 plants per sub-plot. There were three blocks each with 12 sub-plots per block. Each sub-plot measured 4×3m, and each block measured 55m x 3m separated by 3m path, 0.5m and 1m within sub-plots and main plots respectively. Nitrogen fertilizers in the form of calcium ammonium nitrates (26%N) was applied at the rates of (0.40,80 and 120 kgN/ha) was randomly applied in each of the 12 sub-plots comprising the 3 blocks as a top dress in split, first four weeks after planting then three weeks later. Two weeding were carried out manually using hand hoe at 4 and 8 week after planting respectively. Four weeks after sowing the crop was sprayed three times with lamdacyahalothrin 'Karate' (insecticide) and benomyl (benlate) fungicide at the rates of 2 liters and 1.5 kg/ha respectively at 4, 6 and 8 WAP to protect the plants against insect pests and fungal.

2.5 Data Collection and Analysis

Records for each of the 3 central watermelon plants per each sub-plot treatment were taken on the plant attributes include: days to flowering, sex-expression ratio. Ten weeks after planting (WAP) records were also taken on the following plant attributes: fruit number / plant, fruit weight [kg] / plant using a weighing balance, rind thickness [cm] using a ruler, total soluble solids by use of a refractometer using the fruits flesh near the center of the 3 central watermelon in each sub-plot and firmness using a penotrometer. Petiole analysis for N and P was done at early fruit set, fruit ½ sizes and at first harvest. The data collected for each season were subjected to Analysis of Variance (ANOVA) and mean separated using the THSD at ($P = .05$).

3. RESULTS

3.1 Effect of Nitrogen Levels on Yield Components and Yield of Watermelon

3.1.1 Days to flowering

Use of N fertilizer application had significant effect on the days to flowering of watermelon plants (Table 1). In both seasons, the watermelon plants with N levels applied at the rate of 120 kg N/ha had the earliest flowers to appear followed by 80 Kg N/ha and 40 Kg N/.

3.1.2 Sex expression ratio

Significant effects of N levels were observed on sex ratio of watermelon flowers (Table 1). During the first season, N applied at 120 Kg N/ha had the lowest ratio of male to female flowers an indication of more female flowers while those at 80, 40, and 0 (control) Kg N/ha had the highest sex ratios.

3.1.3 Fruit number per plant

Nitrogen had significant effect on watermelon fruit (Table 1). In both seasons, plants with N applied at the rate of 120 Kg N/ha had more fruit followed by 80 Kg N/ha, while N applied at 40 and 0 (control) Kg N/ha had the lowest number of fruit.

3.1.4 Unit fruit weight

Application of N fertilizer had significant effect on watermelon unit fruit weight (Table 1). In both seasons N applied at the rate of 120 Kg N/ha had the heaviest individual fruit followed by 80 Kg N/ha, 40 Kg N/ha while the lowest unit fruit weight was recorded under the control (0 Kg N/ha).

3.1.5 Yield (t/ha)

Application of N fertilizer had significant effect on watermelon yield (Table 2). In both seasons N applied at the rate of 120 Kg N/ha yielded more followed by 80 Kg N/ha, 40 Kg N/ha while the lowest yield was recorded under the control (0 Kg N/ha).

Table 1. Effect of nitrogen levels on yield components and yield of watermelon plants during production in season 1 (April to Aug. 2012) and 2 (Feb. to Jun. 2013)

Nitrogen levels (Kg N/ha)	Days to flowering (days)	Sex expression ratio	Fruit number (no./plant)	Unit weight (Kg)	Fruit weight (Kg/plant)
		a. Season one			
0	33.8 a*	7.52 a*	1.00 c*	1.44 d*	15.95 d*
40	31.8 b	6.41 a	1.07 c	2.06 c	22.93 c
80	29.6 c	6.48 a	1.59 b	2.45 b	27.19 b
120	28.4 d	4.07 b	4.17 a	2.82 a	31.31 a
		b. Season two			
0	33.0 a	9.91	1.78 c	1.61 d	17.89 d
40	31.1 b	10.10	2.11 c	2.17 c	24.06 c
80	29.2 c	8.47	2.85 b	2.64 b	29.42 b
120	27.9 d	4.65	4.92 a	3.01 a	33.41 a

**Means with no letter or followed by the same letter within a column and a season are not significantly different (P = .05) according to the Tukey's honestly significance difference*

3.2 Effect of Nitrogen Levels on Fruit Quality at Harvest of Watermelon

3.2.1 Rind thickness

Application of N fertilizer had significant effect on watermelon rind thickness (Table 2). In both seasons, watermelon plants with N applied at the rate of 120 Kg N/ha had fruits with thin rinds followed by 80 Kg N/ha, 40 Kg N/ha while fruit with the thickest rind was recorded under the control (0 Kg N/ha).

3.2.2 Total soluble solids

Use of N fertilizer had significant effect on total soluble solids (TSS) of watermelon fruits (Table 2). In both seasons, watermelon fruits with N applied at the rate of 120 Kg N/ha had the highest TSS followed by 80 Kg N/ha, 40 Kg N/ha while fruits under the control the lowest TSS (0 Kg N/ha).

3.2.3 Fruit firmness

Applying N and P fertilizer had significant effect on watermelon fruit firmness (Table 2). In both seasons, watermelon fruits with N applied at the rate of 120 Kg N/ha were more firm followed by 80 Kg N/ha, 40 Kg N/ha while fruits under the control were the least firm (0 Kg N/ha).

3.3 Soil Nutrition Status and Tissue Phosphorous Content

The soil and tissue phosphorous content was lower (Tables 3 and 4) than what is considered as ideal (Table 5). The Soil was slightly acidic (Table 3).

3.4 Effect of Nitrogen Levels on Tissue Analysis of Watermelon

The results from watermelon leaf petiole analysis on total nitrogen indicated that higher nitrogen was observed at 40 Kg N/ha and lowest at 0 Kg N/ha in both seasons (Table 6). In general, nitrogen content on watermelon leaves increased with the increasing levels of nitrogen.

4. DISCUSSION

Since it was apparent that the soil was nitrogen deficient and nitrogen application significantly improved watermelon yield components and yield. Nitrogen is known to promote vegetative growth [14]. More leaves translate to better chlorophyll development and higher stomatal conductance hence enhanced photosynthesis. This therefore leads to more photosynthates being manufactured. It is therefore possible that more photosynthates were translocated to the sinks leading to earlier maturity of watermelon fruits, more and heavier fruits subsequently leading to higher yield. The watermelon sex ratio significantly increased with increase in nitrogen up to a point (more than 80 Kgs N/Ha) when further increase led to a reduction. This is in agreement to the findings from Emilson et al. [15] who found out that cucumber grown in increasing levels of nitrogen fertilizers lead to a significant decrease in their sex ratio signifying an increase in female flowers. The decreasing sex ratio leads to a greater potential fruit yield because of the increase in the number of female flowers per plant which develops into fruits.

Furthermore these results are corroborated by the findings of Agba and Enya [16] who recorded higher cucumber yield when nitrogen was applied at the rate of 180 Kg N/ha compared to the control, and with those of Waseem et al. [17] who reported that high dose of nitrogen (100 kg Nha^{-1}) fertilizer significantly maximizes cucumber fruit length, and vine length. In addition, Ahmed et al. [18] and Jiilani et al. [19] reported that an increase in nitrogen application resulted in maximum cucumber and brinjal fruit length and width respectively. The least yield from the control could be attributed to possible depletion of nitrogen from the soil. Whereas the higher yields results can be attributed to the role of nitrogen in creating plant fresh, dry matter and energy-rich compounds which regulates photosynthesis [20].

Table 2. Effect of Nitrogen Levels on fruit quality during Watermelon Production in Season 1 (April to Aug. 2012) and 2 (Feb. to Jun. 2013)

Nitrogen levels (Kg N/ha)	Rind thickness (cm)	TSS (%)	Firmness (KgF)
	a. Season one		
0	0.95 a*	8.9 d*	3.5 d*
40	0.84 b	10.6 c	5.1 c
80	0.65 c	11.4 b	6.8 b
120	0.45 d	12.4 a	8.4 a
	b. Season two		
0	1.05 a	9.1 d	3.7 d
40	0.93 b	10.8 c	5.3 c
80	0.75 c	11.7 b	6.9 b
120	0.57 d	12.6 a	8.8 a

**Means with no letter or followed by the same letter within a column and a season are not significantly different (P = .05) according to the Tukey's honestly significance difference*

Table 3. Soil results for experimental site production in season 1 (April to Aug. 2012) and 2 (Feb. to Jun. 2013)

	Carbon:Nitrogen	Phosphorous	pH (Cacl2)
	a. Season one		
Before planting	22:0.10	9.3 ppm	5.3
After planting	20:0.9	10.4 ppm	4.8
	b. Season two		
Before planting	23:0.09	9.8 ppm	4.9
After planting	24:0.8	10.5 ppm	4.7

Table 4. Plant tissue analysis for water melons grown in Matuga research station in Season 1 (April to Aug. 2012) and 2 (Feb. to Jun. 2013)

	Carbon:Nitrogen	Phosphorous	pH (Cacl2)
	a. Season one		
Before planting	0.5	9.3 ppm	4.7
After planting	1.3	8.4 ppm	4.8
	b. Season two		
Before planting	0.9	9.8 ppm	4.9
After planting	1.3	8.5 ppm	4.7

Table 5. Soil and plant tissue sufficiency ranges

	Carbon:Nitrogen	Phosphorous	pH (Cacl2)
Soil analysis	20:1	20-30 ppm	
Plant tissue analysis	2.5-4.0	0.25-0.7	5.0-6.8

Table 6. Analysis of Leaf Petiole during production in Season 1 (April to Aug. 2012) and 2 (Feb. to Jun. 2013)

Nitrogen levels (Kg N/ha)	Total Nitrogen (%)	Available Phosphorus (%)
	a. Season one	
0	0.92	0.22
40	1.37	0.25
80	1.23	0.24
120	1.35	0.22
	b. Season two	
0	1.00	0.23
40	1.41	0.25
80	1.41	0.25
120	1.40	0.24

In the present experiment there were no nitrogen levels that lead to a decline in yield. According to Hochmuth and Hochmuth [21] excess nitrogen application causes osmotic stress, which can cause oxidative damage injuring many important cellular components, such as lipids, protein, DNA and RNA leading to reduced growth and eventual yield of plants. The decline in yield at very high high N rates could be explained by the fact that a high concentration of soluble N increases the osmotic potential of the soil solution, causing reduction in water uptake by the plant roots [22]. However, Waseem et al. [17] reported that in deficient soils initial application of N increases yield but further increases in N may result in reduction of yield but this is not in all cases. He explains that in some cases excess N may not lead to reduced yield because some plants have the ability to take up N that is not necessarily transformed into dry matter and hence growth. Nitrates absorbed by the plants are normally assimilated in the roots into nitrites then to ammonium and amino acids through the action of nitrate and nitrite reductase enzymes [19] and not all absorbed nitrate ions are reduced and converted into amino acids but are stored in the plant cells as nitrates as influenced by plant nitrogen use efficiency and the amount applied [22]. Using nitrogen fertilizers significantly influenced quality of watermelon fruits at harvest. Quality of the fruit was enhanced with the increase in nitrogen levels. Quantitative analysis revealed the influence of Nitrogen in determining the watermelon quality. This was confirmed through the fruits which were more firm, thinner rind, with significantly increased total soluble solids (TSS) and low acidity as compared to control treatments. These results are in agreement with Aguyoh et al. [23] who reported that increased nutrient application enhances the quality of watermelon fruits. Furthermore Nitrogen helps in starch formation which during ripening is converted to sugar. Nitrogen and Magnesium are integral part of chlorophyll molecule and at higher level they supported the intake of K, hence resulted in the enhanced sweetness of watermelon fruits. Chloroplast contains protein rich in sulphur, manganese maintain chloroplast structure of stroma and grana and at high level of N and its uptake was supported along with water uptake. This resulted to stable chloroplast hence increased chlorophyll content thus higher net photosynthesis leading to increased TSS in the fruits. Nitrogen and phosphorus plays a role in photosynthesis and at higher levels activated potassium and calcium uptake which are responsible for transporting photosynthates from source (leaves) to the sinks (fruits).This enhanced transport led to a good quality fruits with higher TSS, thinner rind thickness and firmer. Water helps in cell wall and membrane development. It is therefore possible that application of nitrogen would have led to higher total soluble solids and firmer fruits respectively. This may therefore be attributed to higher TSS, and firmer fruits due to nitrogen application which are important qualities on watermelon. The application of N at high level improved sugar content compared to zero application. The improved crop growth in response to N application led to higher rate of photosynthesis and therefore higher amount of sugar in the fruit. Application of N at high levels resulted to optimal stomatal conductance. This lead to efficient carbon dioxide assimilation and water use efficiency hence higher photosynthesis and therefore higher sugars in the fruits.

Nitogen at high levels supports manganese uptake leading to enhanced electron transport system hence higher net photosynthesis thus increased TSS in the fruits. Similar results were observed when compound fertilizer applied at different Similar results were observed when compound fertilizer applied at different levels was used in Kenya {NPK (16:16:8)}. Use of compound fertilizer resulted to watermelon fruits with higher total soluble solids and more firm fruits compared with the control [24].

Application of nitrogen influenced availability of nitrogen and phosphorus on both soil and watermelon tissue. Nitrogen and phosphorus availability increased with increasing levels of nitrogen. Calcium ammonium nitrate fertilizers contain calcium ions, this in turn displaces Hydrogen ions in soil solution to adsorption sites resulting into a pH range that favors both N and P availability in the soil, hence increase in nitrogen rates resulted in increase in P and N availability both in soil and leaves (petioles). Agba and Enya [16] found sufficient ranges of recently matured watermelon leaf petiole when nutrient was applied as compound fertilizer (NPK). Therefore, the nutrient uptake of nitrogen was sufficient.

5. CONCLUSION AND RECOMMENDATION

Kwale soils are Nitrogen deficient. Nitrogen fertilizers levels influenced yield and quality of watermelon. Use of N significantly led to more fruits, more female flowers and subsequently higher yield. Quality of watermelon fruits was also significantly improved with thinner rind thickness, and higher TSS. Therefore, this provides coastal farmers with the best nitrogen fertilizer rates that maybe applied in integrated nutrient management for watermelon production.

Based on this finding, further research may be done using more levels and different sources of Nitrogen fertilizers at different areas since ecological conditions may affect the result.

COMPETING INTERESTS

Authors have declared that no competing interests exist.

REFERENCES

1. Huh YC, Solmaz I, Sari N. Morphological characterization of Korean and Turkish watermelon germplasm. 1 Cucurbitaceae 2008, Proceedings of the IXth EUCARPIA meeting on genetics and breeding of Cucurbitaceae (Pitrat M. ed.), INRA, Avignon (France). May 21-24; 2008.
2. Food and Agriculture Organization. Crop Water management-watermelon. Land and Water Management Division. 2007;145.
3. Guner N, Wehner TC. The genes of watermelon. American Journal of Hort. Science. 2004:39(6):1175-1182.
4. Goreta S, Perica S, Dumicic D, Bucan L. Zanic K. Growth and yield of watermelon on polyethylene mulch with different spacings and nitrogen rates. American Journal of Hortultural Science. 2005;40(2): 366-369.
5. Miles C. Icebox watermelons. In: Crop production, vegetable research and extension. Washingston State University. Vancouver Research & Extension Center, Vancouver, USA; 2004.
6. Warncke DD. Nutrient management for cucurbits: Melons, pumpkin, cucumber, and squash. Indiana CCA. Michigan State University. USA; 2007.
7. HCDA. Fruits and vegetables. Horticultural Crops Development Authority Technical Bulletin; 2006.
8. Waseem K, Kamran QM, Jilani MS. Effect of different levels of nitrogen on the growth and yield of Cucumber (*Cucumis sativus* L.). J. Agric. Res. 2008;46:259-266.
9. Ministry of agriculture. Fruits and vegetables. Agricultural Resource Center, Nairobi, Kenya. 2003;120:.
10. Kanyanjua SM, Keter JK, Okalebo RJ, Verchot L. Identifying Potassium-Deficit Soils in Kenya By Mapping and Analysis of Selected Sites. Soil Science. 2006;171(8): 610-625.
11. FURP. The fertilizer Use Recommendation Project (Phase 1). Main Repoort, Methodology and inventory of existing Information. Fertilizer Trial Documentation (FERDOC). Min. of Agriculture, Nairobi, Kenya; 1988.
12. Simonne E, Dukes M, Hochmuth R, Hochmuth G, Studstill D, Davis W. Long-term effects of fertilization and irrigation recommendations on watermelon yield and soil-water nitrate levels in Florida's sandy soils. Acta. Hort. 2002;627:181-186.
13. Okalebo JR, Gathua KW, Woomer PL. Laboratory methods of soil and plant analysis: A working manual. TSBF: Nairobi, Kenya; 2002.
14. Elmstrom GW, Fiskell JGA, Martin FG. Watermelon yield and quality: Effect of fertilizer rate and placement. Proc. Fla. State Hort. Soc. 1973a;86:196-200.

15. Elmstrom GW, Fiskell JGA, Martin FG. Nutrient distribution in soil and watermelon plant uptake: Effect of fertilizer timing, rate, and placement. Soil Crop Sci. Fla. Proc. 1973b;32:154-158.
16. Agba OA, Enya VE. Responses of cucumber (*Cucumis sativa* L) to nitrogen in Obubra, Cross River State. Global Journal of Agricultural Sciences. 2005:4(2): 165-167.
17. Zohary D, Hopf M. Domestication of plants in the old world. 3rd Ed. Oxford University Press; 2000.
18. Ahmed N, Baloch MH, Haleem A, Ejaz M, Ahmed N. Effect of different levels of nitrogen on the growth and production of cucumber. Life Science. Int. J. 2007;1: 99–102.
19. Jilani MS, Afzaal MF, Waseem K. Effect of different nitrogen levels on growth and yield of brinjal J Agric Res. 2008;46:245-251.
20. Hochmuth G, Hochmuth B. Watermelon response to fertilization with controlled-release nitrogen and potassium fertilizers. Fla. Agr. Expt. Sta. Research Report: Suwannee Valley REC 94-04; 1994.
21. Hochmuth GJ, Hanlon EA. Principles of sound fertilizer recommendations. Fla. Coop. Extension Serv. Cir. SL 315; 2010b.
22. Onyango MA. Effect of nitrogen on leaf size and anatomy in onion (*Allium cepa* L.). East; 2002.
23. Aguyoh JN, Audi W, Saidi M, Gao-Qiong L. Growth, yield and quality response of watermelon (*Citrullus lanatus* [Thunb] Mansf. & Nakai) Cv. Crimson Sweet) Subjected to Different Levels of Tithonia Manure. International Journal of Science and Nature. 2010;1(1):7-11.
24. Gichimu BM, Owuor BO, Dida MM. Agronomic performance of three most popular commercial watermelon cultivars in Kenya as compared to one newly introduced cultivar and one local landrace grown on Dystric Nitisols under sub-humid tropical conditions. ARPN Journal of Agricultural and Biological Science. 2008; 3:65-71.

9

Effect of Drought Stress on Canopy Temperature, Growth and Yield Performance of Cowpea Varieties

J. B. Ndiso[1*], G. N. Chemining'wa[2], F. M. Olubayo[2] and H. M. Saha[1]

[1]Department of Crop Sciences, Pwani University, P.O.Box 195 – 80108, Kilifi, Kenya.
[2]Department of Plant Science and Crop Protection, University of Nairobi, P.O.Box 30197 00100 GPO, Nairobi, Kenya.

Authors' contributions

This work was carried out in collaboration between all authors. All authors read and approved the final manuscript.

Editor(s):
(1) Yong In Kuk, Department of Development in Oriental Medicine Resources, Sunchon National University, South Korea.
Reviewers:
(1) Ilkay Yavaş, Menderes University, Turkey.
(2) Guilherme Moraes Ferraudo, CanaVialis/Monsanto, Campinas, Brazil.

ABSTRACT

Aim: To evaluate the influence of drought stress on canopy temperature, growth and yield performance of cowpea.
Study Design: The experiments were laid out in a split plot design, with main plots arranged into three replications as a randomized complete blocks design. Water stress level was assigned to the main plot while the cowpea variety was assigned to the subplot.
Place and Duration of Study: Study was carried out at Pwani University research farm in July – October 2011/2012 cropping season.
Methodology: The treatments comprised three water stress levels (no water stress, water stress at vegetative stage and water stress at flowering stage) and 11 cowpea varieties: KVU 419, Khaki, K80, Macho, Kaima koko, Nyeupe, KVU 27-1, Nyekundu, M66, Kutambaa and Mwandato. The data collected included: Ground cover, canopy temperature, chlorophyll content, leaf number, days to anthesis, shoot dry matter at maturity, pods per plant, grains per pod, 100-grain weight and grain yield.
Results: Water stress imposed at vegetative growth stage and flowering reduced cowpea growth attributes, ground cover and chlorophyll content, but increased canopy temperature, time to

**Corresponding author: E-mail: j.ndiso@pu.ac.ke, j.ndiso@yahoo.com*

anthesis, harvest index, grain yield and yield components for most varieties. Water stress at vegetative and flowering stages increased time to anthesis by 4 and 7 days, respectively.
Conclusion: The impact of water stress on growth is dependent on the cowpea variety. Moderate stress may be beneficial if cowpea is grown for grain production but not if grown for vegetable production. Cowpea varieties which were superior in yield and high harvest indices under water stress included Nyekundu, KVU 27-1, M66, and KVU 419.

Keywords: Cowpea; varieties; growth; yield; water stress.

1. INTRODUCTION

Cowpea (*Vigna unguiculata* (L) Walp) is an important food legume and multipurpose crop [1]. With its high protein content of 25% [2], cowpea may be regarded as a very nutritious food legume for many ethnic communities who use it in their diets [3]. All the plant parts that are used for food are nutritious, providing proteins, vitamins and minerals [4]. The crop is grown throughout the tropical and subtropical areas of the world, where rainfall resources are characteristically low (300 – 600 mm) and variable [5]. Generally, cowpea is better adapted to drought, high temperatures and other biotic stresses than most other crops [6]. It grows well in a wide range of soil textures, from well drained heavy clays to sandy soils, and grows best in slightly alkaline soils (pH 5.5 – 6.5). Cowpea naturally grows under wide and extreme moisture conditions and, once established, it is fairly drought tolerant [7]. It is often grown under rain-fed agriculture in areas receiving at least 600 mm annual rainfall.

Many cultivars of cowpea are, however, damaged by drought and high temperatures, especially during reproductive development [8]. According to [9], the combination of high temperature, drought and long hours of day can slow down or inhibit floral bud development, resulting in few flowers being produced and substantially reduced cowpea productivity. [10] Reported that cowpea yield reduction ranged from 63% to 98.4% under severe water stress, 42.6% to 65.8% under moderate water stress and 9.5% to 47.2% under mild water stress. Under water deficit conditions, as is often the case in the semi-arid zones, the flowering period is cut short and the seed matures earlier. Moreover, the formation of new floral nodes and flowers are delayed and/or aborted, thus leading to low productivity [11]. In addition, cowpea is also sensitive to drought at different stages of growth [8].

Cowpea response to drought stress varies with variety, economic portion of the crop, stage of growth when stress is imposed and the duration of the stress. Earlier studies indicated that cowpea could maintain seed yield when subjected to drought at vegetative stage, provided subsequent conditions were conducive for flowering and pod set [12]. [13] showed that the crop is highly sensitive to water deficits during flowering and pod filling stages. It has been reported by [14], that water stress has a significant adverse effect on the growth and biological nitrogen fixation of cowpea. [15] reported that a decrease in soil water potential can markedly affect root hair and retard nodule growth and nitrogen fixation.

Stability in yields of agronomically acceptable cultivars is generally regarded as the ultimate goal in cowpea improvement [16]. One way to obtain this is to identify genotypes with adequate levels of resistance to drought, heat and other stresses. There is need for cowpea cultivars, which are more tolerant to water deficit or more efficient in water use [17]. However, progress in breeding cultivars for dry environments has been slow [18]. Cowpea possesses high yield plasticity under diverse environments, and could alleviate the economic hardships of farmers in case of severe drought and heat [3]. The objective of the study was to determine the influence of drought stress on canopy temperature, growth and yield of cowpea varieties in coastal lowland Kenya.

2. MATERIALS AND METHODS

2.1 Study Site

The study was carried out on-station at Pwani University (PU) in coastal lowland (CL) Kenya; located at 60 km north of Mombasa, between latitudes 3° S and 4° S and longitudes 39° E and 40° E. The region receives an average annual rainfall of 600–1100 mm that comes in two seasons [19]. The long rains are received in March/April through August while the short rains are received in October, November and December. The long rains season is the most important cropping season and 75% of the annual rainfall is usually received during this time

[20]. Mean monthly minimum and maximum temperatures are about 22°C and 30°C, respectively, and the mean relative humidity is 80% [21]. According to [19], the soils in coastal lowland Kenya are mostly ferralsols. These soils have low organic matter content, are deficient in essential plant nutrients (especially nitrogen) and prone to leaching, and have a pH ranging between 5 and 7 [22]. The study was conducted during dry seasons of 2011 and 2012.

2.2 Experimental Design, Treatments and Crop Husbandry

The experiments were laid out in a split-plot design, with main plots arranged into three replications as a randomized complete blocks design. The main plots consisted of the water stress level while the subplots consisted of the cowpea varieties. The water stress levels were: Well watered (maintained at field capacity), water stress at vegetative stage and water stress at flowering stage. The sub-plots consisted of seven local and four improved varieties: (i). KVU 419 (improved variety from KALRO Katumani); (ii). Khaki (local variety); (iii). K80 (improved variety for the region); (iv). Macho (local variety); (v). Kaima-koko (local variety); (vi). Nyeupe (local variety); (vii). KVU 27-1 (improved variety from KALRO Katumani); (viii). Nyekundu (local variety); (ix). M 66 (improved variety from Katumani); (x). Kuhambala (local variety); and (xi). Mwandato (local variety). The plant spacing was 60 cm x 30 cm with two seeds per hill, no fertilizer was applied. Weeding was done twice; the first and second weedings were done on the second and fifth week after planting respectively.

Water stress was imposed at vegetative and flowering stages. For the vegetative stage, irrigation was stopped six weeks after planting the late maturing cowpea varieties (Nyeupe, Kutambaa and Mwandato) and three weeks after planting the early maturing varieties (KVU 419, Khaki, K80, Macho, Kaima koko, KVU 27-1, Nyekundu and M66). Water stress was imposed for two weeks. For water stress at flowering stage, the irrigation was stopped eight weeks after planting the late maturing varieties and five weeks after planting the early maturing varieties. The water stress was imposed for two weeks when flowering was 50%. The early maturing varieties were planted three weeks after planting the late maturing varieties to synchronize flowering for the drought to be imposed at the same time [23]. Drip irrigation was applied after every 12 hours for three hours to ensure that soil moisture was maintained at close to field capacity.

2.3 Data Collected

The data collected included chlorophyll content, days to anthesis, grains per pod, number of leaves, percent ground cover, canopy temperature, number of pods per plant, 100-grain weight, pod weight, dry matter and grain yield. Chlorophyll content was measured using a chlorophyll meter before flowering stage. A leaf was selected, put in the leaf chamber of a chlorophyll meter and readings recorded on the screen. Number of days to anthesis was calculated by counting the number of days from planting to 50% flowering. The number of grains per pod was determined by counting the number of grains in each pod at harvest. Ten pods were sampled from each of the ten plants sampled in each plot. Leaf number was determined by counting the number of leaves fortnightly after emergence to flowering stage. Percent ground cover was measured in the net plot (5.76 m^2) at vegetative stage using the string and dot method as described by [24]. A string measuring 10 m length was marked with ink every 15 cm and stretched across both diagonals of the plot. The number of marks lying over or under a living plant part were counted and recorded. Percent ground cover was calculated as:

% Ground cover = [(Number of marks over or under plant part/Total number of marks across plot diagonals) x 100]

Canopy temperature was taken in the middle part of the crop canopy using a canopy temperature meter prior to flowering stage. The number of pods per plant was determined from 10 plants in each plot at harvesting time. Weight of 100 grains was determined by weighing 100 grains of the harvested grains in each plot. Pod weight was determined by weighing 10 pods selected from the 10 plants sampled in each plot at harvest time. Total dry matter was determined by uprooting whole plants (together with the pods) at maturity, oven drying and weighing. Grain yield was taken after harvest from the middle part of the drip lines leaving five hills from each end. The area from which the plants were harvested for grain yield determination was 6.4 m^2.

2.4 Data Analysis

Collected data were analyzed by the general linear model (GLM) procedure for analysis of variance using SAS statistical software [25]. Where the F values were significant, means were compared using the least significant difference (LSD) test with 5% of significance level. Linear regression analyses between grain yield and chlorophyll content, canopy temperature, days to anthesis, number of leaves, % ground cover, pods per plant, 100-grain weight, pod weight, and total dry matter were conducted.

3. RESULTS

3.1 Effect of Water Stress at Vegetative and Flowering Growth Stages on Cowpea Percent Ground Cover and Number of Leaves per Plant

Water stress level, cowpea variety and their interaction significantly affected percent ground cover and leaf number (Table 1). Water stress at both vegetative and flowering stages significantly reduced the percent ground cover and leaf number in most cowpea varieties except Macho, Nyeupe and Mwandato. Plants subjected to water stress during flowering had higher percent ground cover and leaf number than plants subjected to water stress during the vegetative stage for all varieties except Kaima koko, M66 and KVU 27-1. Under no water stress, Kutambaa had significantly the highest percent ground cover (99.4%) followed by Mwandoto (85.8%) and Nyeupe (83.6%), while M66 had significantly the lowest percent ground cover (49.2%). Percent ground cover varied from 49.2% (M66) to 99.4% (Kutambaa), 29.6% (Khaki) to 95.2% (Nyeupe) and 40.3% (M66) to 99.4% (Nyeupe) under no water stress, stress at vegetative and flowering stages respectively. Percent ground cover reduction due to water stress ranged from 2.3% to 46.4% at vegetative stage and 6.29% and 37.0% at flowering. Under no stress, Kutambaa had significantly the highest leaf number while Mwandato and Nyeupe had significantly the highest leaf number under water stress imposed during vegetative and flowering stages. Number of leaves per plant ranged from 62 (M66) to 134 (Kutambaa) under no stress, 34 (Khaki) to 128 (Nyeupe) under stress at vegetative stage and 49.3 (M66) to 134 (Nyeupe) under stress at flowering.

Table 1. Effect of water stress at vegetative and flowering stages on percent ground cover and number of leaves per plant of cowpea

Cowpea variety (V)	Ground cover (%)				Number of leaves per plant			
	Water Stress (WS)			V- mean	Water Stress (WS)			V- mean
	Nws	Vws	Fws		Nws	Vws	Fws	
KVU 419	65.4	39.4	43.6	49.5	85.3	48.0	54.0	62.4
Khaki	55.2	29.6	49.2	44.7	70.7	34.0	62.0	55.6
K 80	58.4	39.4	46.4	48.1	75.3	48.0	58.0	60.4
Macho	57.4	56.1	63.8	59.1	74.0	72.0	83.0	73.3
Kaima-koko	73.6	52.6	46.4	57.5	97.0	67.0	58.0	74.0
Nyeupe	83.6	95.2	99.4	92.7	111.3	128.0	134.0	124.4
KVU 27 – 1	51.9	47.8	48.7	49.5	66.0	60.0	61.3	62.4
Nyekundu	65.9	47.8	55.2	56.3	86.0	60.0	70.7	72.2
M 66	49.2	46.4	40.3	45.3	62.0	58.0	49.3	56.4
Kutambaa	99.4	58.9	66.6	75.0	134.0	76.0	87.0	99.0
Mwandato	85.8	94.5	98.0	92.8	114.7	127.0	132.0	124.6
WS-mean	67.8	55.3	59.8		88.8	70.7	77.2	
p-value (V)	0.0001				0.0001			
p-value (WS)	0.0001				0.0001			
p-value VxWS)	0.0001				0.0001			
$LSD_{0.05}$ V	1.5				2.16			
$LSD_{0.05}$ WS	1.0				1.13			
$LSD_{0.05}$ V x WS	2.6				3.7			
CV *main plot* (%)	1.47				1.23			
CV *subplot* (%)	2.68				2.03			

WS –Water stress, Nws – No water stress, Vws – Vegetative water stress and Fws – Flowering water stress

3.2 Effect of Water Stress at Vegetative and Flowering Stages on Cowpea Chlorophyll Content and Canopy Temperature

Water stress significantly reduced chlorophyll content of cowpea while the main effects of variety and the interaction between cowpea variety and water stress had no effect on this parameter (Table 2). Cowpea plants subjected to water stress during the vegetative stage had lower chlorophyll content than non-water stressed plants and plants subjected to water stress during flowering. Water stress during flowering had no effect on chlorophyll content. There were significant differences in canopy temperature due to water stress and the interaction between water stress and cowpea variety (Table 2). Cowpea variety main effect on canopy temperature was not significant. Water stress at vegetative stage significantly increased canopy temperature in Khaki, Kaima koko, and Mwandato but significantly reduced canopy temperature in Macho and Nyekundu. Water stress at flowering increased canopy temperature in K80, M66, Kutambaa and Mwandato. Canopy temperature varied from 20.70°C (Mwandato) to 25.23°C (Macho) under no stress, 20.20°C (Kutambaa) to 24.83°C (Kaima-koko) under water stress imposed at vegetative stage and 22.77°C (KVU 27-1) to 25.60°C (K80) under stress imposed at flowering stage.

3.3 Effect of Water Stress at Vegetative Stage and Flowering on Cowpea Number of Days to Anthesis

Cowpea variety and water stress had significant effects on the number of days to anthesis (Table 3), but their interaction had no significant effect on this attribute. Water stress at vegetative and flowering stages significantly increased the number of days to anthesis. There was no significant difference between water stress at vegetative and at flowering stage in the number of days to anthesis.

Mwandato had significantly the highest number of days to anthesis while KVU 419, Nyekundu and M66 had significantly lower number of days to anthesis than most of the varieties. Kutambaa had a lower number of days to anthesis than all varieties except Mwandato. The number of days to anthesis varied from 44.3 (Nyekundu) to 70.3 (Mwandato). Water stress at vegetative and flowering stages increased time to anthesis by 4 and 7 days respectively.

Table 2. Effect of water stress at vegetative and flowering growth stages on cowpea chlorophyll content and canopy temperature

Cowpea variety (V)	Chlorophyll content index				Canopy temperature (°C)			
	Water Stress (WS)			V- mean	Water Stress (WS)			V- mean
	Nws	Vws	Fws		Nws	Vws	Fws	
KVU 419	54.70	46.43	50.00	50.38	22.67	23.10	23.17	22.98
Khaki	53.37	51.47	56.23	53.69	21.97	24.43	22.93	23.11
K 80	54.43	50.67	50.13	51.74	22.53	23.50	25.60	23.87
Macho	56.13	46.23	55.67	52.68	25.23	21.50	23.27	23.33
Kaima-koko	51.20	47.00	50.77	49.66	22.47	24.83	23.23	23.51
Nyeupe	53.77	51.17	52.23	52.39	22.30	23.10	23.77	23.06
KVU 27 – 1	53.87	48.67	53.63	52.06	22.53	24.37	22.77	23.22
Nyekundu	52.13	45.27	49.53	48.98	24.96	20.47	23.03	22.82
M 66	53.23	46.20	50.20	49.88	21.63	22.33	24.67	22.87
Kutambaa	54.47	48.73	52.47	51.89	20.97	20.20	25.37	22.17
Mwandato	50.77	47.47	54.07	50.77	20.70	23.20	23.83	22.58
WS-mean	53.46	48.12	52.27		22.54	22.82	23.79	
p-value (V)	0.264				0.561			
p-value (WS)	0.0001				0.007			
p-value VxWS)	0.856				0.0003			
$LSD_{0.05}$ V	Ns				Ns			
$LSD_{0.05}$ WS	1.91				0.72			
$LSD_{0.05}$ V x WS	Ns				2.39			
CV *main plot* (%)	2.79				13.65			
CV *subplot* (%)	7.95				7.72			

WS –Water stress, Nws – No water stress, Vws – Vegetative water stress and Fws – Flowering water stress

Table 3. Effect of water stress at vegetative and flowering growth stages on cowpea number of days to anthesis

Cowpea variety (V)	Number of days to anthesis			
	Water Stress (WS)			V- mean
	Nws	Vws	Fws	
KVU 419	42.67	43.00	48.0	44.56
Khaki	44.67	47.33	46.00	46.00
K 80	44.67	45.67	48.00	46.11
Macho	46.67	49.67	48.67	48.33
Kaima-koko	47.67	48.33	49.00	48.33
Nyeupe	58.33	59.67	58.33	58.78
KVU 27 – 1	44.67	46.00	47.00	45.89
Nyekundu	40.00	44.33	44.67	43.00
M 66	42.67	45.67	42.67	43.67
Kutambaa	61.00	62.33	70.00	64.44
Mwandato	63.00	70.33	67.33	66.89
WS-mean	48.70	51.10	51.80	
p-value (V)	0.0001			
p-value (WS)	0.0001			
p-value VxWS)	0.086			
$LSD_{0.05}$ V	2.51			
$LSD_{0.05}$ WS	1.31			
$LSD_{0.05}$ V x WS	Ns			
CV *main plot* (%)	9.81			
CV *subplot* (%)	5.31			

WS –Water stress, Nws – No water stress, Vws – Vegetative water stress and Fws – Flowering water stress

3.4 Effect of Water Stress at Vegetative and Flowering Growth Stages on Cowpea Number of Pods per Plant and Grains per Pod

Cowpea variety, water stress and the interaction between cowpea variety and water stress significantly affected the number of pods per plant (Table 4). Water stress at vegetative and flowering stages significantly increased the number of pods per plant in cowpea varieties. The number of pods per plant varied from 4.3 (Kutambaa) to 10.3 (M66) under no water stress, 6.3 (Mwandato) to 11 (K80) under water stress at vegetative stage and 7.3 (Mwandato) to 12.3 (Macho) under water stress at flowering. Water stress increased the number of pods per plant by 5 and 5.7 at vegetative and flowering stages respectively. Plants subjected to water stress during flowering stages had higher number of pods per plant than water stress at vegetative stage. The varieties which had significantly higher number of pods per plant under water stress at flowering stage were Macho, Nyeupe and M66. Cowpea varieties, water stress and their interactions had no significant effect on the number of cowpea grains per pod (Table 4). The average number of grains per pod for plants under no water stress, water stress at vegetative stage and water stress at flowering was 13.8, 14.2 and 13.8, respectively. Number of grains per pod varied from 11.7 (Nyeupe) to 15.7 (Kaima Koko), 13.0 (Macho) to 15.3 (KVU 419), and 12.7 (Nyeupe and M66) to 16.0 (Khaki) under no stress, stress at vegetative stage and stress at flowering, respectively.

3.5 Effect of Water Stress at Vegetative and Flowering Growth Stages on Cowpea 100-Grain Weight and Pod Weight

Cowpea variety, water stress and interaction between cowpea variety and water stress significantly affected cowpea 100-grain weight (Table 5). Water stress at vegetative and flowering stages significantly reduced 100-grain weight of cowpea varieties. Plants subjected to water stress during flowering stages had a lower 100-grain weight than water stress at vegetative stage. Weight of 100-grains varied from 12.9 g (Nyekundu) to 19.3 g (Nyeupe), 12.0 g (Nyekundu) to 18.9 g (Nyeupe) and 11.7 g (Nyekundu) to 18.6 g (Nyeupe) under no water stress, stress at vegetative and flowering stages, respectively. Variety Nyeupe had significantly higher 100-grain weight than other varieties under all water stress levels. Variety and interaction between variety and water stress significantly reduced cowpea pod weight (Table 5). Main effects of water stress levels had no significant effect on cowpea pod weight.

Table 4. Effect of water stress at vegetative and flowering growth stages on number of pods per plant and grains per pod of cowpea

Cowpea variety (V)	Number of pods per plant				Number of grains per pod			
	Water Stress (WS)			V- mean	Water Stress (WS)			V- mean
	Nws	Vws	Fws		Nws	Vws	Fws	
KVU 419	10.0	9.0	10.0	9.7	13.7	15.3	14.3	14.4
Khaki	7.3	10.7	9.0	9.0	15.3	15.0	16.0	15.4
K 80	9.0	11.0	7.7	9.2	14.7	15.0	14.0	14.6
Macho	10.0	10.0	12.3	10.8	12.3	13.0	13.7	13.0
Kaima-koko	9.0	10.0	9.3	9.4	15.7	14.3	13.0	14.3
Nyeupe	6.0	8.0	11.7	8.6	11.7	13.3	12.7	12.6
KVU 27 – 1	8.0	9.0	8.7	8.6	12.0	14.7	14.7	13.8
Nyekundu	10.0	10.0	9.3	9.8	15.0	13.0	14.3	14.1
M 66	10.3	10.7	11.7	10.8	15.3	14.0	12.7	14.0
Kutambaa	4.3	9.3	8.7	7.4	14.0	14.0	13.0	13.7
Mwandato	5.3	6.3	7.3	6.3	12.3	15.0	13.3	13.6
WS-mean	8.1	9.5	9.6		14.0	14.2	13.8	
p-value (V)	0.0001				0.141			
p-value (WS)	0.0003				0.549			
p-value VxWS)	0.006				0.625			
$LSD_{0.05}$ V	1.46				Ns			
$LSD_{0.05}$ WS	0.76				Ns			
$LSD_{0.05}$ V x WS	2.39				Ns			
CV *main plot* (%)	37.25				12.15			
CV *subplot* (%)	17.52				13.73			

WS –Water stress, Nws – No water stress, Vws – Vegetative water stress and Fws – Flowering water stress

Table 5. Effect of water stress at vegetative and flowering growth stages on cowpea 100-grain weight and pod weight (t/ha)

Cowpea variety (V)	100-grain weight (g)				Pod weight (t/ha)			
	Water Stress (WS)			V- mean	Water Stress (WS)			V- mean
	Nws	Vws	Fws		Nws	Vws	Fws	
KVU 419	13.6	13.1	12.7	13.1	11. 67	6.07	6.97	8.23
Khaki	13.6	13.1	12.8	13.2	9.50	3.97	8.17	7.21
K 80	13.5	13.4	13.3	13.4	10.17	6.07	7.57	7.93
Macho	16.4	15.8	15.6	15.9	10.00	9.70	11.33	10.34
Kaima-koko	13.5	13.5	12.9	13.3	13.43	8.90	7.57	9.97
Nyeupe	19.3	18.9	18.6	18.9	15.60	18.07	19.00	17.56
KVU 27 – 1	18.0	17.3	16.4	17.2	8.80	7.90	8.10	8.27
Nyekundu	12.9	12.0	11.7	12.2	11.77	7.90	9.47	9.71
M 66	14.6	14.4	13.5	14.2	8.17	7.57	6.27	7.33
Kutambaa	14.9	14.1	12.3	13.8	19.00	10.27	11.93	13.73
Mwandato	13.7	13.3	13.1	13.4	16.10	17.93	18.70	17.58
WS-mean	14.9	14.5	13.9		12.20	9.49	10.46	
p-value (V)	0.0001				0.0001			
p-value (WS)	0.0001				0.0001			
p-value VxWS)	0.0001				0.0001			
$LSD_{0.05}$ V	0.09				0.32			
$LSD_{0.05}$ WS	0.05				0.17			
$LSD_{0.05}$ V x WS	0.16				0.56			
CV *main plot* (%)	1.2				15.47			
CV *subplot* (%)	2.20				42.16			

WS –Water stress, Nws – No water stress, Vws – Vegetative water stress and Fws – Flowering water stress

Water stress at vegetative stage significantly increased pod weight of KVU 27-1 but had no effect on the pod weight of the rest of the varieties. In contrast, water stress at flowering significantly increased pod weight of Nyeupe, Kutambaa and Mwandato and significantly reduced pod weight of KVU 419; but had no effect on the rest of the varieties. Under no water stress, KVU 419 and Mwandato had significantly higher and lower pod weight, respectively, than most varieties. Under water stress at vegetative stage, Mwandato and KVU 27-1 had significantly lower and higher pod weight, respectively, than most varieties. Nyeupe and Mwandato had significantly higher and lower pod weight, respectively, than most varieties under stress at flowering stage.

3.6 Effect of Water Stress at Vegetative and Flowering Growth Stages on Cowpea above Ground Dry Matter at Maturity and Grain Yield

Variety, water stress and interaction between variety and water stress significantly affected cowpea shoot dry matter and grain yield (Table 6). Water stress at vegetative and flowering stages significantly lowered the shoot dry matter in cowpea varieties by 56.2% and 36.2% respectively. At vegetative stage the reduction in shoot dry matter ranged from 26.4% in Macho to 86.8% in Mwandato, while, at flowering stage it ranged from 15.3% in KUV 27-1 to 78.3% in Mwandato. At vegetative stage Nyeupe and Mwandato had higher dry matter reduction of 76.3% and 86.8%, respectively while at flowering stage Mwandato had the highest reduction of 78.3% in shoot dry matter. Water stress at vegetative stage significantly increased grain yield in KVU 27, Nyekundu, Kaima Koko, K80 and M66 by 121.1, 102.2, 55.8, 52.8 and 52.4% respectively; but it significantly reduced grain yield in Nyeupe by 44.0%. Water stress at flowering significantly increased grain yield in Nyekundu and Kutambaa by 53.3 and 119.6%, respectively, but had no significant effect on grain yield in the rest of the varieties. KVU 419 had significantly higher grain yield than Kutambaa, Mwandato and Khaki under no stress but its yield was not significantly different from the rest of the varieties. Under water stress at vegetative stage, KVU 27-1, Nyekundu and Kaima koko had significantly higher grain yield than most of the other varieties. Under water stress at flowering, no major differences were noted among the varieties except that Mwandato

Table 6. Effect of water stress at vegetative and flowering growth stages on cowpea above ground dry matter yield at maturity and grain yield

Cowpea variety (V)	Dry matter yield (t/ha)				Grain yield (t/ha)			
	Water Stress (WS)			V- mean	Water Stress (WS)			V- mean
	Nws	Vws	Fws		Nws	Vws	Fws	
KVU 419	20.49	11.67	14.92	15.69	1.34	1.25	1.04	1.21
Khaki	22.67	9.50	12.56	14.91	0.68	0.90	1.03	0.87
K 80	22.41	10.20	16.25	16.29	0.89	1.36	1.23	1.16
Macho	9.18	10.00	12.48	10.55	1.09	0.89	1.41	1.13
Kaima-koko	19.78	13.40	16.27	16.48	1.13	1.76	1.23	1.37
Nyeupe	15.60	3.60	8.11	9.10	1.16	0.65	1.54	1.12
KVU 27 – 1	25.19	8.80	14.44	16.14	0.90	1.99	1.17	1.35
Nyekundu	23.54	11.70	14.89	16.71	0.92	1.86	1.41	1.40
M 66	21.33	8.20	18.06	15.86	1.05	1.60	1.12	1.26
Kutambaa	19.00	7.21	10.34	12.18	0.56	0.74	1.23	0.84
Mwandato	16.00	2.12	3.48	7.20	0.48	0.38	0.65	0.50
WS-mean	19.86	8.69	12.67		0.93	1.22	1.19	
p-value (V)	0.0001				0.0001			
p-value (WS)	0.136				0.0002			
p-value VxWS)	0.0003				0.0001			
$LSD_{0.05}$ V	0.41				0.27			
$LSD_{0.05}$ WS	Ns				0.14			
$LSD_{0.05}$ V x WS	0.70				0.47			
CV *main plot* (%)	1.62				4.92			
CV *subplot* (%)	4.48				28.4			

WS –Water stress, Nws – No water stress, Vws – Vegetative water stress and Fws – Flowering water stress

had significantly lower grain yield than most varieties while Nyeupe had higher grain yield than Mwandato, KVU 419 and Khaki. Average grain yield across all water stress levels varied from 0.5 t/ha (Mwandato) to 1.40 t/ha (Nyekundu).

3.7 Effects of Water Stress at Vegetative and Flowering Growth Stages on Cowpea Harvest Index

Cowpea variety, water stress and the interaction between cowpea variety and water stress significantly affected the harvest index (Table 7). Water stress at vegetative stage significantly increased harvest indices of all cowpea varieties except Macho, Nyeupe and Kutambaa. Water stress at flowering enhanced cowpea harvest indices for only K80, Kaima-koko, Nyekundu, and Kutambaa varieties. Mwandato had significantly the highest harvest indices under all the stress levels. Under water stress at vegetative stage, Kutambaa and Nyeupe had significantly the highest indices than all other varieties. Harvest indices varied from 2.93% (Mwandato) to 12.8% (M66) under no stress, 2.13% (Mwandato) to 25.3% under stress at vegetative stage and 3.47% (Mwandato) to 18% (M66) at flowering.

4. DISCUSSION

4.1 Effects of Variety and Water Stress on Cowpea Chlorophyll Content, Canopy Temperature and Time to Anthesis

Water stress at vegetative stage significantly reduced cowpea chlorophyll content. This observation has been reported in previous studies. [26] Reported 100% reduction in the chlorophyll content of cowpea genotypes under both moderate and severe water stress. Chlorophyll content is among the morphological, biochemical and physiological traits for drought screening in cowpea [27]. In the current study, the response to drought in terms of chlorophyll content was, however, not dependent on the cowpea variety. Chlorophyll content could not therefore be used to identify drought tolerant cowpea varieties. Water stress during flowering had no effect on chlorophyll content.

Water stress significantly increased canopy temperature for varieties such as Khaki and Kaima-koko but reduced the canopy temperature of Macho and Nyekundu. [28] reported that water stress significantly increased canopy temperature of cowpea plants. They attributed this to the fact that cowpea plants Stomata

Table 7. Effects of water stress at vegetative and flowering growth stages on cowpea harvest index

Cowpea variety (V)	Harvest index (%)			
	Control (no stress)	Vegetative stage	Flowering stage	V-mean
KVU 419	11.53	20.53	14.87	15.64
Khaki	7.13	22.77	12.70	14.20
K 80	8.70	22.53	16.27	15.83
Macho	10.93	9.23	12.40	10.85
Kaima-koko	8.37	19.70	16.27	14.78
Nyeupe	7.43	3.57	8.13	6.38
KVU 27 – 1	10.20	25.30	14.90	16.80
Nyekundu	7.80	23.70	14.93	15.48
M 66	12.87	21.17	18.00	17.35
Kutambaa	2.97	7.20	10.33	6.83
Mwandato	2.93	2.13	3.47	2.84
WS-Means	8.26	16.17	12.93	
p-value (V)	0.0001			
p-value (WS)	0.0001			
p-value VxWS)	0.0001			
$LSD_{0.05}$ V	3.52			
$LSD_{0.05}$ WS	1.84			
$LSD_{0.05}$ V x WS	6.10			
CV main plot (%)	7.37			
CV subplot (%)	30.00			

closure is the first responsive event of plants to water deficiency [29]. There was a positive linear relationship between canopy temperature and grain yield, suggesting that stomatal closure resulted in increased water use efficiency. [18] reported that stomata closure under water stress resulted in increased water use efficiency (WUE). Canopy temperatures of Nyekundu and Macho decreased when water stress was imposed suggesting that they did not depend on stomata adjustment as a strategy to deal with water stress. Water stress at vegetative and flowering stages increased time to anthesis by 4 and 7 days, respectively. [10] reported that the onset of and dates to full flowering of cowpea were significantly delayed under high moisture stress. The interaction between variety and water stress did not affect the time to anthesis.

4.2 Effects of Water Stress on Cowpea Leaf Number, Percent Ground Cover and Shoot Dry Matter Yield at Maturity

Water stress caused reduction in cowpea leaf number, percent ground cover and dry matter yield at maturity. Reduction in leaf production and/or increase in leaf senescence and abscission due to water stress have been reported in previous studies [30]. [31] and [32] reported that post-flowering water stress reduced the cowpea total dry matter. Reduction in leaf and plant growth has been attributed to decrease in cellular expansion resulting from a decrease in plant water content [10], reduction in leaf formation and increased abscission of lower leaves [33]. The reduction in growth parameters under water stress conditions could also be attributed to decline in photosynthesis [34]. The effects of water stress on leaf number, percent ground cover and dry matter yield were dependent on cowpea genotypes [26].

4.3 Effects of Water Stress on Harvest Index, Grain Yield and Yield Components

Water stress at both vegetative and flowering stages significantly increased the number of pods per plant, pod weight, grain yield and harvest index for some varieties. Earlier studies indicated that cowpea could maintain seed yield when water stress was subjected at vegetative stage provided subsequent conditions were conducive for flowering and pod set [12]. However, [10] showed that cowpea is highly sensitive to water deficits during flowering and pod filling stages which lead to reduced grain yields. In the current study, there was a positive linear regression relationship between grain yield and number of pods per plant, 100-grain weight and pod weight. Grain yield in cowpea is determined by the product of the number of pods per plant that reach maturity, the average number of grains per pod and 100-grain weight [10]. Under water stress conditions, cowpea closes their stomata to avoid dehydration [28]. This reduces water loss [27] and increases water use efficiency (WUE) [18]. Cowpea varieties with high pod weight under water stress conditions could be making use of the additional photosynthetic capacity of their pods [35]. [3] reported that water stress increased cowpea harvest index. Most of the varieties in the current study apparently responded to water stress by partitioning more photosynthates to the grain relative to the shoot, thus leading to grain yield increases. This suggests that short-term moderate drought stress during vegetative growth and flowering enhances grain yield of some cowpea varieties but reduces shoot growth. Cowpea varieties which were superior in yield and high harvest indices under water stress included Nyekundu, KVU 27-1, M66, and KVU 419. Under no water stress, the cowpea varieties which had high harvest indices were M66, KVU 419, Macho and KVU 27-1.

5. CONCLUSION

Water stress imposed at vegetative and flowering reduced growth parameters and chlorophyll content, but enhanced grain yield and yield components of some varieties. The impact of water stress on growth is dependent on the cowpea variety. Moderate stress may be beneficial if cowpea is grown for grain production but not if grown for vegetable production. Cowpea varieties which were superior in yield and high harvest indices under water stress included Nyekundu, KVU 27-1, M66, and KVU 419.

ACKNOWLEDGEMENTS

Pwani University Research and Innovation Committee and National Commission for Science, Technology and Innovation are appreciated for providing funds for research.

COMPETING INTERESTS

Authors have declared that no competing interests exist.

REFERENCES

1. Sanginga NG, Tholta P, Dashiell K. Effectiveness of rhizobia nodulating recent promiscuous soybean selections in the moist savannah of Nigeria. Soil Biology and Biochemistry. 2002;32:127–133.
2. Quin FM, Introduction. In Singh BB, Mohan Raj DR, Dashiell KE, Jackai LEN, eds. Advances in cowpea research. Co-publication of International Institute of Tropical Agriculture (IITA), Ibadan, Nigeria and Japan International Research Centre for Agricultural Sciences (JIRCAS), Sayce Publishing, Devon, UK. 1997;ix–xv.
3. Dadson RB, Hashem FM, Javaid I, Joshi J, Allen AL, Devine TE. Effects of water stress on the yield of cowpea (*Vigna unguiculata* (L.) Walp) genotypes in the Delmarva Region of the United States. Journal of Agronomy and Crop Science. 2005;191:210–217.
4. Abebe G, Hattar B, Al-Tawaha ARM. Nutrient availability as affected by manure application to cowpea (*Vigna unguiculata* L. Walp) on calcareous soils. Journal of Agriculture and Social Sciences. 2005; 1(10)1–6.
5. Fussel LK, Bidinger FR, Bieler P. Crop physiology and breeding for drought tolerance: Research and development. Field Crops Research. 1991;27:183–199.
6. Onuh MO, Donald KM. Effects of water stress on the rooting nodulation potential and growth of cowpea (*Vigna unguiculata* (L) Walp). Science World Journal. 2009; 4(3):31–34.
7. Gaiser T, Graef F. Optimization of a parametric land evaluation method for cowpea and pearl millet production in semi-arid regions. Agronomic. 2001;21: 705–712.
8. Abdelshakoor HS, Faisal EA. Effects of water potentials on growth and yield of cowpea (*Vigna unguiculata* (L.) Walp). Research Journal of Agriculture and Biological Science. 2010;6(4):401–410.
9. Ahmed FE, Hall AE, DeMason DA. Heat injury during floral bud development in cowpea (*Vigna unguiculata* (L.) Walp). American Journal Botany. 1992;79: 784–791.
10. Abayomi YA, Abidoye TO. Evaluation of cowpea genotypes for soil moisture stress tolerance under screen house conditions. African Journal of Plant Science. 2009; 3(10):229-237.
11. Turk KJ, Hall AE. Drought adaptation of cowpea III. Influence of drought on plant growth and relation with seed yield. Agronomy Journal. 1980;72:428–433.
12. Singh BB, Chambliss OL, Sharma B. Recent advances in cowpea breeding. In: Singh BB, Mohan Raj DR, Dashiell KE, Jackai LEN, (eds). Advances in cowpea research, Co-publication of International Institute of Tropical Agriculture (IITA), Ibadan, Nigeria and Japan International Research Center for Agriculture Science (JIRCAS), Sayce Publishing, Devon, UK. 1997;114–128.
13. Akyeampong E. Seed yield and water use efficiency of cowpea in response to drought stress at different development stages. Ph.D Thesis, Connell University; 1985.
14. Marino D, Frendo P, Gonzalez EM, Puppo A, Arrese C. NADPA recycling systems in oxidative stressed nodules; a key role for the NADP (+) dependent isocitrate dehydrogenase. Planta. 2007;225: 413–421.
15. Hsiao TC, Xu LK. Sensitive of growth of roots versus leaves to water stress: Biophysical anslysis and relation to water transport. Journal of Experimental Botany. 2000;51:1595–1616.
16. Oghiakhe S, Jackai LLE, Makanjuola WA. Evaluation of cowpea genotypes for field resistance to legume pod borer, Maruca testulalis, in Nigeria. Crop Protection. 1995;14:389-394.
17. Anyia AO, Herzog H. Water use efficiency, leaf area and leaf gas exchange of cowpeas under mid-season drought. European Journal of Agronomy. 2004;20: 327–437.
18. Hall AE, Cisse N, Thiaw S, Ismail AM, Ehlers JD. Water use efficiency and drought adaptations of cowpea. In Singh BB, Mohan Raj DR, Dashiell KE, Jackai LEN, eds. Advances in cowpea research. Co-publication of International Institute of Tropical Agriculture (IITA), Ibadan, Nigeria and Japan International Research Center for Agricultural Sciences (JIRCAS), Sayce Publishing, Devon, UK. 1997;87–98.
19. Sombroek WG, Braun HMH, Van der Pouw BJ. Exploratory Soil Survey Report No. E1, Kenya Soil Survey, Nairobi; 1982.
20. Saha HM. Improving resource use under maize – green legume systems in coastal lowland Kenya. Ph.D Thesis, Jomo

Kenyatta University of Agriculture and Technology; 2007.
21. Jaetzold R, Schmidt H, Hornetz B, Shisanya C. Farm management handbook of Kenya. Vol. II/ Part C, Sub-Part C2, Coast Province. Ministry of Agriculture, Kenya, in corporation with the German Agency of International Cooperation (GIZ); 2012.
22. Mureithi JG, Tayler RS, Thorpe W. Productivity of alley farming with leucaena (*Leucaena leucocephala* Lam. de Wit) and Napier grass (*Pennisetum purpureum* K. Schum) in coastal lowland Kenya. Agroforestry Systems. 1995;31:59-78.
23. Ndiso JB, Kibe AM, Mugo S, Pathaka RS. Influence of drought stress on growth, yield and yield components of selected maize genotypes in coastal lowland Kenya. International Journal of Agricultural Sciences. 2012;2(6):178–185.
24. Sarrantonio M. Methodologies for screening soil improving legumes. Rodale Institute Research Centre, Kutztowon, USA. 1991;310.
25. SAS Institute. SAS/STAT users' guide. Release 6.08. Ed. SAS Cary. NC; 1993.
26. Hayatu M, Mukhtar FB. Physiological responses of some drought resistant cowpea genotypes (*Vigna unguiculata* (L.) Walp) to water stress. Bayero Journal of Pure and Applied Sciences. 2010;3(2): 69-75.
27. Souza RP, Machado EC, Silva JAB, Lagoa AMMA, SIlveira JAG. Photosynthetic gas exchange, chlorophyll fluorescence and some associated metabolic changes in cowpea (*Vigna unguiculata*) during water stress and recovery. Environment and Experimental Botany. 2004;5:45–56.
28. Hamidou F, Zombre G, Braconnier S. Physiological and biochemical responses of cowpea genotypes to water stress under glasshouse and field conditions. Journal of Agronomy and Crop Science. 2007;193: 229-237.
29. Lisar SYS, Motafakkerazad R, Hossain MM, Rahman IMM. Water stress in plants: Causes, effects and responses. In: Rahman IMM, ed. Water Stress. In Tech China. 2012;1-15.
30. Abidoye TO. Effects of soil moisture content on growth and yield of cowpea (*Vigna unguiculata* (L) Walp). B. Agric. Dissertation, University of Ilorin, Nigeria; 2004.
31. Okon JE. Effect of water stress on some growth aspects of two varieties of cowpea, *Vigna unguiculata* (L.) Walp Fabaceae. Bulletin of Environment, Pharmacology and Life Sciences. 2013;2(5):69-74.
32. Samson H, Helmut H. Drought effect on yield, leaf parameters and evapotranspiration efficiency of cowpea. Conference of International Agricultural Research for Development, University of Kassel-Witzenhause and University of Gotteingen; 2007.
33. Waseem M, Ali A, Tahir1 M, Nadeem MA, Ayub1 M. Tanveer A, Ahmad R, Hussain M. Mechanism of drought tolerance in plant and its management through different methods. Continental Journal of Agricultural Science. 2011;5(1):10–25.
34. Chaves MM, Flexas J, Pinheiro C. Photosynthetic under drought and salt stress; regulation mechanisms from whole plant to cell. Annals of Botany. 2009;103: 551–560.
35. Ahmed FE, Suliman AS. Effects of water stress applied at different stages of growth on seed yield and water use efficiency of cowpeas. Agriculture and Biology Journal of North America. 2010;1(4):534–540.

10

Quantifying the Occurrence and Ameliorating the Properties of Non-responsive Soils by Inorganic and Organic Fertilizers

Onesmo F. Assenga[1], Tileye Feyissa[1] and Patrick A. Ndakidemi[1*]

[1]School of Life Sciences and Bioengineering, The Nelson Mandela African Institution of Science and Technology, P.O.Box 447, Arusha, Tanzania.

Authors' contributions

This work was carried out in collaboration between all authors. Author OFA prepared the review draft. Authors TF and PAN read and corrected the manuscript. All authors read and approved the final manuscript.

Editor(s):
(1) Rocky Lemus, Department of Plant and Soil Sciences, Mississippi State University, USA.
(2) Marco Trevisan, Institute of Agricultural Chemistry and Environmental Research Centre Biomass, Faculty of Agriculture, Catholic University of the Sacred Heart, Italy.
(3) Radim Vacha, Deputy Director of Research and Development, Research Institute for Soil and Water Conservation, Czech Republic.
Reviewers:
(1) C. N. Mbah, Ebonyi State University, Abakaliki, Nigeria.
(2) Vijay Singh Meena, Vivekananda Institute of Hill Agriculture, India.

ABSTRACT

Conventionally, crop production in sub-Saharan Africa (SSA) depends primarily on natural soil nutrients. Application of inorganic and organic fertilizers is an important way for maximizing agriculture in SSA that targets the full utilization of the agricultural lands for food to feed African population and to obtain surplus produces. However, it has been reported that several areas of agricultural soils fail to respond to application of fertilizers, but the specific reason remain unknown. This review paper aims at exploring techniques that can be used to amend non-responsive soils and evaluate methods of ameliorating such soils through the proper utilization of nutrients supplied from different organic and inorganic sources.

Corresponding author: E-mail: ndakidemipa@gmail.com

Keywords: Non-responsive soils; soil nutrient amelioration; organic fertilizers; inorganic fertilizers; Sub-Saharan Africa (SSA).

1. INTRODUCTION

Agriculture is the main source of livelihood and income for two-thirds of Africa's population [1]. Furthermore, it has been reported that crop production should be improved in order to meet the demand of food and industrial resources for a fast growing global and as well as to obtain surplus crops. Current trend around worldwide shows that the normal consumption of food is beyond the crops produced [2]. However, it is documented that the major limitations in organic matter and other key nutrients largely constrain agricultural productivity to realize an optimum yield [3,4]. Also, it has been known that vigorous cultivation without soil replenishment is among the major driver of soil nutrient degradation in SSA. Population pressure and climate change exacerbate the condition of soils in the region. The depleted soil has caused average yields of grain crops to stagnate at around one ton per hectare since the 1960s while fertilizer use across Africa has remained at around 8 kg ha^{-1} of cultivated land over the past 40 years [3].

There are several reasons for the low level for agricultural productivity and these include availability of inputs, infrastructure and policies that affect the high cost and low accessibility of mineral fertilizers [4]. The studies also revealed that the overriding factor is that most smallholder farms have soils depleted of nutrients and soil organic carbon, following years of nutrient removal by crop harvests with minimal return of crop residues or additions of nutrients through mineral fertilizers or organic inputs [5].

All crops, particularly maize uses substantial amounts of nitrogen (N), phosphate (P_2O_5), and potash (K_2O) and relatively small amounts of secondary nutrients and micronutrients [6]. The nutrients taken up by the plants must be supplied either from soil reserves or by adding nutrients. A deficiency of any of these nutrients can reduce yields [7]. Maize production in SSA faces several production constraints which limit productivity. Poor soil fertility, drought, and parasitism all of this can reduce on-farm yield by over 70% even with the use of high-yielding varieties [8]. It was reported that Nitrogen is the nutrient most deficient in the soils and limits maize yield [9].

Despite the great importance, the growth and productivity of crops such as Soybean and Maize has been slow as a result of physical and chemical environments. The continuous intensive cultivation that mines soil organic matter and nutrients has contributed to the steady decline in soil fertility [10,11]. Soil fertility is linked to soil organic matter, whose status depends on inputs such as biomass management and outputs such as mineralization, erosion and leaching [11].

The need for intensification of agriculture in Sub-Saharan Africa (SSA) has recently gained support, because of the growing recognition that enhanced farm productivity in this region [12]. Various literatures indicated the effect of soil chemical and physical barriers that may hinder crop growth and performance [13,14] and ultimately that may negatively affect crop production. There are huge beneficial impacts from amelioration on yield, particularly those associated with such practices like incorporating with deep nutrients or liming materials [14-16]. Nevertheless, it was reported that these positive impacts may vary with the type soil at the particular location [15,17], but there is opportunity that may make studies that can bring positive results from amelioration, and also their effects may last for several seasons. In this review, some important selected soil nutrients (ameliorants) will be discussed to broadly explore the information of their contribution in the productivity of crops, and whether if the yield could be improved and sustained by using good cultural practices such as deep tillage and soil nutrients additions to the unresponsive soils.

2. OCCURRENCE OF NON-RESPONSIVE SOILS IN TANZANIA

The response of soils to management can be classified into three general categories, which are: (1) soils that may potentially respond to the applied fertilizers and (2) soils that show totally very little response to the applied fertilizers due to other constraints besides the nutrients contained in the fertilizer or non-responsive soils. Hence these soils does not respond well to the recommended doses and types of fertilizers, may be due to deficiencies in essential nutrients [12], and (3) soils that receive large amounts of organic inputs each year, where crops respond little to fertilizer as the soils are partly fertile. It has been reported that, soil fertility status varies significantly from one farm to another and from varying landscape levels in Africa particularly for the smallholder farmers, leading to variation in yield and performance of crops as they respond differently to the inorganic and organic nutrient (fertility) amendments [18]. Some challenges do exist to restore agricultural productivity for such non-responsive or degraded soils as they commonly fail to respond to NPK and other fertilizers. Deficiency in essential or major nutrients like nitrogen, potassium and phosphorus has been identified as a major problem affecting crop productivity [19]. Continuous intensive cropping with insufficient or no fertilizer input is a major contributor to progressive decline in these soil nutrients, resulting in poverty for many Tanzanian farmers [20,21]. In this aspect, the existence of the non-responsive soils in Tanzania and their underlying reasons are discussed, as well as their impact when some management practices are done. In this view, the study on fertility improvement are examined where the selected ameliorants application are studied in detail hence to come up with good common approach to improve its productivity. Soil fertility management practices on unresponsive soils in smallholder farmers are rarely studied in detail, and hence necessitate the use of combination of approaches to overcome the problems associated with some soils that fail to respond adequately to the applied nutrients.

3. CHARACTERISTICS OF NON-RESPONSIVE SOILS IN TANZANIA

Cereal crop production levels in Tanzania are reported to be below the optimum level, exemplified by the low yields of 0.905 t ha^{-1} maize and 0.458 t ha^{-1} for beans which are common to smallholder farmers [22]. It is apparent that this trend of low yield levels is associated with poor soil fertility status as well as low nutrient replenishment [23]. The work of [22] indicated that most of Tanzanian cultivated soils are prone to nutrients loss averaging the rate of 27, 9 and 21 kg N, P_2O_5 and K_2O respectively ha^{-1} in a year. It was projected that there will be an increase to 32, 12 and 25 kg N, P_2O_5 and K_2O per ha per annum respectively in the following year especially if the soil management will not be improved. These nutrient losses are generally related to crops that are harvested, the removal of the crop remains or those occurring during soil erosion activities.

It has been reported that major constraints of acrisols (soils found in in Mara, Tabora, Singida, Dodoma, Tanga, Kilimanjaro, Morogoro, Iringa, Mbeya, Ruvuma, Lindi and Mtwara regions) include low level of major and minor plant nutrients, low to moderate capacity to retain applied nutrients and high leaching losses [24]. Other problems include moderate to high fixation rates of phosphorus fertilizers, low pH and high risk of aluminium toxicity [25]. Furthermore, large areas of the Tanzanian highlands are predominated by fine-textured, P-fixing Oxisols also known as Ferralsols [26], hence making this element to become deficient to the crops grown in these regions. Despite the fact that soil fertility status of no-responsive soils in most areas of East Africa is not well known, there is no adequate research conducted to develop a multi-approach of soil fertility amelioration on unresponsive soils and hence this impose a great need for such initiatives.

3.1 Chemical Characteristics of Soils that Cause Unresponsiveness

The important indicators on soil fertility that have been reported mostly are soil pH, soil organic matter (SOM); cation exchange capacity (CEC); exchangeable bases (EB), salinity and sodicity status and the amount of extractable N, P, K, Mg, Ca and Na [4], (Table 1), [27]. These indicators are realistic in predicting plant growth and development. Hence there is a need for regular and systematic evaluation to establish their levels in the soil so as to achieve sustainable productivity in cropping systems. It has been reported that loss of Soil Organic Matter (SOM); nitrogen (N) and phosphorus (P) deficiencies, and soil acidification are considered as major factors limiting plant growth and crop production in sub-Saharan Africa [28].

According to various studies, there are established minimum levels of nutrients required to support crop growth. For example, the critical value for Mg in most crops was proposed to be 2 cmol kg^{-1} [29] and for K is recommended to be 0.20 cmol kg^{-1}, whereas the proposed critical value of Mg in most crops was 2 cmol kg^{-1} [30] and the proposed critical level of Ca for a majority of crops was 5.0 cmol kg^{-1}. Further, it is suggested that soils having total N below 2.0 g kg^{-1} and C.E.C of 6.0-12.0 cmol kg^{-1} are of poor fertility and soil organic carbon (SOM) less than 20.0 g kg^{-1} has been reported to be of poor quality [31]. Further, based on the current soil fertility recommendations, the critical P critical levels of soil P range widely depending upon soil type, extractant used, and sample timing. Published critical levels of soil P include 6 to 12 mg kg^{-1} with a Mehlich-1 extractant [32], 12 mg kg^{-1} with a Mehlich-3 extractant [33], and 14 to 32 mg kg^{-1} with a bicarbonate extractant [34]. On the other hand, for micronutrients, it has been reported that the recommended critical level (DTPA-extractable) in soil for Cu is 0.2 mg kg^{-1}, for Fe to be from 0.3 to 10 mg kg^{-1} and for Mn optimum level ranged from 2.0 to 5.0 mg kg^{-1}, whereas Zn level analyzed in the soil was found to be 0.4-0.6 mg kg^{-1} and values greater than 10-20 mg kg^{-1} were regarded as excess [35,36]. Despite this, little information is known on specific soil chemical characteristics that cause unresponsiveness and if this remain unchecked, their limitation may result in complex mineral elements imbalances, consequent poor soil quality, and decline in soil productivity and crop yields. In this regard, more studies are required on the specific and complex soil chemical limiting factors.

Table 1. Soil characteristics in two selected nonresponsive soils from sisal farms in Tanga, Tanzania

Soil depth	Bamba estate	Kwamdulu estate
	0-20	0-20
Clay (%)	35	52
Silt (%)	6	8
Sand (%)	59	40
pH(water)1:25	5	4.6
pH(Kcl)1:2.5	3.9	4
Organic C (%)	1.4	1.8
Total N (%)	0.09	Na
Available P (Bray1) mg kg^{-1}	3	3
CEC (NH4O Ac pH 7) mmol kg^{-1}	76	89
Ca mmol$_c$kg^{-1}	16	7
Mg mmolkg^{-1}	5	5
K mmolkg^{1}	1	1
Base saturation (%)	29	14
Al mmolkg^{-1}	11	8
Al saturation (% CEC)	29	37

Physical and chemical properties of selected horizons of Rhodic Ferralsols [37].

3.2 Biological Characteristics of Soils that are Responsible for the Soil Responsiveness

The importance of soil micro-organisms in nutrient cycling and maintenance of soil physical properties is well appreciated [38]. In particular, soil micro-organisms are responsible for key ecosystem functions such as decomposition of organic matter and mineralization and cycling of nutrients, humus synthesis, and aggregate stabilization, nitrogen fixation and the biological control of soil-borne pests and diseases [38-40]. It has been reported that improved soil organic matter contents through organic amendments under Integrated Soil Fertility Management (ISFM) practices enhance nutrient availability, support diverse soil biota, and improve soil aggregation and crop production [12,41]. It has been established that an acre of living topsoil contains approximately 408 kg of earthworms, 1088 kg of fungi, 680 kg of bacteria, 60 kg of protozoa, 403 kg of arthropods and algae, and even small mammals in some cases [42]. Bacteria and fungi are important microfauna in soil that live on organic matter or on living plants. Most bacteria and fungi perform useful functions such as the decomposition of plant residues, release of nutrients, and formation of aggregates. Some bacteria such as rhizobia provide nitrogen to plants [43]. Some fungi live in symbiosis with plant roots, facilitating the uptake of immobile nutrients such as phosphorus and potassium [44-46]. Studies have demonstrated that technologies such as green-manuring (biomass transfer) with shrubs or trees or crop residues increase crop yields [47,48]. These organic residues play an important role in soil fertility management through their short term effects on nutrient supply and longer term contribution to soil organic matter (SOM). Thus, the addition of organic materials to agricultural soil (with or without chemical fertilizers) is important for replenishing the annual C losses and for improving both the biological and chemical properties of the soils [49]. Although the effect of chemical and physical properties on crop response is well explored and known, a direct link between biological properties and crop response is not adequately established in nonresponsive soils of East Africa. Hence, better exploitation of soil-plant-microbe interactions for plant nutrition in non-responsive soils is crucial for enhanced agricultural productivity.

3.3 Physical Characteristics of Soils that Cause Unresponsiveness

Soil texture and structure exerts important influences on the edaphic conditions and hence crop production [5]. Soil texture is a basic soil property, which influences infiltration and moisture retention [50], and thus the availability of water and nutrients to the plant. It is well established that sandy soils cannot support good plant growth and this is attributed in part to poor moisture retention and the leaching of nutrients from the rooting zone [51]. Moreover, [52] have reported that most sandy soils are nonresponsive to nutrient additions. Sandy soils are easily and freely draining, warm in hot days, but susceptible to lack of water in dry periods while clay soils are sticky and plastic when wet and prone to drainage problems, but hard when dry. On the other hand, loams that comprise sand, silt and clay sized particles generally makes good agricultural soils. Silts are also good soils for cultivation but it is associated to structural problems like [16]. It was pointed out that aeration and impedance to root growth through soil compaction are among the dominant factors affecting plant growth [53]. Soil compaction destroys the quality of the soil because it restricts rooting depth and decreases pore size. Among the effects of soil compaction are water-filled pores that become less able to absorb water, increasing runoff and hence erosion [54].

Cropping pattern has immense effect on soil physical and chemical properties and thereby on crop productivity. For example, soil fertility may often change in response to land use, cropping patterns and land management practices [55,56]. Intensive cropping promotes high levels of nutrient extraction from soils without natural replenishment [26]. In addition, the repeated use of inappropriate tillage practices in soils with slopes, and the high intensity of precipitation are key factors for the occurrence of the soil erosion and degradation processes that may reduce the quality of the soil in terms of physical as well as chemical fertility status [57].

Therefore, improvements targeted at such limiting physical conditions of the non-responsive soils can make it possible to optimize crop production in diverse range of farming communities such as those found in East Africa.

4. IMPORTANCE OF SOIL NUTRIENT AMELIORATION TO NON-RESPONSIVE SOILS

It is well documented that the soils in Sub Saharan Africa are inherently infertile as they have been used for agricultural production for many decades with little or no addition of nutrient resources, leading to declining soil fertility [25,58]. Therefore, enriching such soils through the addition of agricultural inputs is key to tackling hunger in the continent [59]. This can be achieved through developing and disseminating appropriate soil management recommendations, detailed up-to-date and spatially explicit information about the condition and trend of soil fertility and health [60,61].

Several studies on advantages of different amelioration practices in enhancing crop yields have been reported [62-66]. For example, [60] reported that restoration of soil fertility through balanced fertilization and organic matter additions was necessary to achieve high crop productivity. Studies conducted in areas where crop growth was limited by the presence of poor physical structure such as hard pan were ameliorated through ripping and addition of nutrients such as gypsum and resulted into improved plant [15,66]. Other studies conducted on the depleted soils across Sub Saharan Africa, it was also revealed that applying N and P increased yields significantly [18]. Similarly, addition of organic soil amendments in poor soils in various forms [18,23,26,64] has increased the retention of nutrients and water, and hence created a better synchrony between the soil nutrients. According to [67] a well balanced soil (with Ca/Mg, Ca/K, and Mg/K ratios) contain the important components that enhance proper plant growth and hence yield and plant health sustenance at an adequate manner. Thus, plants grown in a soil whose cations exchange complex is not balanced, and does not contain the specified cation ratios, may lead to reduced crop yield [68]. The five most abundant exchangeable cations in the soil are calcium, magnesium, potassium, sodium and aluminium. It was reported that cations are held by negatively charged particles of clay and humus called colloids [69]. Cation Exchange Capacity (CEC) varies according to the type of soil, where as sandy soils rely heavily on the high CEC of organic matter for the retention of mineral elements as the OM has the highest CEC value, while the clay has a great capacity to attract and hold cations because of its chemical structure [70]. Cation Echange Capacity can be improved in weathered soils by adding lime and raising the pH [43] or by adding organic matter which is the most effective way of improving the CEC of the soil [71]. It was reprted that CEC of about 15 meq/100 g^{-1} has a relatively high capacity to hold cations, such as Ca^{+2}, Mg^{+2}, K^{+}, NH^{+}, Cu^{+2}, Fe^{+2}, Mn^{+2}, and Ni^{+2}[72]. Soils that are high in clay generally have higher CEC values, although the type of clay can substantially affect CEC [70]. Inadequate information is available in SSA on the importance of soil nutrient amelioration to non-responsive soils and specifically targeting spatial variability's within and between farms in different farming systems found in Africa.

5. STRATEGIES TO IMPROVE NON-RESPONSIVE SOILS USING VARIOUS FERTILIZERS: NPK

It is well established that in soils with exhausted fertility status, strategic fertilizer application with incorporation of crop residues in long period of time would be necessary to improve attainable yields over considerable period of time. This may be accompanied by alternative organic nutrient sources including animal manures and compost, and these may also play an important role in replenishing soil fertility and physical traits in Africa where inorganic fertilizers are mostly not affordable particularly to small holder farmers [73,28]. Strategic use of fertilizer to variable soil fertility conditions may combine incorporating crop residues, animal manures, and crop rotation particularly those involving leguminous crops are crucial for beneficial fertilizer use in smallholder farming systems in SSA [74-76]. In Tanzania, NPK are amongst the major limiting nutrients in the soils, in that order [77]. Nitrogen fertilizers make up the bulk of imports, where 61 percent of all fertilizers are imported [78]. Among the specific types of fertilizer imported, urea constitutes the largest portion. Nitrogen, Phosphorous and Potassium (NPK) blends make the second most common type of fertilizer imported, with 21 percent of fertilizer imports. Diammonium Phosphate (DAP) is the principal phosphate fertilizer, accounting for almost 90 percent of all such fertilizers and 8.5 percent of all fertilizer imports, while Minjingu Rock Phosphate is only P fertilizer manufactured in Tanzania [77]. Therefore, in order to attain sustainable crop production, studies on optimum fertilizer application (NPK) to increase the crop production in unresponsive soils are essential.

5.1 Improving Non-responsive Soils Fertility by Proper Application of N

All crops require essential nutrient elements in order to carry out various physiological processes. The proposed the critical level of nitrogen in the soil for most crops in Tanzania as 2.0 g kg^{-1} [79]. A study by [80] indicated that the percentage total nitrogen in the soils ranged from 0.03 to 0.06 and these values were rated as very low. Any soil with N values below the critical levels provided above may be subjected to lower yield levels and hence the amelioration of the problem through fertilization (with Urea, sulphate of Ammonia, Calcium ammonium Nitrate, and Farm Yard Manure) is of paramount importance. All of non-responsive soils have N levels far below the critical level and hence meeting N requirements through various inorganic and organic sources is recommended.

Numbers of studies have reported optimum fertilizer rates and means of their application in deficient environments [81-83]. The study by [84] reported that optimal fertilizer requirements depend on the productive potential of the cultivar, the previous cropping history and the general fertility of the fields used. In general, the fertilizer requirement for maize and most of cereal crops in tropical conditions ranges about 100-120 kg N ha^{-1}. Generally, in order to attain an optimum crop growth and development, and hence yield, it is required to supply an adequate amount of nitrogen in form of organic or inorganic sources.

According to various studies, nitrogen fertilization plays a significant role in improving soil fertility and increasing crop production [85-87]. For instance, N fertilization in maize increased the grain yield by between 43- 68 percent [88], its biomass (biological yield) between 25-42 percent [89] and boosted soil N stock by a range of 18-34 percent [85,86]. Urea ($NH_2CO\ NH_2$) is the most used fertilizer nitrogen in Africa and contains about 46% N. When urea is applied as a top dressing fertilizer, it undergoes several processes before the crop can take it up. Such processes include hydrolysis to form ammonium carbonate which is further oxidized to form nitrates [82]. During hydrolysis, ammonia gas may also be generated and lost into the air and this takes place when the urea is not fully incorporated into the soil. Therefore, due to this processes, most crops do not respond quickly to fertilizers applications and hence need proper management and timing during application [90].

It is therefore important to replenish unresponsive soils with adequate amount of N that will restore fertility by using nitrogenous fertilizers especially in areas where this nutrient has been depleted in order to attain the optimum yields.

5.2 Improving Non-responsive Soil Fertility by Application of P

Phosphorus (P) is second to nitrogen as an essential mineral fertilizer for crop production [91]. Many non-responsive soils are inherently poor in available phosphorus content, and hence proper management will contribute significantly to sustainable crop production. Studies have indicated that the sufficient P levels by Bray 1 method is above 15.0 mg kg^{-1} [89]. It was revealed that the plant available phosphorus (Olsen's P) ranging from 14.24 to 17.16 mg P kg^{-1} soil rated as medium [80].

In non-responsive soils where yield may be limited because of inherently low P contents, application of P fertilizers is the only way to enhance soil available P status to sustain reasonable plant development and finally achieving higher crop yield [91,92]. Since P is a non- renewable resource, frequent applications are necessary to maintain P levels optimally to sustain crop yields. This can only be achieved by replacing the removed P from the crop harvest through the application of organic and inorganic fertilizers [93].

In summary, steady supply of P fertilizers through inorganic sources, incorporation of plant residues and application of farm yard manures can enhance P supply in the soils and contribute to sustainable crop production in areas such as those involving non-responsive soils.

5.3 Improving Non-responsive Soils Fertility by Addition of K

To achieve or maintain maximal crop yields, supplemental K fertilization is often required, particularly on soils testing low for available soil K to provide enough K to meet crop needs [94]. It is established that the critical level below which yield are severely depressed is less than 0.4 centimole per kilogram, (cmol kg^{-1}) [95]. It has been reported that integrating K^+ in soils has a positive roles in physiological processes, such as those lead to profound effects on crop growth and development [96,97].

Many researchers have reported crop yield increases in response to K fertilization [97-99]. For example, it was reported that maize yield increase and reduced lodging with K fertilization [94,100].

Also an increased ear size with K fertilization, may also contribute to the grain yield increases in maize [94,98]. Whereas the study [98] reported K fertilization increased the stover dry matter at maturity for maize, [101] found that K fertilization did not improve leaf weight or stalk weight at silking stage in maize. Increased stalk weight or stover weight in response to K fertilization may help explain the reduced stalk lodging observed with K fertilization [94].

The presence of K in soil solution is vital for plant growth and development because K is known to be an enzyme activator that promotes metabolism and also assists in regulating the plant's use of water by controlling the opening and closing of leaf stomata's of crops such as the cereals [102]. On the other hand K promotes the translocation of photosythates (sugars) for plant growth or storage in fruits or roots. Potassium has been shown to improve disease resistance in plants, improve the size of grains and seeds of maize, sorghum, and legumes such as soybeans [94], and improve the quality of fruits and vegetables such as eggplant, tomatoes and carrots [103,104]. Plants generally absorb the majority of their potassium at an earlier growth stage. Experiments on potassium uptake by maize showed that 70-80 percent was absorbed by silking time, and 100 percent was absorbed three to four weeks after silking [105].

In agricultural systems, K supply is normally delivered in the form of inorganic fertilizers such as potassium chloride, potassium sulphate, potassium magnesium sulphate and potassium nitrate and organic forms such as compost, wood ash and farm yard manure. Although considerable data exist on the response of K on plant growth and development, little evidence is available on the effects of K on non-responsive soils such as those found in East Africa, hence need for further studies on effects of K on the performance of crop plants grown in the non-responsive soils.

5.4 Contribution of N_2 Fixation in Improving Non-responsive Soils

Biological nitrogen fixation (BNF) can play an essential role in crop establishment and yield, since no N fertilizer is applied and it fulfils most of plants need for nitrogen [106,107]. Study by [108] observed that biological N_2 fixation can contribute as much as 112 kg N ha^{-1} in a season. Generally, the response to Rhizobial inoculation in soils is controlled by the population of indigenous strains and the environmental factors. Therefore, understanding these limitations in non-responsive soils is important in realizing useful results.

Research findings have revealed that *Rhizobium* inoculation of legumes during sowing of seeds and the subsequent process could have positive effects on N_2 fixation, plant growth, uptake of macro and micronutrients and finally enhancing crop yields [109,110]. For example, it was revealed that there was a significant increase in the uptake of macro and micronutrients following inoculation with *Bradyrhizobium*in in common beans, cowpeas and soybeans [74,75,111].

Because the *Rhizobia* depends on the host plant for the energy required in nitrogen fixation, therefore anything that limits normal plant growth and development will affect nitrogen fixation. Symbiotic nitrogen fixation requires the balanced soil nutrients, oxygen, so well-aerated soils with good host plant growth provide the optimum environment [112]. Poorly drained, water-logged soils are detrimental to host plant root growth and to the *Rhizobia* bacteria. *Rhizobia* are adversely affected by very acid soils, and soil pH should be properly monitored and maintained for optimum legume production [113].

However, little attention has been paid to study the role of fixation of N_2 specifically on non-responsive soils in SSA. Hence, a need for detailed studies to tap into this cheaper alternative source of nitrogen for legumes and other companion crops.

5.5 Effects of Supplying FYM to Non-responsive Soils

According to [114] a judicious fertilization of soils with Farm Yard Manure (FYM) was reported to reduce fertilizer requirement, increase soil fertility and then crop production. It is further reported that application of farmyard manure (FYM) with inorganic nutrients may contribute to soil fertility improvement. When FYM is applied at recommended amounts, they may play an important role in replenishing the macro and micro nutrients that are deficient in soils, and in turn contributes to yield increase [115]. Applications of FYM is not only complementary but also bring synergistic effects in soils such as increasing organic inputs and also have beneficial effects beyond their nutrient content [116]. It was reported that organic materials improved soil moisture storage, decrease soil erosion and then minimize leaching losses of nutrients especially N, contribute to P availability, stimulate soil biological activities [117], (Table 2), enhance chelation and bioavailability of micronutrient elements to both plants and soil microorganisms [118,119].

It has been well established that about 84 and 78% of maize grain yield was produced in 2003 and 2004 in degraded soil incorporated with FYMbiomass respectively [120].

Other study showed that incorporation of FYM contributed 65 to 71% of the total N content to a succeeding maize crop (equivalent of 90 to 125 kg N ha^{-1}) and this suggests that the contents supplied in these organic manures has a tendency of releasing its nutrients slowly and can last in the soil for several cropping seasons [121].

Although FYM has been in use for long time, there is a need to explore more information on its effects when applied to non-responsive soils to ensure adequate nutrient supply and benefits from other aspects.

5.6 Supplying Selected Micro Nutrients (Zn, Mo and Bo) to Non-responsive Soils

It is well known that although micronutrients are required in relatively smaller quantities for plant growth, they are important to enhance crops in accomplishing their growth and production cycles. Moreover, it is well established that if one of essential element is not available in the soil or not in adequate amount or balance with other nutrients, plant growth may be limited or even may lead to complete crop inhibition [95]. Deficiencies of micronutrients in fields can be replenished with fertilizers that contain one or combination of these elements. Micronutrients often act as cofactors in enzyme systems and play an important role in reduction reactions, and in addition to having several other vital functions in plants. It is further reported that, micronutrients are involved in the key physiological processes of plants including the photosynthesis and respiration activities [95,96]. Deficiency of these nutrients may severely restrict such vital physiological processes thus limiting growth and yield gain in most of crops. For example, boron (B) deficiency can lead to yield reduction considerably in wheat and lentil crops [122,123], while in rice and maize zinc (Zn) deficiency can be a major hidden yield-limiting feature in most of the soils [124].

Therefore, although they are required in small quantities, these trace elements are important to be applied in the soils, as they are continuously being taken up by plants without being replenished [125]. While numbers of studies have targeted the relationship of macronutrients and plants growth, there is limited information for the role trace elements on non-responsive soils that might cause their deficiency and lead to poor plant nutrition, growth and production and hence need for more studies in this aspect.

5.7 Application of Zinc to Non-responsive Soils

It has been reported that significant zinc deficiencies occur in soils for different regions, and especially in sub-Saharan Africa [126]. Zinc is essential element for the proper plant growth and it has a role in important physiological processes such as the reproduction activities in most of plants. It is well documented that when the supply of zinc is inadequate in soils, crop yields are reduced and the qualities of crop products are impaired [19,127-129]. Zinc is reported to have an important key role(s) in plants particularly those related to structural constituent or regulatory co-factor of different enzymes that regulates many important biochemical pathways [130]. Losses of maize yield of 40% or more in many zinc deficient soils have a major economic impact on the farmer due to the reduced income as a result of lost yield [131]. Poor growth and small brown spots on leaves are common symptoms in rice and maize plants grown on Zn deficient soils [132].

According to studies done by [133-135], the critical level of HCl-extractable Zn and DTPA extractable Zn in the soil ranged from 0.2 to 1.0 $mg \cdot kg^{-1}$ and hence in this respect it is inadequate for plant growth.

Studies on the effect of zinc fertilization on growth and yield of many plants such as alfalfa, wheat, maize, barley and potato have been reported increased yield with zinc application [136-138]. Also zinc fertilization of maize and alfalfa increased herbage, hay, dry matter crude protein yields and zinc concentration of alfalfa [136]. However, there are limited studies in SSA that have reported the effect of Zn on improving crop productivity in the non-responsive soils. Hence, there is necessity for detailed studies on the effects of Zn on yield of commonly grown crops such as maize and soybean.

5.8 Application of Boron to Non-responsive Soils

Boron (B) is required for all plant to enhance growth [139]. It has been reported that adequate B nutrition is critical for high yields and quality of crops [140]. Deficiencies of B may result in many anatomical, biochemical and physiological changes in most crops [141,142]. It is well documented that Boron deficiency is among the factors that lead to low crop yield, due to interferences of activities involving B in the metabolism of food reserves such as carbohydrates and protein, and cell structural components synthesis [143,144]. For most of crops, B deficiency symptoms are related to interference in flowering and fruiting processes [145] and hence ultimately poor yields, which may be having poor grain or fruit qualities [146]. Also it has been reported that B deficiency symptoms vary among different species of crops. For instance, in soybean, B deficiency may cause empty or hollow heart seeds, while in black gram no symptoms of B deficiency may be visible in seeds, yet grain yield may be reduced considerably [147]. It was documented that the critical limit of B on maize was below 95 ppm [148]. Studies have revealed that Boron deficiency occurring in more than 30% of agricultural soils is even more widespread in SSA, affecting approximately 50% of the soils [126,149]. In order to attain optimum yields for any crop, nutrient stress should effectively be minimized. Therefore, a further investigation on supplying Boron in non-responsive soils is important to address the problems of such limiting selected micronutrients in tropical *soils.*

Table 2. Effects of FYM on improvement of the soil properties

Treatment	Organic matter	Soil bulk density	Available water holding capacity of soil
FYM	31.841*	0.487*	2332.168*
N	1.266 ns	0.002 ns	24.78 ns
P	0.477 ns	0.004 ns	4.121 ns
FYM × N	1.199 ns	0.003 ns	9.402 ns

** indicates statistical significance at $P \leq 0.05$ and ns indicates non statistical significance.*
This trend in table shows positive effects of the organic manures on the soil properties [150]

5.9 Application of Molybdenum to Non-responsive Soils

Studies indicate that, Mo is an important element because it is responsible in nitrogen nutrition as it helps plants to use nitrates that are absorbed from the soil [151,152]. It is well established that legumes and other plants with no or low access to Mo can lead to poor growth, particularly, those field crops that are supplied with inadequate amounts of Mo. In such cases, nitrates tend to fail to be assimilated into protein products. In legumes, Mo plays the crucial role to help root nodule bacteria to fix atmospheric N and convert it into a usable form [153]. By counting all this merits of Mo, simple techniques such as seed coating may play an effective role in improving crop performance. However, some reports indicate that Mo can have toxic effects on bacterial strains used for inoculation in seeds of legumes coated with this element. Therefore, it is important that the efficacy of Mo seed coating with bacterial strains to be evaluated before using Mo seed coating.

Also, the study indicated that there was a significant response in yield and other yield components of *P. vulgaris L.* supplied with Mo at 0, 6 and 12 g kg-1 of seeds [42]. Further it was reported that significant interactive effect was between *Rhizobium* inoculation and the Mo on the number of seeds pod^{-1}, seed weight, and grain yield [111,154]. *Rhizobium* inoculated treatments in combination with the highest rate of Mo gave good results suggesting that significant additive results by mixing the Mo and *Rhizobium* and this corroborates with other studies that stated that Mo has a crucial role in N_2 fixation [104].

Currently there is limited research in SSA focusing on the position of Mo in non-responsive soils and hence need for detailed studies to establish the usefulness of this micronutrient in supporting plant growth in heterogeneous soil types found in Africa.

6. THE NEED FOR USE OF COMBINATION OF SOIL AMELIORANT'S TO IMPROVE SOIL RESPONSIVENESS

It is well established that crop yields can only be produced at an adequate or optimum levels, when the proper management of the soil fertility are observed and implemented in Africa [26]. Proper soil fertility replenishment in Africa requires utilization of both inorganic fertilizers and organic manures at the desirable levels [40]. In order to make pace in this, it is desirable that the sufficient levels of nutrients to be applied for the purpose of making crops to have more yields and at the desired level, thus this can enhance maximum production and hence creating the desirable profit to the farmers [4], (Table 3).

It was reported that the combined application of boron with molybdenum or zinc resulted in higher yields for some crops such as rapeseed and other grain crops yield and quality than the application of boron without molybdenum or zinc alone, and the seed yield of the B+Mo+Zn treatment was the highest in all treatments [155].

Other studies have revealed positive interactions between some inorganic minerals particularly zinc and organic manure (FYM) in cereal crops. That can be contributed by synergistic activity of the two nutrient sources. For example, the study by [156] documented an increased Zn uptake, as well as grain and stover yields with the use of 120 kg N, 10 t FYM, and 5 kg Zn ha^{-1} in the study involving maize. Further, it was also observed that combined use of 5 tons animal manure and 16 kg $ZnSO_4$ ha^{-1} resulted in higher maize yield components specifically the stems and seed yields. This effect was associated with more uptake of the Zn that resulted in increased yields. Similarly, the use of FYM had capacity to improve the uptake of Zn and other elements such as NPK uptake in plants hence enhancing positively the fertility of soils [157].

Table 3. Effects of combination of nutrient management on maize yield, and yield components

Treatment	plant height cm	grain yield t ha^{-1}	Biological yield t ha^{-1}	harvest index %	Oil contents %
Control	102.5c	0.72c	2.7d	24.29c	3.3c
Recommended dose offertilizer (200-120-125 kg NPK ha^{-1})	02.5c	0.72c	26.7c	24.29c	3.3c
Single spray of multnutrients (1.2 L ha^{-1})	154.9b	4.13b	13.75ab	30.44b	4.6b
Recommended dose NPK and 1 spray multi-nutrients	176.9a	5.78a	15.73a	36.63a	4.96ab

Means followed by similar letter(s) in a column are not significantly different. Source: [158]

Results from the research showed that there are dual mutual effects of phosphorous and zinc. It was indicated that the minimum leaf phosphorous content (0.262 mg kg^{-1}) was obtained where there was zinc, but the maximum leaf phosphorous content (0.369 mg kg^{-1}) was achieved at the application where no zinc was applied [159]. This shows that P and Zn had antagonistic effects. Similar results were also reported by [66]. A study by [160] found that maize production increased considerably by applying N, P, and Cu at rates of 120, 80, and 10 kg ha^{-1}, respectively. Similarly, it was reported that the highest grain yield (5.84 t ha^{-1}) was produced when all micronutrients (B, S, Mn, Mo and Zn) were applied in combination with NPK fertilizers at 120:60:40 kg ha^{-1} followed by those which did not receive molybdenum (Mo) at the same level of fertilizers (5.26 kg ha^{-1}) [161].

Integrated use of plant nutrients aim at combined use of inorganic and organic sources of plant nutrients to improve efficiency of applied nutrients, reduce environmental hazards and improve crop productivity [162,163]. It was also reported that the shoot and root dry matter of maize increased with increasing rate of the applied fertilizers [7]. Integrated use of organic manures and mineral fertilizer is reported to reduce the cost and amount of fertilizer required by crops [28]. Also, other study [164] reported that proper soil fertility management and sustainable agriculture can be achieved with the use of both mineral fertilizer and organic manure. It was further suggested that integrated nutrient management through combined use of organic materials and chemical fertilizers can be an effective approach to combat nutrient depletion, increase yield and promote sustainable crop productivity [165].

Study by [166] showed that when the required essential elements were applied at right time on sandy soils, the dry weight of maize was highest (5.18 t ha^{-1}) for the NPK treatment, followed by the K treatment, P treatment, N treatment, and lowest (2.54 t ha^{-1}) for the control treatment. The study revealed that dry weight for the NPK treatment was significantly higher than those for the N, P and control treatments. It was further found that, the dry weight of maize for the manure treatment was higher than that of the NPK treatment, suggesting the positive effect of organic matter application in addition to inorganic fertilizer application. In separate studies involving legumes, foliar application of nitrogen, phosphorous, potassium, and sulphur at seed filling stage [167] reported greater influence on the growth and yield of soybean. In another study involving soybean, highest seed yield of 4.21 t ha^{-1} was recorded in a combination involving cattle manure with rock phosphate at the rate of 50 kg ha^{-1} and control treatment gave 4.10 t ha^{-1} [168].

This trend indicates that the use of combined treatments of major elements such as NPK; trace elements such as Zn, Bo and Mo as well as organic manures that are readily available particularly FYM may enhance replenishment of lacking nutrients in the soil and ultimately enhance crop growth and productivity and reward farmers efforts on using inputs.

7. CONCLUSION

In conclusion, to overcome soil constraints this can be done by adapting techniques that will discover the limiting nutrients in the soil. This is important approach to maximize yield in SSA, where particularly the non-responsiveness of the soils is a common problem. The adverse soil conditions that contribute to these problems should be explored well to enhance crop productivity as ultimately the soils will respond well to the applied nutrients. Mutual scientific efforts by optimizing specific limiting nutrient use and nutrient cycling to minimize external inputs and maximize their use efficiency is a suitable approach to enhance more crop productivity.

Therefore, knowledge of soil nutrient status and the specific constraints is necessary to plan for the soil fertility management options in non-responsive soils. Therefore, it is necessary to undertake detailed analysis of soil fertility by considering spatial variability between and within sites. Amelioration of the soil fertility constraints can be achieved by supplying appropriate doses of macro and micro nutrients through organic and inorganic fertilizer formulations.

ACKNOWLEDGEMENTS

This work was generously funded by the Nelson Mandela African Institution of Science and Technology.

COMPETING INTERESTS

Authors have declared that no competing interests exist.

REFERENCES

1. Dittoh S, Madhusudan A, Margaret A, Hanjra MA. Micro irrigation-based vegetable farming for income, employment and food security in West Africa. Global Food Security. 2013;178-199.
2. Imhoff ML, Bounoua L, Ricketts T, Loucks C, Harriss R, Lawrence WT. Global patterns in human consumption of net primary production. Nature. 2004;429:870-873.
3. Monreal C, Dinel H, Schnitzer M, Gamble D, Biederbeck V. Impact of carbon sequestration on functional indicators of soil quality as influenced by management in sustainable agriculture. Advances in soil sciences: soil processes and the carbon cycle. Lal, R. 1997;435-457.
4. Larson BA, Frisvold GB. Fertilizers to support agricultural development in sub-Saharan Africa: What is needed and why. Food Policy. 1996;21(6):509-525.
5. Smaling E, Stoorvogel J, Windmeijer P. Calculating soil nutrient balances in Africa at different scales. Fertilizer Research. 1993;35(3):237-250.
6. Vitosh M, Johnson J, Mengel D. TH-state fertilizer recommendations for corn, soybeans, wheat and alfalfa; 1995.
7. Vanotti M, Bundy L. An alternative rationale for corn nitrogen fertilizer recommendations. Journal of Production Agriculture. 1994;7(2):243-249.
8. Oikeh S, Carsky R, Kling J, Chude V, Horst W. Differential N uptake by maize cultivars and soil nitrate dynamics under N fertilization in West Africa. Agriculture, Ecosystems & Environment. 2003;100: 181-191.
9. Carsky R, Iwuafor E. Contribution of soil fertility research and maintenance to improved maize production and productivity in sub-Saharan Africa. In strategy for sustainable maize production in West and Central Africa. Proc. Regional Maize Workshop, IITA-Cotonou, Benin Republic; 1997.
10. Mafongoya P, Bationo A, Kihara J, Waswa BS. Appropriate technologies to replenish soil fertility in southern Africa. In Advances in Integrated Soil Fertility Management in sub-Saharan Africa: Challenges and Opportunities. Springer. 2007;29-43.
11. Roose E, Barthes B. Organic matter management for soil conservation and productivity restoration in Africa: A contribution from Francophone research. Nutrient Cycling in Agroecosystems. 2001; 61(1-2):159-170.
12. Vanlauwe B, Kihara J, Chivenge P, Pypers P, Coe R, Six J. Agronomic use efficiency of N fertilizer in maize-based systems in sub-Saharan Africa within the context of integrated soil fertility management. Plant and Soil. 2011;339:35-50.
13. Chartres C, Cresswell H, Murphy B, Geeves G, Little I, Gessler P. Distribution of physical and chemical subsoil constraints and their prediction in landscapes. In National workshop on subsoil constraints to root growth and high soil water and nutrient use by plants.'Tanunda, SA; 1992.
14. Belford R, Dracup M, Tennant D. Limitations to growth and yield of cereal and lupin crops on duplex soils. Animal Production Science. 1992;32(7):929-945.
15. Hamza M, Anderson W. Responses of soil properties and grain yields to deep ripping and gypsum application in a compacted loamy sand soil contrasted with a sandy clay loam soil in Western Australia. Crop and Pasture Science. 2003;54(3):273-282.
16. Buol SW, Southard RJ, Graham RC, McDaniel PA. Soil genesis and classification. John Wiley & Sons; 2011.
17. Holloway RE, Bertrand I, Frischke A, Brace D, McLaughlin MJ, Shepperd W. Improving fertiliser efficiency on calcareous and alkaline soils with fluid sources of P, N and Zn. Plant and Soil. 2001;236:209-219.
18. Zingore S, Murwira H, Delve R, Giller K. Influence of nutrient management strategies on variability of soil fertility, crop yields and nutrient balances on smallholder farms in Zimbabwe. Agriculture, Ecosystems & Environment. 2007;119:112-126.
19. Akinrinde E. Issues of optimum nutrient supply for sustainable crop production in tropical developing countries. Pakistan Journal of Nutrition. 2006;5(4):377-387.
20. Tittonell P, Giller KE. When yield gaps are poverty traps: The paradigm of ecological intensification in African smallholder agriculture. Field Crops Research. 2013; 143:76-90.
21. Ngoze S, Riha S, Lehmann J, Verchot L, Kinyangi J, Mbugua D, Pell A. Nutrient constraints to tropical agroecosystem productivity in long-term degrading soils. Global Change Biology. 2008;14:2810-2822.

22. Kaihura F, Stocking M, Kahembe E. Soil management and agrodiversity: A case study from Arumeru, Arusha, Tanzania. In Proceedings of the Symposium on Managing Biodiversity in Agricultural Systems; 2001.
23. Ndakidemi P. Agronomic and economic potential of Tughutu and Minjingu phosphate rock as alternative phosphorus sources for bean growers. Pedosphere. 2007;17(6):732-738.
24. Haule K, Solonchaks P. 6.8 The extent and management of problem soils in Tanzania. World Soil Resources Reports. 1985;58.
25. Nandwa SM, Bekunda M. Research on nutrient flows and balances in East and Southern Africa: State-of-the-art. Agriculture, Ecosystems & Environment. 1998;71(1):5-18.
26. Sanchez PA, Shepherd KD, Soule MJ, Place FM, Buresh RJ, Izac A-MN, Mokwunye AU, Kwesiga FR, Ndiritu CG, Woomer PL. Soil fertility replenishment in Africa: An investment in natural resource capital. Replenishing Soil Fertility in Africa. 1997;1-46.
27. Brady NC, Weil RR. The nature and properties of soils. Prentice-Hall Inc; 1996.
28. Jones C, Jacobsen J. Plant nutrition and soil fertility. Nutrient Management Module. 2005;2:11.
29. Joachim H, Makoi J, Ndakidemi PA. Selected chemical properties of soil in the traditional irrigation schemes of the Mbulu district, Tanzania. African Journal of Agricultural Research. 2008;3(5):348-356.
30. Marx E, Hart JM, Stevens RG. Soil test interpretation guide. Oregon State University Extension Service Oregon; 1996.
31. Kemper WD, Koch EJ. Aggregate stability of soils from Western United States and Canada: Measurement procedure, correlations with soil constituents. Agricultural Research Service, US Department of Agriculture; 1966.
32. Cope J. Relationships among rates of N, P, and K, soil test values, leaf analysis and yield of cotton at 6 locations. Communications in Soil Science & Plant Analysis. 1984;15(3):253-276.
33. Cox F, Barnes J. Peanut, corn, and cotton critical levels for phosphorus and potassium on Goldsboro soil. Communications in Soil Science and Plant Analysis. 2002;33(7-8):1173-1186.
34. Duggan B, Gaff N, Singh D, Yeates S, Constable G. Cotton yield response to sources, rates and placement of P fertilizers in tropical Australia. In Proc. Beltwide Cotton Conf., Nashville, TN. 2003;6-10.
35. Mbila MO, Thompson ML, Mbagwu JS, Laird DA. Distribution and movement of sludge-derived trace metals in selected Nigerian soils. Journal of Environmental Quality. 2001;30:1667-1674.
36. Karamanos R, Kruger G, Stewart J. Copper deficiency in cereal and oilseed crops in northern Canadian prairie soils. Agronomy Journal. 1986;78(2):317-323.
37. Hartemink AE, Wienk JF. Sisal production and soil fertility decline in Tanzania. Outlook On Agriculture-ICI Limited Then Pergamon Press. 1995;24:91-91.
38. Bremer C, Braker G, Matthies D, Reuter A, Engels C, Conrad R. Impact of plant functional group, plant species, and sampling time on the composition of nirK-type denitrifier communities in soil. Applied and Environmental Microbiology. 2007;73: 6876-6884.
39. Coleman DC, Reid C, Cole C. Biological strategies of nutrient cycling in soil systems. Advances in Ecological Research. 1983;13:1-55.
40. Stanford G, Smith S. Nitrogen mineralization potentials of soils. Soil Science Society of America Journal. 1972;36(3):465-472.
41. Ayuke F, Brussaard L, Vanlauwe B, Six J, Lelei D, Kibunja C, Pulleman M. Soil fertility management: Impacts on soil macrofauna, soil aggregation and soil organic matter allocation. Applied Soil Ecology. 2011;48:53-62.
42. Pimentel D, Harvey C, Resosudarmo P, Sinclair K, Kurz D, McNair M, Crist S, Shpritz L, Fitton L, Saffouri R. Environmental and economic costs of soil erosion and conservation benefits. Science-New York Then Washington. 1995;1117-1117.
43. Bambara S, Ndakidemi PA. Changes in selected soil chemical properties in the rhizosphere of *Phaseolus vulgaris* L. supplied with Rhizobium inoculants, molybdenum and lime. Scientific Research and Essays. 2010;5(7):679-684.
44. Cwala Y, Laubscher C, Ndakidemi P, Meyer A. Mycorrhizal root colonisation and the subsequent host plant response of soil less grown tomato plants in the presence

and absence of the mycorrhizal stimulant. Mycotech. Afr. J. Microbiol. Res. 2010;4: 414-419.
45. Joachim HM, Patrick AN. The agronomic potential of vesicular-arbuscular mycorrhiza (VAM) in cereals–legume mixtures in Africa. African Journal of Microbiology Research. 2009;3(11):664-675.
46. Fließbach A, Oberholzer H-R, Gunst L, Mäder P. Soil organic matter and biological soil quality indicators after 21 years of organic and conventional farming. Agriculture Ecosystems & Environment. 2007;118:273-284.
47. Makoi JH, Ndakidemi PA. Reclamation of sodic soils in northern Tanzania, using locally available organic and inorganic resources. African Journal of Biotechnology. 2007;6(16).
48. Palm CA, Myers RJ, Nandwa SM. Combined use of organic and inorganic nutrient sources for soil fertility maintenance and replenishment. Replenishing Soil Fertility in Africa. 1997;replenishingsoi:193-217.
49. Goyal S, Chander K, Mundra M, Kapoor K. Influence of inorganic fertilizers and organic amendments on soil organic matter and soil microbial properties under tropical conditions. Biology and Fertility of Soils. 1999;29:196-200.
50. Jury W, Gardner W, Gardner W. Soil Physics, Wiley, New York; 1991.
51. Sullivan DM, Brown B, Shock CC, Horneck DA, Stevens R, Pelter G, Feibert EBG. Nutrient management for onions in the Pacific Northwest. [Covallis, Or.]: Oregon State University Extension Service; 2001.
52. Glaser B, Lehmann J, Zech W. Ameliorating physical and chemical properties of highly weathered soils in the tropics with charcoal– A review. Biology and Fertility of Soils. 2002;35(4):219-230.
53. Letey J. Relationship between soil physical properties and crop production, in advances in soil science. Springer. 1985; 277-294.
54. Hillel D. Fundamentals of Soil Physics. Academic Press, Inc.(London) Ltd; 1980.
55. Alam MK, Salahin N, Rashid M, Islam A, Hossain M. Effect of tillage depths and cropping systems on soil physical properties in grey terrace soils; 2013.
56. Salam MA, Alam M, Rashid M. Effects of different tillage practices and cropping patterns on soil physical properties and crop productivity. Journal of Tropical Resources and Sustainable Sciences. 2013;1(1):51-61.
57. Ramos M, Martínez-Casasnovas J. Impact of land levelling on soil moisture and runoff variability in vineyards under different rainfall distributions in a Mediterranean climate and its influence on crop productivity. Journal of Hydrology. 2006; 321(1):131-146.
58. Bationo A, Kihara J, Vanlauwe B, Waswa B, Kimetu J. Soil organic carbon dynamics, functions and management in West African agro-ecosystems. Agricultural Systems. 2007;94:13-25.
59. Clover J. Food security in sub-Saharan Africa. African Security Studies. 2003; 12(1):5-15.
60. Hofmockel M, Callaham MA, Powlson DS, Smith P. Long-term soil experiments: Keys to managing Earth's rapidly changing ecosystems. Soil Science Society of America Journal. 2007;71:266-279.
61. Busse MD, Sanchez FG, Ratcliff AW, Butnor JR, Carter EA, Powers RF. Soil carbon sequestration and changes in fungal and bacterial biomass following incorporation of forest residues. Soil Biology and Biochemistry. 2009;41:220-227.
62. Zingore S, Murwira H, Delve R, Giller K. Influence of nutrient management strategies on variability of soil fertility, crop yields and nutrient balances on smallholder farms in Zimbabwe. Agriculture, Ecosystems & Environment. 2007;119:112-126.
63. Hamza M, Anderson W. Improving soil physical fertility and crop yield on a clay soil in Western Australia. Crop and Pasture Science. 2002;53(5):615-620.
64. Vanlauwe B, Bationo A, Chianu J, Giller K, Merckx R, Mokwunye U, Ohiokpehai O, Pypers P, Tabo R, Shepherd K. Integrated soil fertility management operational definition and consequences for implementation and dissemination. Outlook on Agriculture. 2010;39:17-24.
65. Graham R, Turner N, Ascher J. Evidence for subsoil constraints and potential benefits from amelioration. In National workshop on subsoil constraints to root growth and high soil water and nutrient use by plants'. Tanunda. (Ed. GRDC). (GRDC); 1992.
66. Ellington A. Effects of deep ripping, direct drilling, gypsum and lime on soils, wheat

growth and yield. Soil and Tillage Research. 1986;8:29-49.
67. Major J, Rondon M, Molina D, Riha SJ, Lehmann J. Maize yield and nutrition during 4 years after biochar application to a Colombian savanna oxisol. Plant and Soil. 2010;333:117-128.
68. Sparks D, Page A, Helmke P, Loeppert R. Cation exchange capacity and exchange coefficients; 1996.
69. Van de Graaff R, Patterson RA. Explaining the mysteries of salinity, sodicity, SAR and ESP in on-site practice. In Proceedings of On-site'01 Conference: Advancing Onsite Wastewater Systems; 2001.
70. Brady N, Weil R. Soil organic matter. The nature and properties of soils. Prentice Hall, Upper Saddle River, New Jersey. 1999;446-490.
71. Loveland P, Webb J. Is there a critical level of organic matter in the agricultural soils of temperate regions: A review. Soil and Tillage Research. 2003;70(1):1-18.
72. Muchena F, Hilhorst T. Nutrients on the move: Soil fertility dynamics in African farming systems; 2000. IIED.
73. Place F, Barrett CB, Freeman HA, Ramisch JJ, Vanlauwe B. Prospects for integrated soil fertility management using organic and inorganic inputs: Evidence from smallholder African agricultural systems. Food Policy. 2003;28:365-378.
74. Nyoki D, Ndakidemi PA. Influence of *Bradyrhizobium japonicum* and phosphorus on micronutrient uptake in cowpea. A case study of zinc (Zn), iron (Fe), copper (Cu) and manganese (Mn). American Journal of Plant Sciences. 2014; 5(04):427.
75. Tairo EV, Ndakidemi PA. *Bradyrhizobium japonicum* inoculation and phosphorus supplementation on growth and chlorophyll accumulation in soybean (*Glycine max* L.). American Journal of Plant Sciences. 2013;4:2281.
76. Ojiem J, De Ridder N, Vanlauwe B, Giller K. Socio-ecological niche: A conceptual framework for integration of legumes in smallholder farming systems. International Journal of Agricultural Sustainability. 2006;4:79-93.
77. Msolla M, Semoka J, Borggaard OK. Hard Minjingu phosphate rock: an alternative P source for maize production on acid soils in Tanzania. Nutrient Cycling in Agroecosystems. 2005;72(3):299-308.
78. Benson T, Kirama SL, Selejio O. The supply of inorganic fertilizers to smallholder farmers in Tanzania; 2012.
79. Ndakidemi P, Semoka J. Soil fertility survey in western Usambara Mountains, northern Tanzania. Pedosphere. 2006; 16(2):237-244.
80. Kajiru G, Mrema J, Mbilinyi B, Rwehumbiza F, Hatibu N, Mowo J, Mahoo H. Assessment of soil fertility status under rainwater harvesting systems in the Ndala River catchment Northwest Tanzania: Farmers' versus scientific approaches; 2005.
81. Cooke G. Fertilizer for maximum yield, Granada Publishing Ltd., London; 1980.
82. Msarmo K, Mhango W. Yield of maize as affected by fertiliser application practices. Bunda College of Agriculture; 2000.
83. Vanlauwe B, Wendt J, Diels J. Combined application of organic matter and fertilizer. Sustaining soil fertility in West Africa. 2001;sustainingsoilf:247-279..
84. Yayock J, Lombin G, Owonubi J. Crop science and production in warm climates. Macmillan Publishers; 1988.
85. Dang T-H, Cai G-X, Guo S-L, Hao M-D, Heng L. Effect of nitrogen management on yield and water use efficiency of rainfed wheat and maize in Northwest China. Pedosphere. 2006;16:495-504.
86. Habtegebrial K, Singh B, Haile M. Impact of tillage and nitrogen fertilization on yield, nitrogen use efficiency of< i> tef</i>(< i> Eragrostis tef</i>(Zucc.) Trotter) and soil properties. Soil and Tillage Research. 2007;94(1):55-63.
87. Binder DL, Sander DH, Walters DT. Maize response to time of nitrogen application as affected by level of nitrogen deficiency. Agronomy Journal. 2000;92(6):1228-1236.
88. Khan A, Jan MT, Marwat KB, Arif M. Organic and inorganic nitrogen treatments effects on plant and yield attributes of maize in a different tillage systems. Pak. J. Bot. 2009;41:99-108.
89. Ogola J, Wheeler T, Harris P. Effects of nitrogen and irrigation on water use of maize crops. Field Crops Research. 2002; 78(2):105-117.
90. Doran JW, Jones AJ. Measurement and use of pH and electrical conductivity for soil quality analysis; 1996.
91. Drew M. Comparison of the effects of a localised supply of phosphate, nitrate, ammonium and potassium on the growth of the seminal root system, and the shoot,

in barley. New Phytologist. 1975;75(3): 479-490.

92. Damodar Reddy D, Subba Rao A, Sammi Reddy K, Takkar P. Yield sustainability and phosphorus utilization in soybean–wheat system on Vertisols in response to integrated use of manure and fertilizer phosphorus. Field Crops Research. 1999;62:181-190.
93. Hedley M, Stewart J, Chauhan B. Changes in inorganic and organic soil phosphorus fractions induced by cultivation practices and by laboratory incubations. Soil Science Society of America Journal. 1982;46(5): 970-976.
94. Pettigrew WT. Potassium influences on yield and quality production for maize, wheat, soybean and cotton. Physiologia Plantarum. 2008;133(4):670-681.
95. Mengel K, Kosegarten H, Kirkby EA, Appel T. Principles of plant nutrition. Springer;l 2001.
96. Marschner H, Marschner P. Marschner's mineral nutrition of higher plants. Academic Press. 2012;89.
97. Ebelhar S, Varsa E. Applications in sustainable production: Tillage and potassium placement effects on potassium utilization by corn and soybean. Communications in Soil Science & Plant Analysis. 2000;31(11-14):2367-2377.
98. Heckman J, Kamprath E. Potassium accumulation and corn yield related to potassium fertilizer rate and placement. Soil Science Society of America Journal. 1992;56(1):141-148.
99. Mallarino AP, Bordoli JM, Borges R. Phosphorus and potassium placement effects on early growth and nutrient uptake of no-till corn and relationships with grain yield. Agronomy Journal. 1999;91(1):37-45.
100. Kant S, Kafkafi U, Pasricha N, Bansal S. Potassium and abiotic stresses in plants. Potassium for sustainable crop production. Potash Institute of India, Gurgaon. 2002;233-251.
101. Ebelhar S, Kamprath E, Moll R. Effects of nitrogen and potassium on growth and cation composition of corn genotypes differing in average ear number. Agronomy Journal. 1987;79(5):875-881.
102. Kaya C, Kirnak H, Higgs D. Enhancement of growth and normal growth parameters by foliar application of potassium and phosphorus in tomato cultivars grown at high (NaCl) salinity. Journal of Plant Nutrition. 2001;24(2):357-367.
103. Fawzy Z, El-Nemr M, Saleh S. Influence of levels and methods of potassium fertilizer application on growth and yield of eggplant. Journal of Applied Sciences Research. 2007;3(1):42-49.
104. Ravikovitch S, Porath A. The effect of nutrients on the salt tolerance of crops. Plant and Soil. 1967;26(1):49-71.
105. Leigh R, Wyn Jones R. A hypothesis relating critical potassium concentrations for growth to the distribution and functions of this ion in the plant cell. New Phytologist. 1984;97(1):1-13.
106. Hungria M, Vargas MA. Environmental factors affecting N< sub> 2</sub> fixation in grain legumes in the tropics, with an emphasis on Brazil. Field Crops Research. 2000;65(2):151-164.
107. Zahran HH. Rhizobium-legume symbiosis and nitrogen fixation under severe conditions and in an arid climate. Microbiology and Molecular Biology Reviews. 1999;63(4):968-989.
108. Ojiem J, De Ridder N, Vanlauwe B, Giller K. Socio-ecological niche: A conceptual framework for integration of legumes in smallholder farming systems. International Journal of Agricultural Sustainability. 2006; 4:79-93.
109. Rodelas B, González-López J, Martinez-Toledo M, Pozo C, Salmeron V. Influence of Rhizobium/Azotobacter and Rhizobium/Azospirillum combined inoculation on mineral composition of faba bean (*Vicia faba* L.). Biology and Fertility of Soils. 1999;29:165-169.
110. Rengel Z, Batten G, Crowley Dd. Agronomic approaches for improving the micronutrient density in edible portions of field crops. Field Crops Research. 1999; 60(1):27-40.
111. Ndakidemi PA, Bambara S, Makoi JH. Micronutrient uptake in common bean (*Phaseolus vulgaris* L.) as affected by rhizobium inoculation, and the supply of molybdenum and lime; 2011.
112. Chianu JN, Nkonya EM, Mairura F, Chianu JN, Akinnifesi F. Biological nitrogen fixation and socioeconomic factors for legume production in sub-Saharan Africa: a review. Agronomy for Sustainable Development; 2010.
113. Tran YT. Response to and benefits of rhizobial inoculation of soybean in the

south of Vietnam. in The 4th International Crop Science Congress; 2004.
114. Roy RN, Misra RV, Montanez A. Decreasing reliance on mineral nitrogen-yet more food. AMBIO: A Journal of the Human Environment. 2002;31(2):177-183.
115. Bekeko Z. Effect of enriched farmyard manure and inorganic fertilizers on grain yield and harvest index of hybrid maize (bh-140) at Chiro, eastern Ethiopia. African Journal of Agricultural Research. 2014;9: 663-669.
116. Shah Z, Ahmad MI. Effect of integrated use of farm yard manure and urea on yield and nitrogen uptake of wheat. Journal of Agricultural and Biological Science. 2006; 1(1):60-65.
117. Horst W, Härdter R. Rotation of maize with cowpea improves yield and nutrient use of maize compared to maize monocropping in an alfisol in the northern Guinea Savanna of Ghana. Plant and Soil. 1994;160(2):171-183.
118. Sylvia DM, Fuhrmann JJ, Hartel P, Zuberer DA. Principles and applications of soil microbiology. Pearson Prentice Hall New Jersey; 2005.
119. Marinari S, Masciandaro G, Ceccanti B, Grego S. Influence of organic and mineral fertilisers on soil biological and physical properties. Bioresource Technology. 2000; 72:9-17.
120. Getachew M, Belete K. Yield related traits and yield of maize (*Zea mays* L.) as affected by green manuring and nitrogen levels at Mizan Teferi, South-west Ethiopia. International Journal of Agronomy and Plant Production. 2013; 4(7):1462-1473.
121. Gill H, Meelu O. Studies on the substitution of inorganic fertilizers with organic manure and their effect on soil fertility in rice—wheat rotation. Fertilizer Research. 1982; 3(4):303-314.
122. Srivastava S, Bhandari T, Yadav C, Joshi M, Erskine W. Boron deficiency in lentil: yield loss and geographic distribution in a germplasm collection. Plant and Soil. 2000;219:147-151.
123. Rerkasem B, Jamjod S. Boron deficiency in wheat: A review. Field Crops Research. 2004;89(2):173-186.
124. Atique-ur-Rehman, Farooq M, Cheema ZA, Wahid A. Role of boron in leaf elongation and tillering dynamics in fine-grain aromatic rice. Journal of Plant Nutrition. 2013;36:42-54.
125. Fageria N, Baligar V, Clark R. Micronutrients in crop production. Advances in Agronomy. 2002;77:185-268.
126. Godfray HCJ, Beddington JR, Crute IR, Haddad L, Lawrence D, Muir JF, Pretty J, Robinson S, Thomas SM, Toulmin C. Food security: the challenge of feeding 9 billion people. Science. 2010;327:812-818.
127. Cakmak I. Enrichment of cereal grains with zinc: agronomic or genetic biofortification? Plant and Soil. 2008;302(1-2):1-17.
128. Passioura JB. Review: Environmental biology and crop improvement. Functional Plant Biology. 2002;29(5):537-546.
129. Yilmaz A, Ekiz H, Torun B, Gultekin I, Karanlik S, Bagci S, Cakmak I. Effect of different zinc application methods on grain yield and zinc concentration in wheat cultivars grown on zinc-deficient calcareous soils. Journal of Plant Nutrition. 1997;20:461-471.
130. Alloway BJ. Zinc in soils and crop nutrition. International Zinc Association Brussels, Belgium; 2004.
131. Timperley M, Brooks R, Peterson P. The significance of essential and non-essential trace elements in plants in relation to biogeochemical prospecting. Journal of Applied Ecology. 1970;429-439.
132. Sharma CP. Plant micronutrients. Science publishers Enfield, NH; 2006.
133. Kparmwang T, Chude V, Raji B, Odunze A. Extractable micronutrients in some soils developed on sandstone and shale in the Benue Valley, Nigeria. Nigerian Journal of Soil Research. 2000;1:42-48.
134. Landon J, Manual BTS. A Handbook for Soil Survey and Agricultural Land Evaluation in the Tropics and Subtropics, Longman Scientific and Technical, Harlow; 1991.
135. Sakal R, Singh A, Tandon H. Boron research and agricultural production. Micronutrient research & agricultural production. 1995;1-31.
136. Mousavi SR, Galavi M, Ahmadvand G. Effect of zinc and manganese foliar application on yield, quality and enrichment on potato (*Solanum tuberosum* L.). Asian Journal of Plant Sciences; 2007.
137. Shaheen R, Samim M, Mahmud R. Effect of zinc on yield and zinc uptake by wheat on some soils of Bangladesh. Journal of Soil Nature. 2007;1(1):07-14.
138. Kinaci G, Kinaci E. Effect of zinc application on quality traits of barley in

semi arid zones of Turkey. Plant Soil and Environment. 2005;51(7):328.
139. Gupta UC, MacLeod J, Sterling J. Effects of boron and nitrogen on grain yield and boron and nitrogen concentrations of barley and wheat. Soil Science Society of America Journal. 1976;40(5):723-726.
140. Rashid A, Rafique E, Bughio N. Diagnosing boron deficiency in rapeseed and mustard by plant analysis and soil testing. Communications in Soil Science & Plant Analysis. 1994;25(17-18):2883-2897.
141. Pilbeam D, Kirkby E. The physiological role of boron in plants. Journal of Plant Nutrition. 1983;6(7):563-582.
142. Shelp BJ, Marentes E, Kitheka AM, Vivekanandan P. Boron mobility in plants. Physiologia Plantarum. 1995;94:356-361.
143. Dell B, Huang L. Physiological response of plants to low boron. Plant and Soil. 1997; 193(1-2):103-120.
144. Tanaka M, Fujiwara T. Physiological roles and transport mechanisms of boron: perspectives from plants. Pflügers Archiv-European Journal of Physiology. 2008; 456(4):671-677.
145. Farooq M, Wahid A, Siddique KH. Micronutrient application through seed treatments- a review. Journal of Soil Science and Plant Nutrition. 2012;12(1): 125-142.
146. Shorrocks VM. The occurrence and correction of boron deficiency. Plant and Soil. 1997;193(1-2):121-148.
147. Keerati-Kasikorn P, Panya P, Bell R, Loneragan J. Effect of boron on peanut grown in north-east soils: 1986; 1988.
148. Adhikary BH, Shrestha J, Baral BR. Effects of micronutrietns on growth and productivity of maize in acidic soil; 2003.
149. Moura J, Prado R, Benvindo R, Chaves Alencar L. Applying boron to coconut palm plants: effects on the soil, on the plant nutritional status and on productivity boron to coconut palm trees. Journal of Soil Science and Plant Nutrition. 2013;13:79-85.
150. Tadesse T, Dechassa N, Bayu W, Gebeyehu S. Effects of farmyard manure and inorganic fertilizer application on soil physico-chemical properties and nutrient balance in rain-fed lowland rice ecosystem. American Journal of Plant Sciences. 2013;4:309.
151. Foy CD. Soil chemical factors limiting plant root growth, in Limitations to plant root growth. Springer. 1992;97-149.
152. Evans HJ. Role of molybdenum in plant nutrition. Soil Science. 1956;81(3):199-208.
153. Campo RJ, Albino UB, Hungria M. Importance of molybdenum and cobalt to the biological nitrogen fixation, in nitrogen fixation: From molecules to crop productivity. Springer. 2000;597-598.
154. Valenciano J, Boto A, Marcelo V. Response of chickpea *(Cicer arietinum L.)* yield to zinc, boron and molybdenum application under pot conditions. Spanish
155. Rong DYQLX, Huang Zhi Yao H, Yuan JHZXH, Ting C. Effect of boron and molybdenum on nitrogen metabolism of peanut [J]. Acta Agronomica Sinica. 2001; 5:008.
156. Glaser B. Prehistorically modified soils of central Amazonia: a model for sustainable agriculture in the twenty-first century. Philosophical Transactions of the Royal Society B: Biological Sciences. 2007; 362(1478):187-196.
157. Sharma S, Sharma S. Effect of different combinations of inorganic nutrients and farmyard manure on the sustainability of a rice-wheat-mungbean cropping system. Acta Agronomica Hungarica. 2006;54(1): 93-99.
158. Saracoglu A, Saracoglu KT, Aylu B, Fidan V. Influence of integrated nutrients on growth, yield and quality of maize (*Zea mays* L.). American Journal of Plant Sciences. 2011;2:63.
159. Raboy V, Dickinson DB. Effect of phosphorus and zinc nutrition on soybean seed phytic acid and zinc. Plant Physiology. 1984;75(4):1094-1098.
160. Lisuma JB, Semoka J, Semu E. Maize yield response and nutrient uptake after micronutrient application on a volcanic soil. Agronomy Journal. 2006;98(2):402-406.
161. Reddy TP, Umadevi M, Rao PC. Effect of fly ash and farm yard manure on soil properties and yield of rice grown on an inceptisol. Agricultural Science Digest. 2010;30(4):281-285.
162. Altieri MA. Agroecology: The science of natural resource management for poor farmers in marginal environments. Agriculture, Ecosystems & Environment. 2002;93(1):1-24.
163. Hobbs PR, Sayre K, Gupta R. The role of conservation agriculture in sustainable agriculture. Philosophical Transactions of the Royal Society B: Biological Sciences. 2008;363(1491):543-555.

164. Celik I, Ortas I, Kilic S. Effects of compost, mycorrhiza, manure and fertilizer on some physical properties of a Chromoxerert soil. Soil and Tillage Research. 2004;78(1):59-67.
165. Khaliq A, Abbasi MK, Hussain T. Effects of integrated use of organic and inorganic nutrient sources with effective microorganisms (EM) on seed cotton yield in Pakistan. Bioresource Technology. 2006;97(8):967-972.
166. Yanai J, Nakata S, Funakawa S, Nawata E, Tulaphitak T, Katawatin R, Kosaki T. Re-evaluation of fertility status of sandy soil in northeast Thailand with reference to soil-plant nutrient budgets. Management of Tropical Sandy Soils for Sutainable Agriculture. FAO Regional Office for Asia and the Pacific. 2005;368-372.
167. Odeleye F, Odeleye O, Dada OA. The performance of soybean *(Glycine max* (L.) Merrill*)* under varying weeding regimes in south western Nigeria. Notulae Botanicae Horti Agrobotanici Cluj-Napoca. 2007; 35(1):27-36.
168. Piraveena S, Seran TH. The effect of cattle manure enriched with ERP fertilizer on seed yield of soybean (*Glycine max*) in sandy regosol. Journal of Food and Agriculture. 2013;3(1-2):24-30.

11

Number of Nodes and Part of Vine Cutting Effect on the Growth and Yield of Sweetpotato *(Ipomoea batatas (L.)* Lam) in Transitional Zone of Ghana

M. E. Essilfie[1*], H. K. Dapaah[1], J. W. Tevor[1] and K. Darkwa[1]

[1]College of Agriculture Education, University of Education, Winneba P.O.Box 40, Mampong-Ashanti, Ghana.

Authors' contributions

This work was carried out in collaboration between all authors. Author HKD designed the study, wrote the protocol, analyzed data and edited manuscript. Author MEE was involved in the design of the study, carried out the field work and wrote the first draft. Author KD collected and compiled data. Author JWT managed the literature searches and critically reviewed the first draft. All authors read and approved the final manuscript.

Editor(s):
(1) Marco Trevisan, Institute of Agricultural Chemistry and Environmental Research Centre BIOMASS, Faculty of Agriculture, Catholic University of the Sacred Heart, Italy.
Reviewers:
(1) Shelley Gupta, P G Moze College of Engineering, Pune, India.
(2) Mohammad Arif, Kansas State University, Kansas, USA.

ABSTRACT

Two field experiments were conducted at the Multipurpose crop nursery of the University of Education, Winneba, Mampong campus from May to September, 2013 and June to October, 2014 respectively to investigate into the effect of number of nodes (4, 5 and 6) and vine part (apical and semiwoody) on the growth and yield of sweetpotato. The results from both seasons showed that many of the characters measured increased with increase node number up to 6 nodes and with apical vine part. The apical vine cuttings established earlier than semiwoody cuttings in both seasons. Node numbers did not have significant effect on any of the yield characters measured although the 5 and 6 node cuttings gave higher marketable tuber weight per plot, tuber length and marketable tuber number per plot than the 4 node cuttings during 2014 growing season. The 5 node apical cuttings produced significantly higher vegetative biomass than 5 node semiwoody and the 4 and 6 node apical and semiwoody cuttings.

**Corresponding author: E-mail: kwabenadarkwa@gmail.com*

Keywords: Sweetpotato; apical vine cutting; number of nodes; semiwoody vine cutting.

1. INTRODUCTION

Among the root and tuber crops grown in the world, sweetpotato ranks second after cassava [1]. Sweetpotato has one of the highest dry matter productivity rates among crops [2]. The crop produces 152 MJ ha^{-1} day^{-1} calories as compared to 121 MJ ha^{-1} day^{-1} in cassava, 151 MJ ha^{-1} day^{-1} in rice, 135 MJ ha^{-1} day^{-1} in wheat and 159 MJ ha^{-1} day^{-1} in maize [3]. The carbohydrate rich root is used as subsidiary food after boiling or baking. In some countries the vine tops are used as vegetables. In previous studies in Ghana, consumers accepted and utilized sweetpotato leaves as food through modified and culturally-acceptable traditional recipes [4]. Vines form an excellent source of green fodder for cattle. The crop is efficient in the production of carbohydrates, proteins, vitamins and cash income per unit area of land and time [5]. The average storage root yield of sweetpotato varies between 10 and 28 t ha^{-1} [3]. When compared with other root and tuber crops, sweetpotato yields considerably high amount of energy per unit area such that it comes third to the potato (*Solanum tuberusom*) and cassava (*Manihot esculenta*) in production across the world [6]. The Crop Research Institute of the Council for Scientific and Industrial Research (CSIR-CRI) in Ghana has developed the *Okumkom* variety of sweetpotato which has a yield capacity of over 19 t ha^{-1} and are for human consumption as *'ampesi'* and chips. Important as it is, the crop has hitherto received little attention from consumers in Ghana. Sweetpotato appears to be the least recognized by both farmers and consumers. This is probably because the crop is considered poor man's food and also as a dessert rather than as a staple food even though it is comparable to the other root crops in yield and quality. Yields of sweetpotato could be improved by the use of good planting materials in terms of number of nodes and part of vine used.

Vine cuttings of sweetpotato are better planting materials than sprouts from vine cutting free from soil-borne diseases [7]. Yield of sweetpotato could be improved by the use of good planting materials. Cuttings from the apical portion of the shoot are regarded as a better planting material than middle or basal vine cuttings [8,9]. Apical cuttings may ensure better rooting and establishment and faster shoot growth and therefore early closure for weed suppression [10,8]. The apical and middle portion vine cuttings are also regarded to be the best planting materials for getting higher root yield [11]. Basal vine cuttings are usually thick and woody, sometimes fail to establish and there is a greater chance of weevil incidence due to proximity with the crown portion where sweetpotato weevil multiplies [12]. Cuttings from apical shoot grow more vigorously and produce large storage roots compared with cuttings from middle and basal portions. Tewe et al. [13] reported that apical cuttings of sweetpotato had significantly higher marketable and total tuber yield than the middle followed by the basal cuttings. Cuttings from the shoot apex give more branches due to the fact that they establish quickly in the soil by initiating more roots and thereby encourage subsequent production of branches and greater number of root tubers than semi hardwood and hardwood cuttings [14].

The number of nodes on cuttings used as planting material may be an important aspect of yield variability. Increase in node number might result in more nodes buried and more roots initiated from the nodes for early establishment. In higher node number cuttings, early rapid growth, tuber initiation and bulking began earlier than in lower node number cuttings. This translates into higher tuber yield in the higher node number cuttings. Onwueme and Sinha [15] indicated that tuber yield tends to increase with increase in the length of the vine cutting used and a length of about 30 cm is recommended. Cuttings of greater length than this tend to be wasteful of planting material, while shorter cuttings establish more slowly, and give poor yields. Farmers in Ghana generally use long lengths of vine as planting material obtained from established sweetpotato field. This method uses quite a lot of plant material that if taken from older portions of the vine may be a potential source of insect pests and diseases [16]. The length of vine and the part used may affect the growth and yield of sweetpotato favourably or unfavourably. Most farmers plant the crop using the vine cuttings irrespective of the number of nodes and part of vine cutting. The agronomy of sweetpotato is well documented. Nevertheless, understanding the effect of number of nodes and part of vine cutting on growth and development is important for improving the yield and quality of sweetpotato. The objective of this study was to assess the effect of number of nodes and part of vine cutting on the growth and yield of *Okumkom* variety of sweetpotato in the transitional zone of Ghana.

2. MATERIALS AND METHODS

Two field experiments were conducted at the Multi-purpose crop nursery of the University of Education, Winneba, Mampong-Ashanti campus from May to September, 2013 and June to October, 2014 respectively. The soil type is the savanna ochrosol formed from the Voltaian sandstone of the Afram plains. Texturally, the soil is friable with a thin layer of organic matter and is deep and brown-sandy loam and well-drained. It however has a good water-holding capacity. The soil has been classified by FAO / UNESCO legend as Chronic Luvisol and locally as the Bediesi series with a pH range of 4.0-6.5.The weather conditions during the experimental periods are summed up in Tables 1 and 2.

The planting material used for the study was vine cuttings of *Okumkom* variety of sweetpotato obtained from Crop Research Institute of CSIR at Fumesua near Kumasi. The experimental design was a 2 x 3 factorial arranged in randomized complete block design (RCBD) with three replicates made up of three number of nodes (4, 5 and 6) and two parts of vine cuttings (apical and semiwoody) was assigned to each block. Each treatment plot measured 4.0 m x 3.0 m. Ridges were constructed 1.0 m between rows and planted at 0.3 m within row plants.

Vine cuttings of sweetpotato variety *Okumkom* of node numbers 4, 5 and 6 and topmost apical and semiwoody sections were planted after preparation of ridges. Number of branches were counted on three plants from the two central rows and three plants were also randomly sampled for vine length, vine diameter, fresh and dry matter accumulation. Data were collected from four weeks after planting and at two weeks interval for twelve weeks. Percentage crop establishment was estimated by counting the number of vine sprouted and established from the two central rows and the percentage estimated for each treatment plot. Fresh vine weight and tuber dry weight at harvest, number of marketable and unmarketable tubers, marketable and unmarketable tuber weight, tuber yield, and yield components including tuber length, tuber diameter, percentage pest infested tuber were estimated from the two central rows. Data analysis was done using ANOVA and Genstat Statistical Package. Least Significant Difference (LSD) was used to separate means at 5% level of probability.

3. RESULTS AND DISCUSSION

3.1 Percentage Crop Establishment

The percentage crop establishment during the 2013 and 2014 growing seasons range from (79.6 – 98.1) and (72.2-92.6), respectively. There was no significant difference between node numbers (4, 5 and 6) in percentage crop establishment during the 2013 and 2014

Table 1. Climate data for 2013 rainy season for experiment one (1)

Month	Total rainfall (mm)	Mean relative humidity (%)		Mean temperature (℃)	
		06:00 h	15:00 h	Min	Max
May	207.4	97	62	22.7	31.6
June	114.9	96	67	22.6	30.0
July	138.0	97	72	21.9	48.3
August	6.0	95	70	21.5	27.7
September	219.8	97	69	22.9	29.6
Total	686.1				

Meteorological Department – Mampong Ashanti, (2013)

Table 2. Climate data for 2014 rainy season for experiment two (2)

Month	Total rainfall (mm)	Mean relative humidity (%)		Mean temperature (℃)	
		06:00 h	15:00 h	Min	Max
June	268.0	97	67	23.0	30.4
July	71.6	96	71	22.2	28.7
August	56.4	96	73	21.8	28.0
September	113.2	95	72	22.4	29.2
October	180.1	97	64	22.5	31.0
Total	689.3				

Meteorological Department – Mampong Ashanti, (2014)

growing seasons although 6 node cuttings produced higher percentage establishment during the 2013 growing season than in the 2014 growing season (Table 3). This might be due to the fact that with increase in node number, the number of buds increased and these acted as sources of hormones for rooting and sprouting. The increase in percentage establishment with increase in number of nodes per cutting is in agreement with Amoah [17] that the 5- node and 7- node cuttings took significantly less number of days to achieve 100 percent establishment than the 3- node cuttings. The apical vine part had significantly higher percentage crop establishment than the semiwoody vine part. This might be due to actively growing portion of the vine part coupled with reduction in weed competition for good plant growth later during the season [10]. There was no significant difference between node numbers in percentage crop establishment. This might be attributed to the fact that with increasing node number, vine cutting use for planting tends to be wasteful while shorter cutting establish more slowly [15].

Table 3. Percentage crop establishment during 2013 and 2014 experimental periods

Treatment	2013	2014
4 nodes + apical vine part	94.4	92.6
4 nodes + Semi Woody part	79.6	72.2
5 nodes + apical vine part	90.7	88.9
5 nodes + Semi Woody part	85.2	83.3
6 nodes + apical vine part	98.1	87.0
6 nodes + Semi Woody part	83.3	74.1
Mean	88.6	83.0
LSD (P=0.05) node	NS	NS
LSD (P=0.05) Vine part	6.74	NS
LSD (P=0.05)Node + Vine part	NS	NS
% CV	9.73	11.11

A substantially higher percentage crop establishment was produced during the 2013 growing season than during the 2014 growing season. This might be due to initial higher rainfall experienced during the 2013 growing season (Table 3).

3.2 Vegetative Growth

3.2.1 Vine fresh weight, vine dry weight and root dry weight

Vine fresh weight, vine dry weight and root dry weight with the different node numbers (4,5 and 6) and vine part (apical and semiwoody cuttings) did not differ significantly from treatments in both 2013 and 2014 growing seasons (Figs. 1, 2 and 3). The lack of significant response showed by the different node numbers and vine part might be attributed to the innate characteristics of the sweetpotato variety, *Okumkom* used for the study. *Okumkom* can colonize or establish on marginal soils in spite of the prevailing climatic conditions of the growing area. It readily produces adventitious roots, trailing vines and exhibits aggressive ground cover [18].

3.3 Vine Length

Vine length of sweetpotato grown during the 2013 growing season with semiwoody vine part was significantly higher than the apical vine cutting (Fig. 4). The 4 node apical cutting produced higher vine length compared with 5 and 6 node cuttings at 86 days after planting (DAP) in both growing seasons. This result however, contradicts the report of Ray et al. [6] who reported that increasing the number of nodes increased vine length. Apical and semiwoody cuttings with 4, 5 and 6 nodes grown during the 2014 growing season produced higher vine length than in 2013. This might be due to better climatic condition experienced during the 2014 growing season than 2013 season. The result is in conformity with the report of Rasco et al. [19] that field and environmental conditions may influence the relationship between cutting length of sweetpotato and the crop improvement.

3.4 Number of Branches

The interaction between number of nodes and vine part was significant in number of vine branches during the 2013 and 2014 growing seasons (Fig. 5). The 5 nodes apical cutting produced higher numbers of branches from 58 DAP to 86 days after planting in both growing seasons than 4 and 6 nodes apical cutting. However, the 4 nodes semiwoody cutting produced higher number of branches from 58 DAPS to 86 days after planting in both growing seasons than 5 and 6 nodes semiwoody cutting (Fig. 5). This might be due to differences in node number and vine part used as planting material. This result is in agreement with Mukhopathyay et al. [11] that apical and semiwoody portions of vine cutting are found to be the best planting materials for getting higher root yield which is as a result of production of higher number of vine branches.

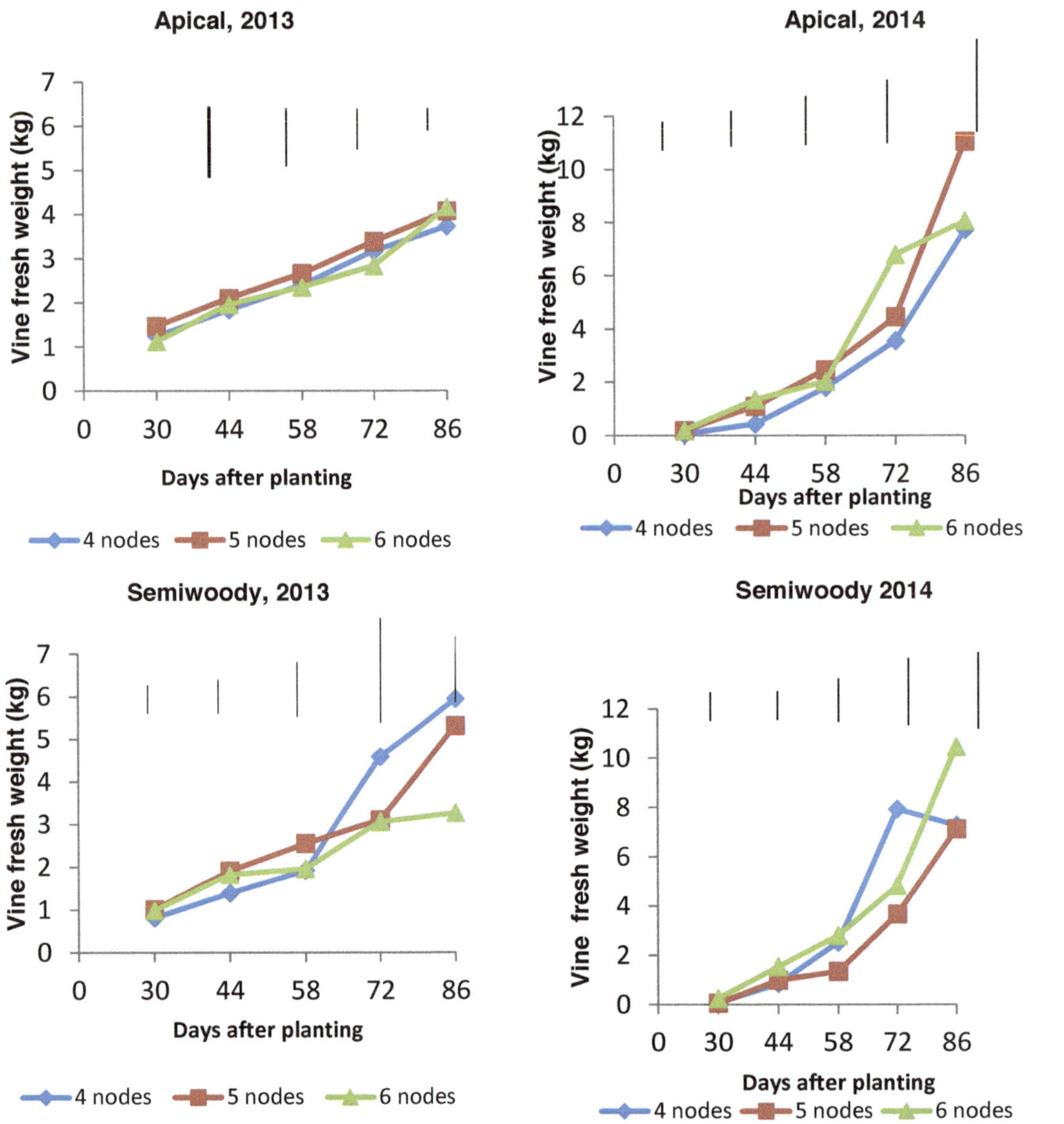

Fig. 1. Vine fresh weight of apical and semiwoody cuttings 30-86 days after planting during 2013 and 2014 growing seasons

There was a significant difference between apical and semiwoody vine cutting in number of branches during both growing seasons (Fig. 5). This result is similar to that found by Choudbury et al. [14] that cuttings from the shoot apex give more branches due to the fact that they establish quickly in the soil by initiating more roots and thereby encourage subsequent production of branches than semi hardwood cuttings. Degras [20] also reported that for better performance of sweetpotato the apical vine cuttings are the best planting material to use.

3.5 Stem Diameter

The 5 nodes apical cutting produced higher vine diameter from 58 DAP to 86 days after planting in both growing seasons than 4 and 6 nodes apical cutting (Fig. 6). The increase in stem diameter with 5 node apical cuttings in both growing seasons might be due to its early establishment than the 4 and 6 node cuttings and this agrees with the findings of [17]. However, the 4 nodes semiwoody cutting produced higher vine diameter from 58 DAP to 86 days after planting with the lowest recorded by the 5 nodes cutting in both growing seasons. This might be due to differences in node number and its interaction with the environment. Semiwoody vine part is also regarded to be one of the best planting material and this is in agreement with the findings by Mukhopathyay et al. [11].

3.6 Yield and Yield Components of Sweetpotato

3.6.1 Number of marketable tubers per plot

Apical vine with 4 and 6 nodes had higher number of marketable tuber per plot for 2013 and 2014 growing season respectively. Apical vine part produced significantly higher number of marketable tubers per plot than semiwoody vine during the 2013 growing period. A significantly higher number of marketable tubers per plot with apical vine part might be due to more vigorous growth and actively growing part of the vine [8, 10 and 13] indicated that apical cutting ensure better rooting and establishment and faster shoot growth and also getting higher marketable root tubers [11]. In the 2013 growing season both apical and semiwoody vines with 4, 5 and 6 nodes had higher number of marketable tubers per plot than during the 2014 season. The lower number of marketable tubers per plot for apical vine part with 4 and 6 nodes during 2014 growing season might be due to higher rainfall during tuber formation stage that resulted in the higher vegetative growth at the expense of tuber formation [21].

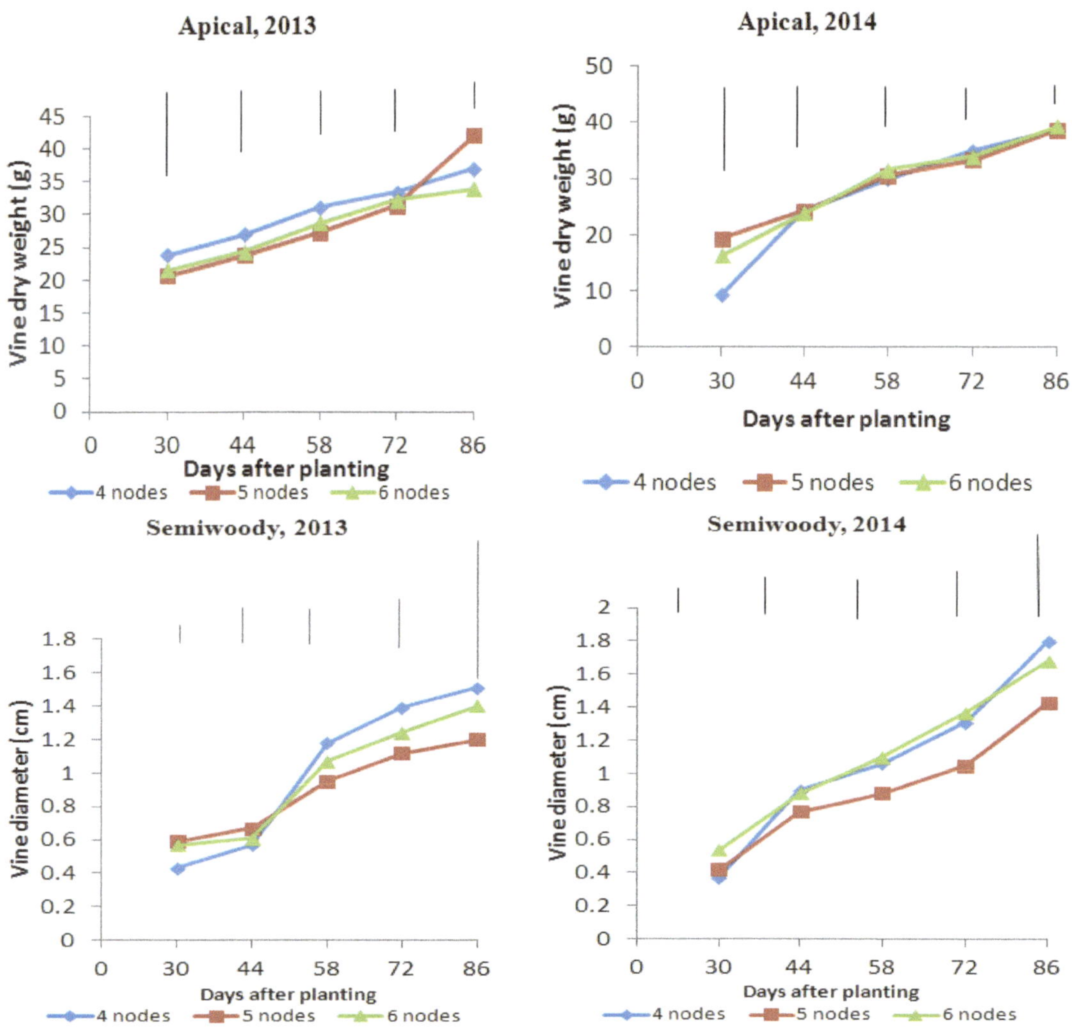

Fig. 2. Vine dry weight of apical and semiwoody cuttings 30-86 days after planting during 2013 and 2014 growing season

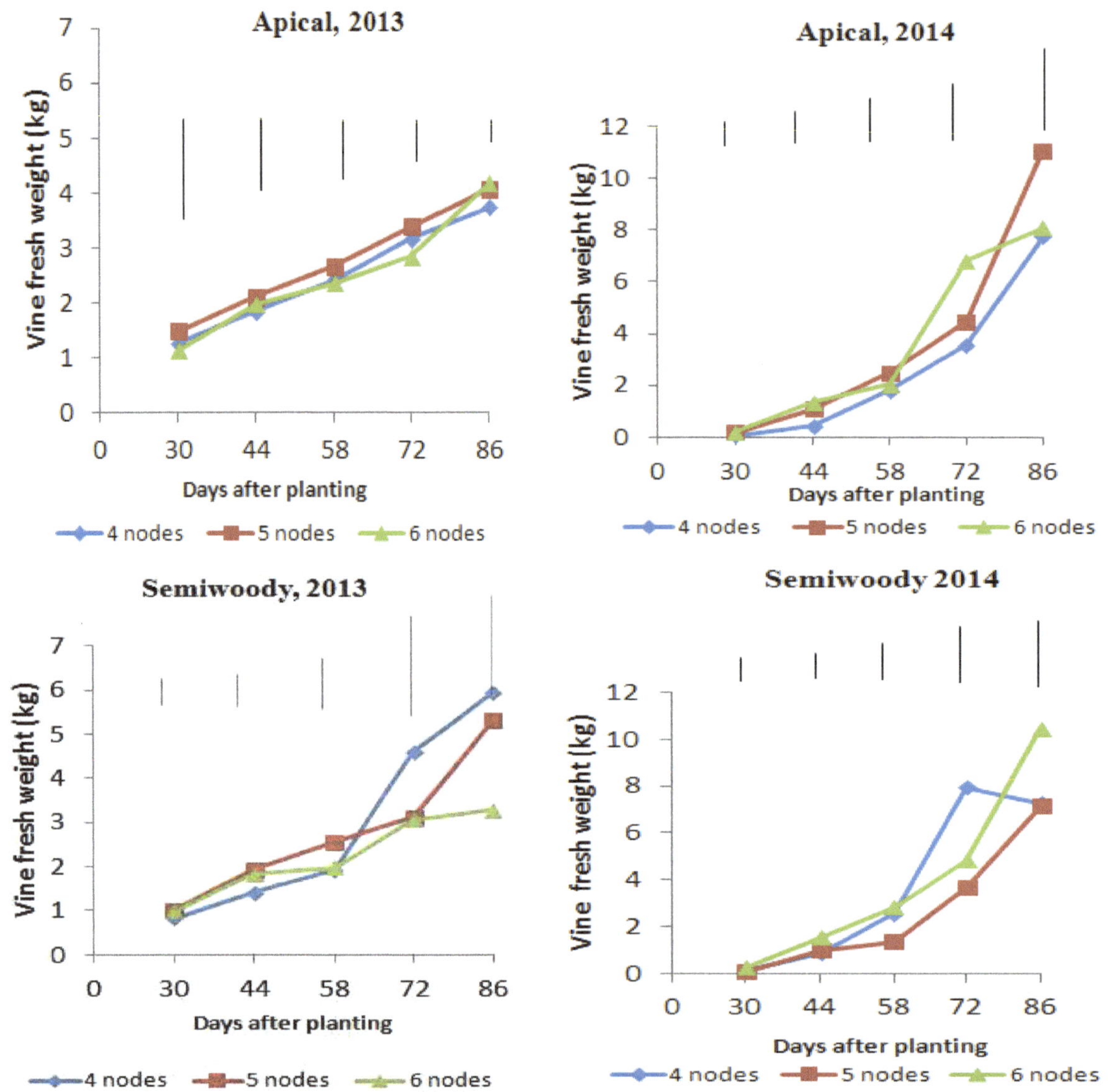

Fig. 3. Root dry weight of apical and semiwoody cuttings 30-86 days after planting during 2013 and 2014 growing seasons

Table 4. Number of marketable tuber per plot during 2013 and 2014 experimental periods

Treatment	2013	2014
4 nodes + apical vine part	52.0	11.0
4 nodes + Semi Woody part	31.0	20.0
5 nodes + apical vine part	50.0	29.0
5 nodes + Semi Woody part	39.0	17.0
6 nodes + apical vine part	45.0	32.0
6 nodes + Semi Woody part	32.0	21.0
Mean	41.4	21.6
LSD (P= 0.05) Node	NS	NS
LSD (P= 0.05) Vine part	9.09	NS
LSD (P= 0.05) Node x Vine part	NS	NS
%CV	46.60	66.44

3.6.2 Number of unmarketable tubers per plot

There was a significantly higher number of unmarketable tuber per plot between apical vine part from semiwoody vine part during 2013 growing season. This might be due to higher number of branches produced by the apical vine cutting than the semiwoody cuttings that resulted in more tuberous root formation than the semiwoody [14]. There was however, no significant difference between node numbers (4, 5 and 6) in unmarketable tuber per plot. In the case of 2014 growing season, there was no significant difference between node number and vine part. In the 2013 growing season higher number of unmarketable tubers was produced than in the 2014 growing season. This might be due to initial higher rainfall during the 2013 growing season. [21] indicated that excessive rain prevent proper root formation and that luxurious growing condition caused by excessive irrigation lead to luxurious vine growth that result in poorly developed roots.

3.7 Tuber Length

There was no significant difference between node numbers (4 ,5 and 6) and vine part (apical and semiwoody) in tuber length during the 2013 and 2014 growing seasons although apical and semiwoody cuttings with node numbers (4, 5 and 6) produced higher tuber length during the 2014 growing season than in the 2013 growing season. The reduction in tuber length during the 2013 growing season might be due to initial drought experienced during tuber formation stage for the growing period. Similarly, the higher rainfall during the 2014 growing season might have contributed to the result obtained. Similar result was found in [22,16] who indicated that irrigation generally increase yields and improve the quality of storage roots of sweetpotato.

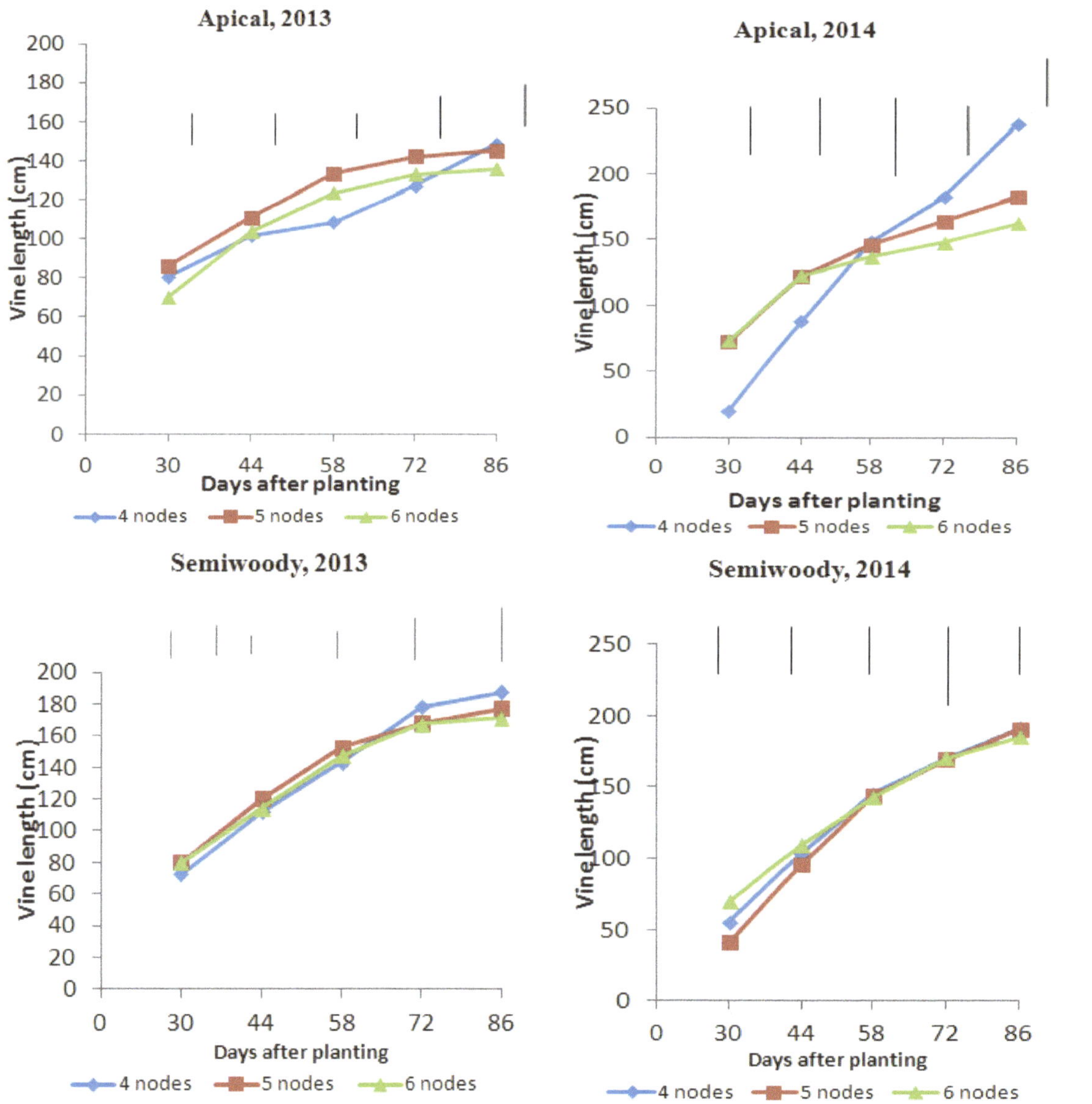

Fig. 4. Vine length of apical and semiwoody cuttings 30-86 days after planting during 2013 and 2014 growing seasons

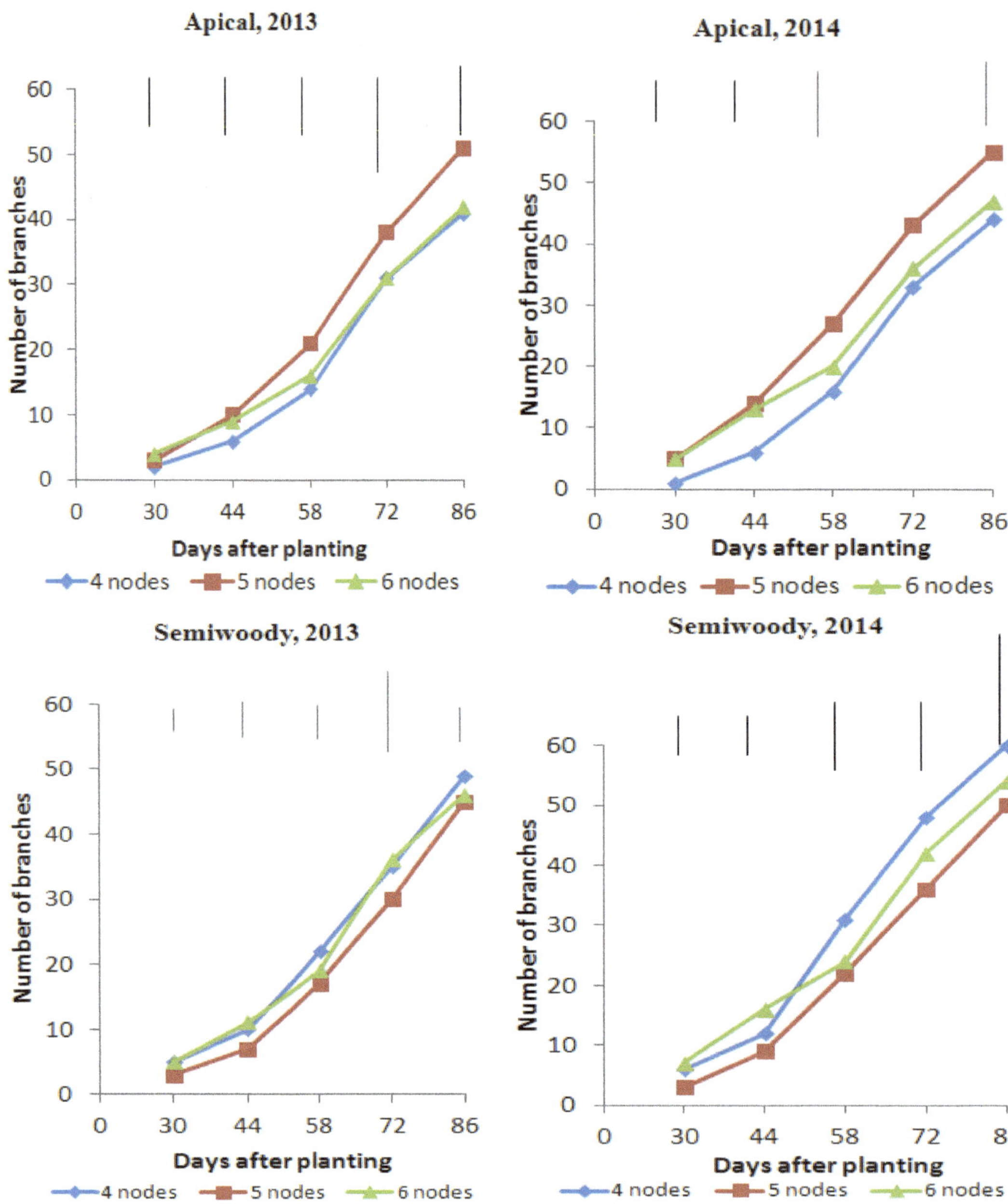

Fig. 5. Number of branches of apical and semiwoody cuttings 30-86 days after planting during 2013 and 2014 growing seasons

3.8 Tuber Diameter

There was no significant difference between node numbers (4, 5 and 6) and vine part in tuber diameter during 2013 and 2014 growing seasons although 5 node apical cutting grown during 2013 growing season produced higher tuber diameter than the 4 and 6 nodes cutting while in 2014 season 4 nodes apical had the highest tuber diameter (Table 7). Tuber diameter increase due to apical vine part used might be due to the fact that vine cuttings from apical shoot grow more vigorously and produce large storage roots compared with cuttings from middle and basal portions. This result is similar to those found by [13] that apical cuttings of sweetpotato had significantly higher tuber size than the middle followed by the basal cuttings.

The higher tuber diameter produced by the 5 node cuttings than the 4 and 6 nodes cutting during the 2013 growing season might be due to increase in root tuber size. Increase in root tuber size occurs through increase in cell number and cell size while the storage root weight increases through accumulation of photosynthates. Wilson [23] reported that storage root growth (bulking) involves increase in size and weight. The higher tuber diameter produced during the 2014 growing season than during 2013 growing season (Table 7) might be due to differences in rainfall experienced during the 2014 growing season than the 2013 season. Increase in cell number and cell size in root tuber is under the control of high rainfall. This is in line with the findings of [16] that irrigation generally improves the grade and quality of storage roots of sweetpotato.

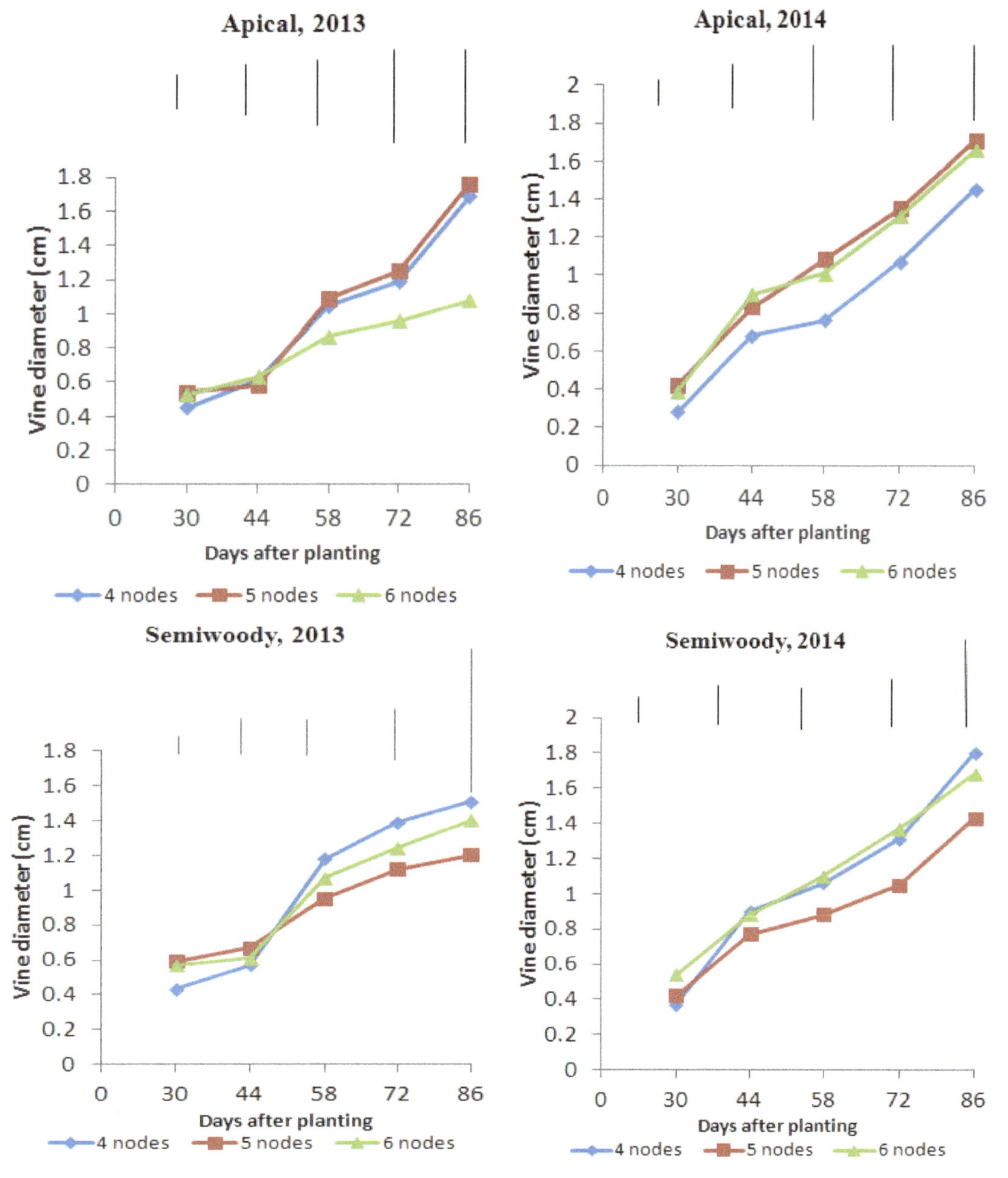

Fig. 6. Vine diameter of apical and semiwoody cuttings 30-86 days after planting during 2013 and 2014 growing seasons

Table 5. Number of unmarketable tuber per plot during 2013 and 2014 experimental periods

Treatment	2013	2014
4 nodes + apical vine part	33.0	9.0
4 nodes + Semi Woody part	12.0	7.0
5 nodes + apical vine part	25.0	13.0
5 nodes + Semi Woody part	21.0	5.0
6 nodes + apical vine part	28.0	11.0
6 nodes + Semi Woody part	18.0	9.0
Mean	22.9	9.0
LSD (P=0.05) node	NS	NS
LSD (P=0.05) Vine part	6.12	NS
LSD (P=0.05) Nodes x vine part	10.61	NS
% CV	36.14	66.44

Table 6. Tuber length during 2013 and 2014 experimental periods

Treatment	2013	2014
4 nodes + apical vine part	14.7	18.3
4 nodes + Semi Woody part	13.5	18.5
5 nodes + apical vine part	14.3	21.0
5 nodes + Semi Woody part	13.6	17.1
6 nodes + apical vine part	14.6	19.4
6 nodes + Semi Woody part	14.6	19.6
Mean	14.2	18.8
LSD (P=0.05) node	NS	NS
LSD (P=0.05) Vine part	NS	NS
LSD (P=0.05) Nodes x vine part	NS	NS
% CV	25.81	12.98

Table 7. Tuber diameter during 2013 and 2014 experimental periods

Treatment	2013	2014
4 nodes + apical vine part	3.2	7.8
4 nodes + Semi woody part	4.3	7.4
5 nodes + apical vine part	4.7	7.3
5 nodes + Semi woody part	4.6	7.1
6 nodes + apical vine part	4.5	7.1
6 nodes + Semi woody part	3.4	6.5
Mean	4.1	7.2
LSD (P=0.05) node	NS	NS
LSD (P=0.05) Vine part	NS	NS
LSD (P=0.05) Nodes x vine part	NS	NS
% CV	25.89	11.80

3.8.1 Marketable tuber weight per plot

There was no significant difference between node numbers (4 ,5 and 6) and vine part (apical and semiwoody) cuttings in marketable tuber weight per plot in both growing seasons although 5 and 6 nodes apical cuttings grown during 2013 and 2014 growing seasons produced higher marketable tuber weight per plot respectively than the 4 node cutting (Table 8). This might be due to initial drought experienced during tuber formation stage with supplementary irrigation that resulted in higher marketable tuber weight per plot during the 2013 growing season. According to [24] supplementary irrigation may significantly improve root quality in terms of tuber weight when the rainfall distribution is erratic. Similarly, yield increase due to increasing node number might be due to increasing node number points for tuber formation. Due to early rapid growth in higher node number cuttings, tuber initiation and bulking began earlier than in lower node number cuttings which translated into higher tuber yield in the higher node number cuttings. The higher tuber weight of the 5 and 6 nodes cuttings is in agreement with the findings of [17] that the 5 and 7 node cuttings produced significantly higher tuber weight than the 3 node cuttings. The interaction between number of nodes and vine part was significant in marketable tuber weight per plot during the 2014 growing season than in 2013.

The high marketable tuber weight per plot during the 2014 growing season might be due to higher fresh vine weight experienced during the 2014 growing season than in 2013 season. This result is similar to that reported by Keutgen et al. [25] that storage root growth depends on the sink strength, the potential of leaves to export photosynthesis and on the photosynthetic efficiency of leaves.

Table 8. Marketable tuber weight per plot (Kg) during 2013 and 2014 experimental periods

Treatment	2013	2014
4 nodes + apical vine part	11.9	3.2
4 nodes + Semi woody part	10.8	6.5
5 nodes + apical vine part	12.2	9.1
5 nodes + Semi woody part	10.6	4.7
6 nodes + apical vine part	11.3	10.4
6 nodes + Semi woody part	9.5	5.7
Mean	11.08	6.6
LSD (P=0.05) node	NS	NS
LSD (P=0.05) Vine part	NS	NS
LSD (P=0.05) Nodes x vine part	NS	4.31
% CV	19.02	54.52

3.8.2 Unmarketable tuber weight per plot

The weight of unmarketable tuber per plot was not significantly affected by either the node number or the part of vine cutting. However, there was a general trend towards an increase in unmarketable tuber weight per plot with apical vine cutting for the different node numbers (4, 5 and 6) in both 2013 and 2014 growing seasons. This might probably be due to higher number of branches in apical vine cutting with more number of unmarketable tubers that led to higher weight of unmarketable tuber per plot than semiwoody vine cutting [14].

Table 9. Unmarketable tuber weight per plot during 2013 and 2014 experimental periods

Treatment	2013	2014
4 nodes + apical vine part	1.5	0.4
4 nodes + Semi woody part	0.6	0.3
5 nodes + apical vine part	1.3	0.7
5 nodes + Semi woody part	1.0	0.2
6 nodes + apical vine part	1.6	0.4
6 nodes + Semi woody part	0.8	0.3
Mean	1.13	0.45
LSD (P=0.05) node	NS	NS
LSD (P=0.05) Vine part	NS	NS
LSD (P=0.05) Nodes x vine part	NS	NS
% CV	42.09	43.20

3.8.3 Vine fresh weight at harvest

The interaction between number of nodes and vine part was significant in vine fresh weight at harvest during the 2013 growing season. The 5 and 6 nodes apical cutting had higher vine fresh weight at harvest than the 4 node apical cutting in both growing seasons. The difference between 4, 5 and 6 nodes cutting was not significant although there was a general vine weight increase at harvest with increasing node number for both 2013 and 2014 growing seasons. This might be due to the higher number of branches recorded and resultant higher number of leaves in higher node number cutting. The vine cutting part treatment did not have significant effect on weight of fresh vegetative biomass, although there was a decreasing vine fresh weight at harvest with 4 and 6 nodes apical vine cutting during the 2014 season. The vine fresh weight at harvest for both apical and semiwoody cuttings with the different node numbers during the 2014 growing season was higher compared with the 2013 growing season. This might be due to initial higher monthly rainfall during the 2014 growing season than in 2013.

3.8.4 Percentage pest infested tuber

Percentage pest infested tuber was not significantly affected by either the node number or the part of vine cutting in both 2013 and 2014 growing seasons although there was a general trend towards an increase in the percentage pest infested tuber for both apical and semiwoody cuttings with different node numbers (4, 5 and 6) during the 2014 growing season (Table 11). The total monthly rainfall values were relatively high during the initial cropping period of 2014 coupled with high temperature (Table 2). This might have resulted in the high incidence of pest infestation. This result is similar to that found by Sowley [26] who reported that wet and warm conditions increase the likelihood of serious pest infestations in sweetpotato. Despite the fewer nodes (4 and 5) and vine part (semiwoody) vine cuttings for root formation, the 4 and 5 nodes semiwoody produced lower number of pest infested tubers than (4, 5 and 6 nodes) apical cuttings during the 2013 growing season while the 4 node apical cutting however, produced lower number of pest infested tubers during the 2014 growing season.

Table 10. Vine fresh weight at harvest during 2013 and 2014 experimental periods

Treatment	2013	2014
4 nodes + apical vine part	17.5	24.20
4 nodes + Semi woody part	27.8	32.07
5 nodes + apical vine part	28.7	34.90
5 nodes + Semi woody part	20.9	28.20
6 nodes + apical vine part	19.8	29.13
6 nodes + Semi woody part	30.1	33.40
Mean	24.05	30.31
LSD (P=0.05) node	NS	NS
LSD (P=0.05) Vine part	NS	NS
LSD (P=0.05) Nodes x vine part	9.01	NS
% CV	27.20	19.93

Table 11. Percentage pest infested tuber during 2013 and 2014 experimental periods

Treatment	2013	2014
4 nodes + apical vine part	2.9	13.3
4 nodes + Semi woody part	2.2	15.6
5 nodes + apical vine part	5.7	23.7
5 nodes + Semi woody part	2.2	17.6
6 nodes + apical vine part	4.5	19.3
6 nodes + Semi woody part	3.7	14.7
Mean	3.5	17.4
LSD (P=0.05) node	NS	NS
LSD (P=0.05) Vine part	NS	NS
LSD (P=0.05) Nodes x vine part	NS	NS
% CV	44.81	45.50

4. CONCLUSION

The results of this study revealed that cuttings used for sweetpotato propagation should come from the apical part with 4, 5 and 6 nodes. Cuttings with 4 nodes apical although do not give high yield but gave high tuber length and tuber diameter in 2013 and 2014 growing seasons respectively. Cuttings with 5 nodes apical gave high vegetative growth in both growing seasons while cuttings with 6 nodes apical gave high yield and marketable tuber during 2014 growing season. However, cuttings with 6 nodes semiwoody gave high vine fresh weight at harvest in both growing seasons. Cutting from semiwoody part with least number of nodes, especially 4 and 5 nodes did not give any significant increase in vegetative biomass and yield and found to be inappropriate planting material.

ACKNOWLEDGEMENTS

The authors are most grateful to the management of the University of Education, Winneba, College of Agriculture Education, Mampong-Ashanti campus, and the entire management of Crop Research Institute-CSIR in Kumasi for the provision of land and planting materials respectively.

COMPETING INTERESTS

Authors have declared that no competing interests exist.

REFERENCES

1. Ray RC, Ravi V. Post harvest spoilage of sweet potato and its control measures. Critical Review of Food Science and Nutrition. 2005;35:623-644.
2. Scott GJ, Rosegrant M, Ringler C. Roots and tubers for the 21st century. Trends, projections and policy for developing countries. Food, agriculture and the environmental discussion paper. International Food Policy Research Institute (IFPRI). Washington D.C. USA; 2000.
3. CTCRI. Tuber crop varieties released the central tuber crops research institute, Technical bulletin series 24 revised and reprinted, central tuber crops research institute. Thiruvananthapuram. 2006;64.
4. Bonsi E, Bonsi C, Zabawa R, Doamekpor P, Kebede E. Concentrated nutritional and economic enhancement of Ghanaian diets using orange-fleshed sweet potato products. Horticulture collabotative research support program. Tuskegee University; 2009
5. Magagula NEM, Ossom EM, Rhykerd RL, Rhykerd CL. Effects of chicken manure on soil properties under sweet potato [*Ipomoea batatas* (L). Lam.] culture in Swaziland. American-Eurasian Journal of Agronomy. 2010;3(2):36-43.
6. Nedunchezhiyan M, Byju G, Jata SK. Sweet potato agronomy, fruit, vegetable and cereal science and biotechnology. Global Science Books. 2012;1-10.
7. Phills BR, Hill WA. Sweet potato propagation. A macro propagation of the sweet potato. The sweet potato for the space mission. Carver Res found. Tuskegee University Tuskegee AL; 1984.
8. Eronica CA, Escalado RG, Trenuela RM. Effect of different portions and length of storage of vine cuttings on the growth and yield of sweet potato. Ann. Trop. Rex 3. 1981;144-149.
9. Schuthesis JR, Cantliffe DJ, Buyan HH. Early plant growth from seed, vegetative cuttings and somatic embryos. J. Am. Soc. Hort. Sci. 1994;119:1104-1115.
10. Hall MR. Shoot duration pre-sprouting enhances sweet potato post production. Hort. Sci. 1987;22:314.
11. Mukhopathyay SK, Sen H, Hana PK. Effect of planting materials on growth and yield of sweet potato. J. of food crops. 1990;16(2):119-222.
12. Nair EM. Agro-techniques and planting material production in sweet potato. In: Byju G, (Ed.) Quality planting material production in tropical tuber crops. Central tuber crops research institute. Thiruvamanthapuram, India. 2006;55-58.
13. Tewe OO, Ojenyi FE, Abu OA. Sweet potato production, utilization and marketing in Nigeria. Social sciences Department. International potato centre (CIP) Lima Peru. 2003;44.
14. Choudbury SH, Ahmed SV, Sharfuddin AFM. Effect of number of nodes in different types of vine cuttings on the growth and yield of sweet potato. Bangladesh Hort. 1986;14(1):29-33.
15. Onwueme C, Sinha TD. Field crop production in tropical Africa. Technical centre for Agriculture and rural co-operation CTA publication, Ede. The Netherlands. 1991;293.
16. Ray CS, Antony E, Singh R, Kar G, Verma HN. Source and relationship in sweet

potato under different irrigation regimes. Journal of Root Crops. 2001;27(1):164-168.

17. Amoah FM. The effect of number of nodes per cutting and potassium fertilizer on the growth, yield and yield components of sweet potato (*Ipomoea batatas* Poir). Ghana Journal of Agriculture Science. 1997;30:53-62.
18. Crop Research Institute. Council for Scientific and Industrial Research, Ghana. Sweetpotato Production Guide. A Resource and Reference Manual. 2004; 12-14.
19. Rasco ET Jr, Amante V. Sweet potato in tropical Asia. Field guides and technical manual. International potato center. Lima, Peru. 2000;266.
20. Degras L. Sweet potato. The tropical Agriculturalist. Macmillan publishers Ltd. Lima, Peru; 2003
21. Maynard AA, Hill EV. How to grow sweet potatoes in Connecticut. The Connecticut Agricultural Experimental Station. New Haven. CT 06504; 2000. Available:http:www.ct.gov\caes
22. Nair GM, Nair VM, Streedhuran C. Response of sweet potato to plastic stress irrigation in summer rice fallows. J of Root Crops. 1996;22(1):45-49.
23. Wilson LA. Tuberization in sweet potato (*Ipomoea batatas* (L.) Lam.) In: Villareal RL, Griggs TD. (eds.), Sweet potato. Proc. Int. Symp. AVRDC, Taiwan, China. 1982;79-93.
24. Ravi V, Indica P. Anatomical studies on tuberization in sweet potato under water deficit stress and stress condition. Journal of Root Crops. 1996;22:105-111.
25. Keutgen M, Kubota F, Saitou K. Effects of exogenous injection of sucrose solution to plant on the carbon distribution to tuberous root and tuberous root production in sweet potato (*Ipomoea batatas* Lam.). Japan. J. Crop Sci. 2001;70:575-579.
26. Sowley ENK. Etiology of storage root of sweet potato (*Ipomoea batatas* (L) Lam) and its control by curing. An Mphil thesis presented to the Department of Crop Science, University of Ghana, Legon. 1999;120.
27. Meteorological Department. Temperature, relative humidity and rainfall distribution Mampong- Ashanti, Ghana; 2013.
28. Meteorological Department. Temperature, relative humidity and rainfall distribution Mampong- Ashanti, Ghana; 2014.

12

Root Architectural Development and Yield Sensitivity of Phosphorus Tolerant Common Bean under Low Soil Phosphorus and Drought

Namugwanya Margaret[1,2], John Stephen Tenywa[1*], Erasmus Otabbong[3] and Twaha Ali Basamba[1]

[1]Department of Agricultural Production, Makerere University, P.O.Box 7062, Kampala, Uganda.
[2]Department of Agriculture, Kyambogo University, P.O.Box 1, Kyambogo, Kampala, Uganda.
[3]Departmentof Soil and Environment, Swedish University of Agricultural Sciences, Ornskoldsvik, Sweden.

Authors' contributions

This study was carried out in collaboration amongst all authors. Author NM conducted this study as part of her Ph.D. studies, she designed, implemented the experiment and analysed data. Author JST co-designed and supervised the study implementation. Authors EO and TAB provided technical input in the implementation of the research. All authors participated in the development of the manuscript and approved it for publication.

Editor(s):
(1) Sławomir Borek, Faculty of Biology, Department of Plant Physiology, Adam Mickiewicz University, Poland.
Reviewers:
(1) Hewan Demissie Degu, Hawassa University, Ethiopia.
(2) Rogerio Faria Vieira, Empresa de Pesquisa Agropecuária de Minas Gerais, Brazil.
(3) Agnaldo Rodrigues de Melo Chaves, Embrapa Semiárido, Brazil.

ABSTRACT

Aims: The study assessed the effect of limited soil phosphorus and drought on yield sensitivity and root architectural development of low phosphorus tolerant common bean *(Phaseolus vulgaris* L.) materials
Study Design: This was a randomised complete block design in which 2*3*5 factorial treatment combinations of drought, P levels and genotypes, respectively.
Place and Duration of Study: The study was carried out in Central Uganda; Nakasongola representing a drought-stress and Mukono Zonal Agricultural Research and Development Institute

**Corresponding author: E-mail: johntenywa@gmail.com*

(MUZARDI) representing non-drought-stress for two rainy seasons (March-June and August - December 2014).

Methodology: In each study site, four low phosphorus tolerant genotypes (AFR703-1, AFR708, JESCA, and MCM2001 using K131 as a local check) were planted in plots treated with 0, 60 and 160 kg P ha^{-1} in the form of Triple Super Phosphate.

Results: The P-treatments neither had significant influences on grain yield nor root response; and no significant interactions with drought and genotypes. Yield did not significantly vary by drought treatments, but by genotypes ($P<0.001$), with AFR708 registering the highest yield of 1122 kg ha^{-1}. Drought-stress induced significant root development, namely adventitious roots, tap and lateral root lengths, and total root lengths, in some genotypes. Genotype AFR703-1 and AFR708 had an edge over the local check; they produced multiple root systems.

Conclusion: Under drought stress, the AFR gene pool was superior in root development, namely number of adventitious and lateral roots, and taproot and lateral root length. In contrast, grain yield of these materials was suppressed by drought stress.

Keywords: Phaseolus vulgaris; drought-stress; multiple root systems; climate changes.

1. INTRODUCTION

Plant-available phosphorus in the soil, together with drought are key abiotic factors severely constraining common bean production, especially in resource-constrained Sub-Saharan African farming systems [1-4]. The two constraints frequently co-exist in soils of the tropics and impact on bean production by well over 60% [1,5-8]. The supply of plant-available P declines along with decreasing soil moisture [9,10], thus the changing climate from wetter to drier conditions aggravates the situation.

Root architecture is a significant component in the performance of common beans grown under low soil P levels and drought conditions. Depending on their architecture, root systems govern the ability of a plant to access nutrients and moisture in a wide range of the soil [11]. Moreover, P is a sparingly mobile nutrient that resides largely in the topsoil layers. Consequently, root systems that dwell more in topsoil explore it and take advantage of P acquisition [12-14]. Conversely, soil moisture is often more abundant in the deeper soil horizons, especially during dry periods; thus, the deeper the root system the more the plant will access moisture [3,11,15].

Based on architectural root traits, selection of genotypes has resulted in improved root systems that perform well either under low soil P availability or drought conditions. Studies show that some plants have abundant diverging patterns of root systems in both top and sub-soil layers, which enhance efficient absorption of P and soil moisture. Suriyagoda, Ryan [7] reported that roots of some Australian native perennial legumes explored extensively, the top and lower layers of soils deficient in both plant-available P and moisture, respectively. A study carried out by Morino, Obrador [6] on Mediterranean shrubs, also revealed that some plant species had sufficient root systems in different soil layers, enabling them to utilise both P and moisture. Thus, improved low P tolerant and drought resistant genotypes have been investigated, but implementation of the results is yet to be promoted [4]. Besides, the response of architectural root traits to the combined effects of low levels of P and moisture deficit in soils has received less attention in common bean because the architectural root traits have been explored independent of the two abiotic constraints. It is not clear whether the yield sensitivity of the P-tolerant genotypes can be stimulated or compromised by the combined effects of low P and drought. It also remains to be investigated the changes in root architecture of these genotypes in response to the combined effects.

It has been observed that common bean genotypes adapt with varying capacities to low levels of P and drought conditions. This may be explained by their different abilities to acquire P and moisture from the soil. Beebe, Rao [8] demonstrated that selection of common bean genotypes to overcome the impact of drought enhanced yield in P limited conditions. A study conducted by Ho, Rosas [11] on root architectural trade-offs for water and P acquisition reported that the shallow-rooted BAT 477 developed a deep root system that increased utilization of P in soils deficient of the nutrient and moisture. Thus, synergism within the root system might be central for adaptation in common bean genotypes to low levels of plant-available P and moisture in the soil. The objective of this study was to assess the effects of limited phosphorus and drought on yield sensitivity and root architectural development of low P-tolerant bean (*Phaseolus vulgaris* L.) materials. The study tested the hypothesis that some low P-tolerant bean genotypes have multiple root systems for both P and soil moisture acquisition.

2. MATERIALS AND METHODS

2.1 Description of Study Sites

A field experiment was carried out in Mukono and Nakasongola districts of central Uganda, during two rainy seasons (March-June and August-December 2014). Mukono District is located at 00° 15' 00" N latitudes and 32° 30' 00" E. It is characterised by bimodal rainfall seasons with a mean annual rainfall of 1100-1600 mm. The district has a mean annual maximum temperature of 25 to 28°C and minimum annual temperature of 15 to 18°C [16]. This site represented the non-drought-stressed environment, herein referred to as non-drought-stress (NDS) treatment. Nakasongola District is located in the cattle corridor and is one of driest districts in Uganda. It lies between 1° 19' 60" N and 1° 19' 58" N, and 32° 30' 00" E. The district receives 500 to 1000 mm of rainfall *annum*$^{-1}$ and experiences a five-month long dry period. The district is characterised by extreme spatial and temporal rainfall variability and experiences high frequent and severe droughts (Kisamba-Mugerwa, 2001 unpublished). This site represented the drought-stressed environment herein referred to as drought-stress (DS) treatment.

2.2 Soil Measurements

For pre-experimental characterisation at each site, soil samples were collected at a depth of 0-20 cm before the plots were prepared. The samples were air-dried and analysed for pH in water (1:1), total N, Bray-1-extractable-P and organic matter. Other properties included exchange cations (K+, Ca^{2+} and Na+). Total N was determined by Kjeldahl digestion process followed by colorimetric procedure and the rest were determined by procedures described by Page, Miller [17]. Laboratory data are presented in Table 1.

2.3 Field Experiment

Treatments included four low P-tolerant common bean (*Phaseolus vulgaris* L.) genotypes; AFR 708, AFR 703-1, JESCA and MCM 2001, tested against K131 as a local check. AFR 708, AFR 703-1, MCM 2001 and K131 are crosses of Mutikired/AFR308, Bean98/ZAA39//ZAA39, IVT831629/BAT1554 and IVT831607/RAB71, respectively. AFR 703-1 and AFR 708 are Andean materials. The rest of the genotypes are of Mesoamerican origin (Dr. Clare Mukankusi Clare, Regional Breeder, East and Central African Bean Network, Kampala Uganda, pers. comm.). The materials were obtained from Bean Research Program of the International Centre for Tropical Agriculture (CIAT) in Uganda. The genotypes were planted in plots treated with P levels of 0, 60 and 160 kg ha^{-1} in the form of Triple Super Phosphate (TSP) in each drought treatment. This was to assess possible drought effects on their tolerance under low P supply. The experiment was arranged in a factorial structural combination of drought treatment, P levels and low P-tolerant common bean genotypes. All treatments were in triplicates and were laid out in a randomised complete block design (RCBD). Plants were spaced at 60 cm between and 20 cm within rows, giving a plant density of 213,333.3 plant ha^{-1}. The experiment was repeated once; in blocks a few meters away from the first blocks in both districts.

Table 1. Physical and chemical properties of the experimental soil (0-20 cm)

Soil characteristics and its unit	Drought-stress (DS)	Non-drought-stress (NDS)	P-value
Soil pH (H_2O)	6.32	6.11	0.45
Organic matter (%)	2.22	2.44	
Total N (%)	0.16	0.14	0.340
Bray-1 P (mg kg^{-1})	8.4	7.1	0.120
Exchangeable bases (cmol. 100 g^{-1}):			
Na^{+}	0.25	0.52	0.060
K^{+}	0.32	0.28	0.277
Ca^{2+}	8.17	9.84	0.045
Soil textural class	Sandy loam,	Loam	

T-test performed and mean differences are based on null hypotheses of H_0: difference = 0

Seeds were inoculated with Rhizobium tropici CIAT899 strain, provided by the Biological N_2 fixation laboratory of the School of Agricultural and Environmental Sciences, Makerere University. Four inoculated seeds were sown manually; and seedlings were thinned to two plants per hill at 14 days after emergency (DAE). Weeding was done manually with hand hoes at 28 and 49 days after planting. Pests for common beans were controlled using Tafgor-40-EC (Dimethoate 40% inflammable) at 600 ml ha^{-1}.

2.4 Data Collection and Analysis

Climate data on the amount of rainfall, number of rainy days, minimum and maximum temperatures were collected from the sites using rain-gauges placed within the study sites. The climate data are presented in Table 2.

For root measurements, destructive sampling was used whereby five plants were carefully uprooted from the middle of each plot, using a random sampling procedure at mid-pod formation stage depending on the genotypes. The plants were bagged in polyethylene before being washed with tap water. They were then left to drip for a 20-30 minutes and placed flat on graph paper. The roots were separated into adventurous, basal and tap roots, and lateral roots on the tap root.

The number of roots, together with their root hairs were counted and root length were measured using a ruler, after which the roots were oven-dried at 80°C for 48 hr and dry weight determined using a digital balance with a precision scale up to +0.001. Specific root length (SRL) was calculated as total length per unit of root dry weight. Root hairs were visually counted under a magnifying lens of X40 resolution. The root to shoot ratio was also determined as a proportion of root biomass to shoot biomass. Pod height was measured at physiological maturity on five sampled plants per plot, using a ruler. The number of pods per plant and seeds per pod were also determined by counting, respectively, pods and seeds on the same plants. Total grain yield was recorded by harvesting the plants in the entire plot, and the weight of the total grains measured using a digital electronic balance. The weight of 100 seeds was determined on seeds randomly collected from all plants harvested per plot.

2.5 Statistical Analysis

Normality for all the variables was tested by Shapiro-Wilk test [18]. All the root variables were normal (P=0.05). Data for the two seasons were tested for homogeneity using Bartlett's test [19] before the combined analysis. According to the Bartlett's test for homogeneity of variance, the adventitious roots, basal root hairs, length of tap root $plant^{-1}$ and total yield were homogeneous (P<0.001). However, the number of adventitious root hairs, basal roots, taps root hairs, lateral roots plant tap $root^{-1}$, length of lateral roots on the tap roots, total root length, stem biomass, shoot biomass, root biomass and root to shoot ratio, were not homogeneous across the two seasons at 5% level of significance. Therefore, to test mean differences, a combined analysis was carried out by a Generalised Linear Model (GLM) with a random statement using StataSe statistical package, 11th edition. A simple correlation was used to establish the relationship between root characteristics and yield; and yield components of low P tolerant bean genotypes.

Table 2. Weather conditions during two rainy seasons of the study at the non-drought-stress (NDS) and drought-stress (DS) sites in Uganda

Season	Drought treatment	Total rainfall (mm $season^{-1}$)	Daily rainfall (mm day^{-1})	Number of rainy days	Daily mean min. temp. (°C)	Daily mean maxim temp. (°C)
March - June (A)	DS	241.0	2.38	29	18.4	30.1
March - June (B)	NDS	587.6	5.82	52	17.9	27.5
Mean Difference (A-B)		-346.6	-3.43 (p=0.007)	-23	0.53 (p=0.000)	3.08 (p=0.000)
August - December (C)	DS	256.1	2.54	28	18.0	29.8
August– December (D)	NDS	542.8	5.38	45	17.4	27.7
Mean Difference (C-D)		-286.7	-2.84 (p=0.014)	-17	0.60 (p=0.001)	2.32 (p=0.001)

T-test was performed and mean differences are based on null hypotheses of H_0: difference = 0. Values in parentheses are probability values

3. RESULTS

3.1 Yield and Biomass

The effect of P and its interactions with bean genotypes, drought or both were not significant (P>0.05) on grain yield and biomass production. Likewise, Drought treatments (DS and NDS) did not affect the grain yields significantly (Table 3), but grain yields from genotypes were significantly (*P*=0.05) different (Table 4).

The interaction between bean genotypes and drought on grain yield was also significant (*P*=0.05). The DS reduced grain yield of AFR 703-1 and AFR 708 by 31 and 42%, respectively, more than the check (Fig. 1); however, the grain yield of MCM 2001 was not reduced significantly.

Contrastingly, the differences in drought treatments neither affected accumulation of the below nor above-ground biomass of the bean plants (Table 3). However, genotype JESCA differed significantly in shoot biomass from the check, while JESCA and MCM 2001 produced lower root: shoot ratio than the check (Table 4).

The interactions between genotypes and drought treatment were significant on shoot and root biomass (Fig. 2). Genotype AFR 708 produced 29% more shoot biomass in the DS than the local check (Fig 2a). AFR 708 and MCM 2001produced 34 and 29%, respectively, more root biomass in the DS than the check (Fig. 2b); the biomass of AFR 703-1 and JESCA were not statistically affected by the interactions.

Table 3. Grain yield, root variables and biomass allocation in low P-tolerant bean genotypes under drought regimes

Root variables	NDS	DS	Difference
Grain yield (kg ha^{-1})	930	816	NS
Adventitious roots (numbers)	13.70	16.40	***
Adventitious root hair (numbers)	153	1181	**
Basal roots (numbers)	5.23	5.13	NS
Basal root hair (numbers)	320	202	**
Tap root hairs (numbers)	44.17	34.97	***
Tap root length (cm)	6.30	8.50	***
Lateral roots on tap root (numbers)	2.68	5.31	***
Length of Lateral roots on tap root (cm)	21.37	49.00	***
Total root length (cm)	173	186	NS
Specific root length (cm g^{-1})	304	300	NS
Shoot biomass (g $plant^{-1}$)	5.07	4.60	NS
Root biomass (g $plant^{-1}$)	0.62	0.69	NS
Root: shoot	0.14	0.15	NS

, **, *, indicate that treatment differences in the same row are significant at P= 0.05, 0.01 and 0.001, respectively; NS = non-significant*

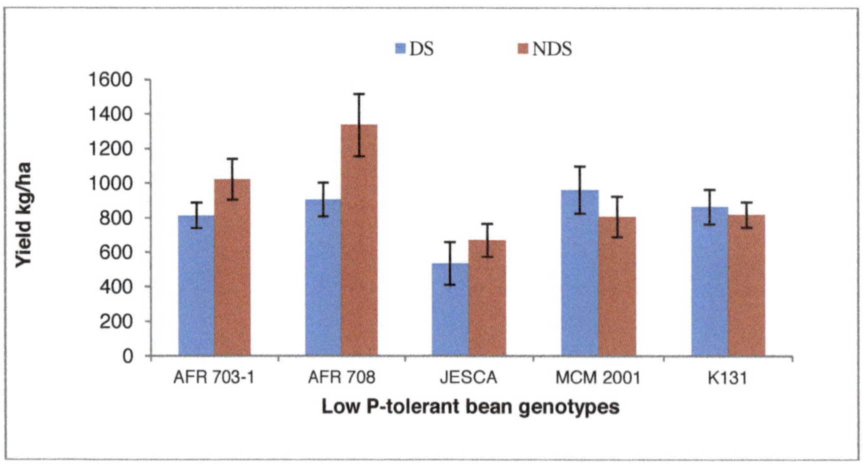

Fig. 1. Interactive effect of drought treatments and bean genotypes on grain yield. (Bars over the mean indicate standard error)

3.2 Root Architectural Responses to P-levels and Drought Treatments

Application of P did not significantly affect bean root characteristics. Also, there were no significant interactions of P and drought treatments, genotypes or both (P>0.05). Common bean plants grown in the DS environment developed deeper taproots, longer and more lateral roots and more adventitious roots, than their counterparts growing in the NDS environment. In contrast, the plants had less root hairs in the DS than in NDS (Table 3). Genotypes significantly influenced the number of adventitious roots, tap root hairs, and tap root length, number of lateral roots and their lengths (Table 5). Genotypes AFR 703-1 and AFR 708 had more adventitious roots, adventurous root hairs, lateral roots on tap root plant, longer lateral roots, total root length and higher specific root length than the check. MCM 2001 had fewer tap root hair than the check, whereas the tap root length plant^{-1} was higher (P=0.05) in AFR 703-1 and lower (P=0.05) in MCM 2001 than the local check.

The interactive effects between drought treatments and bean genotypes on root variables are presented in Fig. 3. There were significant interactions between drought treatments and

Table 4. Influence of low P-tolerant bean genotypes on biomass production and grain yield

Genotypes	*Biomass variables*			*Grain yield (kg ha^{-1})*
	Shoot (g plant^{-1})	Root (g plant^{-1})	Root: Shoot	
AFR 703-1	4.889	0.692	0.147	918
AFR 708	4.682	0.672	0.158	1122*
JESCA	5.390*	0.672	0.123**	603*
K131 [a]	4.363	0.679	0.160	884
MCM 2001	4.832	0.625	0.140*	842

, **, * indicate that the effect of genotype in the same column are significantly different from the check at P = 0.05, 0.01 and0.001, [a] = local check genotype*

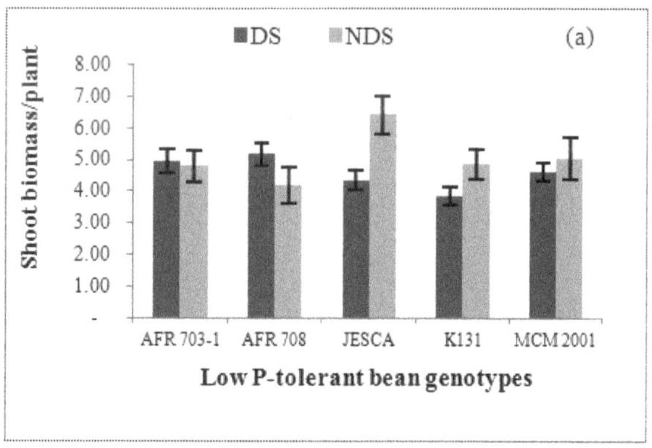

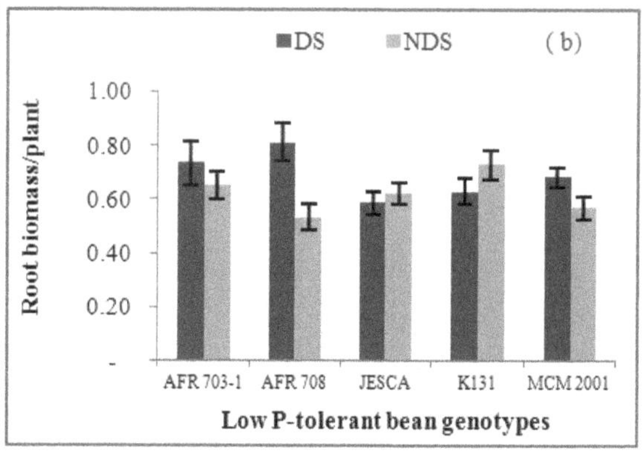

Fig. 2.The effect of drought treatments on (a) shoot biomass and (b) root biomass of the common bean genotypes. (Bars over means indicate standard error)

Table 5. Influence on root variables

Genotypes	Root variables							
	Adventitious roots #	Adventitious root hair #	Tap root hairs #	Tap root length (cm)	Lateral roots on tap root #	Length of lateral	Total root length	SRL
AFR 703-1	18.59 ***	1594*	39.29	8.659*	4.697*	45.14**	220.8***	364.6***
AFR 708	17.21 ***	1574*	39.32	7.912	4.963**	45.40**	206.6***	346.8**
JESCA	13.54	1286	39.64	6.838	3.092	25.47	161.8	286.3
K131 [a]	12.97	1186	43.32	7.371	3.699	30.51	161.4	259.9
MCM 2001	12.96	1141	36.26**	6.214*	3.535	29.41	148.3	253.4

**, ** and ***indicate significant different from a local check at P = 0.05, 0.01 and 0.001. SRL = Specific root length, # = numbers, [a]= local check genotype*

genotypes on adventitious roots (P<0.001), number of lateral roots on the tap root and their lengths (P<0.001). The DS increased the number of adventitious roots (Fig. 3a), tap root lengths (Fig. 3b), number and lengths of lateral roots (Fig. 3c and d) in AFR 703-1 and AFR 708, compared to the local check. Drought stress significantly reduced the number of tap root hairs (P=0.05) in JESCA compared to the local check (Fig. 3e).

3.3 Correlations between Plant Parameters

The correlation between biomass with the grain yield; and yield components of low P-tolerant common bean genotype grown under drought and non-drought stressed conditions are shown in Table 6. Whereas shoot and root biomass were positively associated with grain yield under non-drought stress conditions, the tested biomass variables were not significantly associated with grain yield or yield component under the drought-stress.

The correlations between root variables and grain yield; and yield components were not significantly associated with the grain yield (Table 7). However, the adventitious roots, tap root length, lateral roots lengths, total root length and specific root length were positively associated with weight of 100 seeds under the drought-stress.

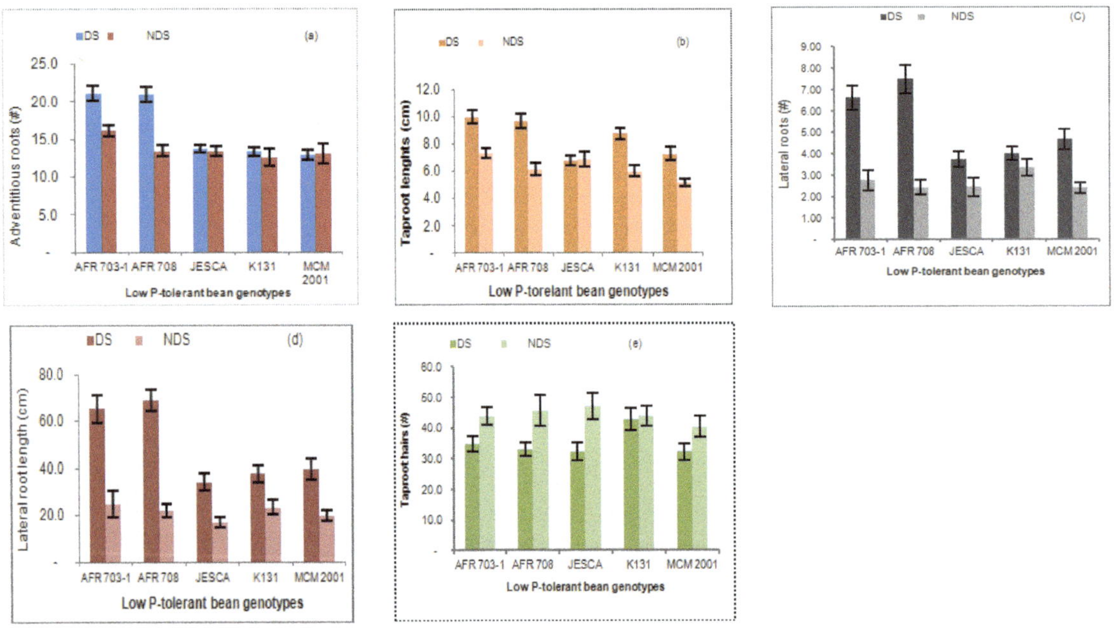

Fig. 3. Drought treatments and genotypes interactions on: (a) adventitious roots (#), (b) length of tap root and (c) Lateral roots (#), (d) length of lateral roots (cm), (e) tap root hairs. (Bars over the mean indicate standard error)

Table 6. Correlation coefficients (r) between biomass with grain yield; and yield components of bean genotypes grown under the drought treatments

Biomass components	Correlation under non-drought-stress (r)				Correlation under drought-stress (r)			
	Grain yield	Weight of 100 seeds	Pods plant^{-1}	Seeds pod^{-1}	Grain yield	Weight of 100 seeds	Pods plant^{-1}	Seeds pod^{-1}
Root biomass	0.40**	-0.17	0.48***	0.03	-0.13	0.19	-0.22	-0.03
Shoot biomass	0.36*	-0.14	0.47***	0.14	-0.14	0.21	-0.07	-0.03
Root: shoot	-0.29	0.06	-0.26	-0.20	0.03	0.03	-0.18	0.04

*The correlation is significant at *P= 0.05, **P =0.01, ***P<0.001*

Table 7. Correlation coefficients (r) between root variables and yield of bean genotypes grown under the drought treatments

Root variables	Correlation under non-drought (r)				Correlation under drought stressed (r)			
	Grain yield	Weight of 100 seeds	Pods $plant^{-1}$	Seeds pod^{-1}	Grain yield	Weight of 100 seeds	Pods $plant^{-1}$	Seeds pod^{-1}
Adventitious roots	-0.03	0.21	-0.13	0.14	0.06	0.72***	-0.02	0.06
Basal roots	0.16	0.22	-0.04	0.033	0.18	0.24	-0.36*	-0.35
Tap root length	0.11	0.18	0.13	0.04	-0.01	0.48***	-0.24	-0.22
Lateral root/tap root	0.07	-0.10	0.05	0.08	0.12	0.55***	0.05	0.05
Length of lateral roots	0.17	0.03	0.19	0.02	0.01	0.59***	-0.07	0.01
Total root length	0.20	0.28	0.10	-0.05	0.07	0.45***	0.04	0.11
Specific root length	-0.20	0.34***	-0.26*	-0.08	0.14	0.20	0.20	0.07

*The correlation is significant at *P= 0.05, **P =0.01, ***P<0.001*

4. DISCUSSION

The grain yield and root development of low P-tolerant genotypes used in the current study did not show significant differences ($P>0.05$) in relation to P application under different drought regimes. Similar findings were reported by Mourice and Tryphone [20], who observed that some low P-tolerant genotypes did not show significant response to increasing P levels in terms of yield. However, this was in disagreement with the study conducted by [12], which showed dominance in yield of P-efficient genotypes under low soil P levels compared to P-insufficient genotypes. Furthermore, the results revealed that low P-tolerant genotypes also have potential to perform equally well under combined different levels of plant-available P and moisture regimes. This is in agreement with [8], who reported that breeding for unfavourable condition did not hinder performance potential of the developed genotypes under favourable condition. This was further in disagreement with a study conducted by [13] that G19833; a low P tolerant genotype performed well in P deficient soils but poorly in P enriched soils. In the same study it was observed that AND696 performed well in P sufficient soil but poorly under low P soils. This suggests that the materials conserve their tolerance to low soil P supply in spite of their exposure to drought conditions.

Phosphorus levels did not influence the root variables that are associated to P acquisition and no significant interactions existed between P levels, drought treatments, or genotypes. However, that pattern was reversed in terms of root numbers and length for drought treatments (Table 3).The adventitious roots, tap root length, number of lateral roots as well as their lengths were significantly greater in the DS than in the NDS. Increased root length and rooting depth allow the crop to explore a larger soil volume and exploit the accessible soil moisture from deeper profiles, after the shallow roots have exhausted all the available moisture at the upper profiles [4]. Specifically, greater root length in the lower soil profile is a significant mechanism for drought resistance in common beans [11]. However, the grain yield in the present study did not significantly correlate with root variables under drought stress, possibly because the adaptations to unfavourable conditions by rooting mechanisms do not necessarily translate into yield [21]. Similar results were reported by Porch, Ramirez [22], who observed that BAT 477 with both shallow and deep rooted genotype did not yield better under interactive stressing conditions.

The low P-tolerant genotypes significantly differed in the number of adventitious, adventitious and basal root hairs; tap root length, number of lateral roots and their length, total and specific root length. Genotypes AFR 703-1 and AFR 708 registered considerably higher values in most of the root variables than the local check (Table 5). They equally possessed both shallow and deep root variables that provided them the capacity to exploit topsoil P and deep moisture resources, respectively. Similar results were reported by [11], that BAT 477 possessed both shallow and deep root systems. Suriyagoda, Ryan [7] too, reported that Cullen species developed both shallow and deep root systems when grown in low P soils under drought conditions. Our results have confirmed the hypothesis that some low P tolerant genotypes have multiple root systems for both P and soil moisture acquisition.

According to Suriyagoda, Ryan [7], root biomass is key for acquisition of soil resources that are limited in supply. Studies have revealed that root: shoot ratio rises with shrinking soil P levels [12,23] and moisture [9]. However, a higher allocation of biomass to roots and a lesser amount of shoots (root: shoot) in response to soil P deficiency and/or moisture deficit results into reduced plant growth, hence reduced grain yield production. This means that allocation of biomass to roots is done at the expense of the shoot biomass. However, in the present study AFR 708 which produced more root biomass in the DS than in the NDS did not do so at the expense of the shoot biomass. According to Ho, Rosas [11], common bean genotypes that are P tolerant and drought resistant produce higher root and shoot biomass and higher yields than the non-efficient genotypes. Therefore, since root and shoot biomass were positively associated with grain yield under NDS, they could be useful traits for exploitation in the selection of low P-tolerant bean genotypes under drought conditions.

5. CONCLUSION

Differences in phosphorus application rates had no effect on root architecture and grain yield of all genotypes. Genotypes AFR 703-1 and AFR 708 outperform the local check (K131) in terms of root development by the ability to develop the multiple root systems, especially under drought conditions. Overall, under drought stress, the AFR gene-pool proved superior in root development, namely number of adventitious and lateral roots, and taproot and lateral root length. In contrast, grain yield of these materials was suppressed by drought stress. The results have significant contributions for common bean breeding as far as improving performance of the crop under limited soil P and exploitation of moisture at different soil depths is concerned.

ACKNOWLEDGEMENTS

This study was funded in part by the Swedish International Development Cooperation Agency (Sida) and Makerere University, Kampala, Uganda.

COMPETING INTERESTS

Authors have declared that no competing interests exist.

REFERENCES

1. Wortmann SC, et al. eds. Atlas of common bean (*Phaseolus vulgaris* L.) production in Africa. International Centre for Tropical Agriculture, CIAT: Cali, Colombia. 2004;33.
2. Beebe ES, et al. Genetic improvement of common beans and the challenges of climate change, in crop adaptation to climate change, Yadav SS, et al. Editors. Blackwell Publishing Ltd: Colombia. 2011; 356-369.
3. Beebe ES, et al. Phenotyping common beans for adaptation to drought. Frontiers in Physiology. 2013;4(35):1-20.
4. Namugwanya M, et al. Development of common bean (*Phaseolus vulgaris* L.) production under low soil phosphorus and drought in Sub-Saharan Africa: A review. Journal of Sustainable Development, 2014; 7(5):128-139.
5. Acosta-Díaz E, et al. Adaptation traits in dry bean cultivars grown under drought stress. Agricultura Técnicaen México. 2009;35(4):416-425.
6. Morino G, et al. Fine root distribution in Dehasas of Central-Western Spain. Plant Soil. 2005;277(1-2):153-162.
7. Suriyagoda DBL, et al. Multiple adaptive responses of Australian native perennial legumes with pasture potential to grow in phosphorus- and moisture-limited environments. Annals of Botany. 2010; 105:755-767.
8. Beebe ES, et al. Selection for drought resistance in common bean also improves yield in phosphorus limited and favourable environments. Crop Science. 2008;48: 582-592.
9. Farooq M, et al. Plant drought stress: Effects, mechanisms and management: A review. Agronomy for Sustainable Development. 2009;29:185-212.
10. Lambers H, et al. Root structure and functioning for efficient acquisition of phosphorus: Matching morphological and physiological traits. Annals of Botany. 2006;98:693-713.
11. Ho DM, et al. Root architectural trade-offs for water and phosphorus acquisition. Functional Plant Biology. 2005;32:737-748.
12. Beebe ES, et al. Quantitative trait loci for root architecture traits correlated with phosphorus acquisition in common bean. Crop Science.2006;46:413-423.
13. Cichy AK, Snapp SS, Blair WM. Plant growth habit, root architecture traits and tolerance to low soil phosphorus in an Andean bean population. Euphytica. 2009; 165:257-268.
14. Namayanja A, et al. Genotypic variation for tolerance to low soil phosphorous in common bean under controlled screen

house conditions. Agricultural Sciences. 2014;5:270-285.
15. Frahm AM, et al. Breeding beans for resistance to terminal drought in the lowland tropics. Euphytica. 2004;136: 223-232.
16. Ntambirweki-Karugonjo B, Barungi J. Agri-food system governance and service delivery in Uganda: A case study of Mukono district. Kampala: ACODE; 2012. Policy research series No: 57.
17. Page R, Miller H, Keeney RD, eds. Methods of soil analysis. Part 2. Agronomy Monograph 2. 2nd ed. American Society of Agronomy, Soil Science Society of America: Madisin, WI, USA. 1965;577.
18. Ghasemi A, Zahediasl S. Normality tests for statistics: A guide for non-statisticians. International Journal of Endocrinology and Metabolism. 2012;10(2):486-489.
19. Bartlett MS. Some examples of statistical methods of research in agriculture and applied biology. Supplement to the Journal of the Royal Statistical Society. 1937;4(2): 137-183.
20. Mourice SK, Tryphone GM. Evaluation of common bean (*Phaseolus vulgaris* L.) genotypes for adaptation to low phosphorus. Agronomy. 2012;1-9.
21. CIAT, Annual report 2008, in Outcome line SBA-1. Improved beans for the developing world, CIAT: Cali, Colombia. 2008;39-65.
22. Porch TG, et al. Evaluation of common bean for drought tolerance in Juana Diaz, Puerto Rico. Journal of Agronomy and Crop Science. 2009;195:328-334.
23. Rubio G, et al. Crop ecology, management and quality: Top foraging and its role in plant competitiveness for phosphorus in common bean. Crop Science. 2003;43: 598-607.

13

Larrea tridentata Uptake of Trace Metals in Mine Tailings within an Arid Region of Nevada, USA

Douglas B. Sims[1*], Christopher J. Collumb[1], Vernon F. Hodge[2], Konstantinos I. Vatalis[3] and George Valakas[4]

[1]Department of Physical Sciences, College of Southern Nevada, North Las Vegas, NV, USA.
[2]Department of Chemistry, University of Nevada, Las Vegas, NV, USA.
[3]Environmental Engineering Department, Technological Institution, Western Macedonia, Kozani, Greece.
[4]School of Mining and Metallurgical Engineering, NTUA, Athens, Greece.

Authors' contributions

This work was carried out in collaboration between all authors. Authors DBS, CJC and VFH designed the study, wrote the protocol and wrote the first draft of the manuscript. Authors DBS and CJC managed the literature searches, analyses of the study samples and ICP-OES analysis. Authors KIV and GV performed the statistical analysis. Authors DBS and CJC performed the experimental process and identified the species of plant. All authors read and approved the final manuscript.

Editor(s):
(1) Junhong Bai, School of Environment, Beijing Normal University, Beijing, China.
Reviewers:
(1) Volodymyr Chernyshenko, Palladin Institute of Biochemistry NAS of Ukraine, Ukraine.
(2) Emil Jose Hernandez-Ruz, Universidade Federal do Para, Brazil.
(3) Anonymous, Centro de Investigaciones Biológicas del Noroeste, S. C. (CIBNOR), Mexico.
(4) Shruti Murthy, Bangalore University, India.
(5) Pavlina Simeonova, Bulgarian Academy of Sciences, Bulgaria.
(6) Rodrigo Crespo Mosca, Sao Paulo University, Brazil.

ABSTRACT

Larrea tridentata (creosote bush) growing in contaminated tailings of the Techatticup Mine in Nelson, Nevada were sampled and analyzed for trace metals. Samples were also collected outside the mine tailings to measure geogenic trace metal levels. These data show that some trace metals (Cu, Pb, Zn, Co, Fe, Li) enter *L. tridentata* through root tissues, but at significantly lower levels than in the tailings area. Trace metals that enter the root are higher in concentration in the outer than in the inner root tissue, possibly due to *L. tridentata* blocking their absorption at the outer root surface.

**Corresponding author: E-mail: Douglas.Sims@csn.edu*

Data further show the plant's ability to block the intake of these toxic trace metals that may adversely affect the plant. Statistical analysis suggests that certain metals, while not in high abundance, may be utilized by the plant for self-defense mechanisms or to aid in plant development. Finally, differences between plant components may be the result of hyper-accumulation of useful trace metals (e.g. B, Cr, Zn) and a blockage of potentially toxic trace metals (e.g. Cd, Pb, V).

Keywords: Mining; Nevada; trace metals; arid; Larrea tridentata; creosote bush.

1. INTRODUCTION

Waste from abandoned mine sites has the potential to impact more than just water and sediment resources. The processes and pathways that affect transport of trace elements in any ecosystem can pose environmental and human health risks, even beyond the vicinity of the site. Hooda [1] states that such risks are contingent on the bioavailability of trace elements, not just on their presence in the environment. Biological impacts include the health and wellbeing of flora, fauna and humans exposed to contaminated water and food resources [2-4]. The possible incorporation of trace metals from abandoned mining areas adjacent to population centers into biomass poses unknown hazards. Tracking how a contaminant moves from sediments to biomass of local plant species, thus potentially entering the food web, is of great importance.

Mining site wastes have been shown to have effects on the abundance of trace metals (e.g. Cu, Ni, Fe, Co, Zn, U, and Pb) in plant and animal tissues in the vicinity of the mining site [5,6]. However, the uptake of trace metals is a complex process of metal-ligand binding and chemical conversions (biocatalysts) involving transfer of chemical species across root membranes to plant cells. In addition to the bioavailability of trace metals, biomass uptake can be limited by other factors such as moisture, nutrients (e.g. N and P), and organic acids [7-10]. The mycorrhiza associated to the plant root systems have been shown to reduce plant tissue exposure to trace metals in contaminated sediments [11,12]. Other studies of trace metal abundance in vegetation have focused on the impact to plants in stream sediments and farming adjacent to tailings dumps [13]. In arid regions, however, contaminated sediments may be redistributed by short duration and high intensity storm-water transport events, exposing drought resistant plants to substantial trace metal loads.

Researchers have shown that the bioavailability and transfer of trace metals from sediment to plant is controlled by sediment moisture, pH, presence of mycorrhiza, and plant physiology [1,14]. Studies have determined sediment pH is a key factor influencing solubility and resultant bioavailability of metals to plants. The presence of certain plant species also plays an important part in determining which trace metals are used or blocked [14]. Some vascular plant species are capable of utilizing metals as a self-defense mechanism from insects, while others are capable of blocking certain trace metals due to their adverse effect on a plant growth [15,16]. A symbiotic relationship of plants with mycorrhiza fungi has proven to be a benefit to both the mycorrhiza fungi and the plant itself, and in particular it has been shown that arbuscular mycorrhiza fungi found in association with the creosote bush (*L. tridentata*) are also capable of protecting plants from adverse trace metal exposure [17,11,12].

This study was performed at the area of the Techatticup Mine, a large-scale mining operation that functioned between 1850 and 1960 located close to Nelson, Nevada [18]. This mine area was chosen due to the presence of heavily contaminated tailings exposed to the open environment [19]. A more detailed overview of the operation of this mine, its ore and tailing processing can be found in Sims [20]. This area is located within the southern edge of the Great Basin, where topography controls regional hydrologic processes resulting in closed valley systems with no significant external drainage [21]. Ephemeral streams in the area run east-west towards Lake Mohave, a regional reservoir. Mine waste (tailings) were dumped into the ephemeral washes during operation. Tailings mounds currently have a variety of sparse desert vegetation growth including *L. tridentata, Ambrosia dumosa,* and *Eriogonum inflatum*.

Among these, the dominant species, *L. tridentata*, was chosen for sampling. This is a perennial plant that can accumulate trace elements and metalloids in its tissues, preserving its leaf structure for extended periods of time, and can extend its main root (tap root) up to 25 m below the surface in search of water resources [22,23]. Studies have documented its uptake of trace metals and metalloids by *L. tridentata* from sediments, even in desert climates [22,24,25]. This research evaluates the uptake, accumulation, and transfer of trace elements and metalloids into particular tissues of *L. tridentata*.

2. MATERIALS AND METHODS

2.1 Sampling Methodologies

Seven (7) plants were sampled within a 20 by 20 meter area of the Techatticup tailings dump (Fig. 1). Three (3) separate *L. tridentata* samples were collected outside of the Techatticup Mine area for background analysis to measure geogenic levels of trace metals. At each sampling location, 60-cm long stems containing leaves were cut from a mature *L. tridentata* that was at least 120 cm in height. Additionally, the main root (tap root) was removed at each bush location for analysis. Sediment samples were collected from the contaminated tailings of the seven bush locations to 30 cm below the surface, composited, and analyzed for trace metals. In addition, dust on the leaves and branches was collected for analysis.

2.2 Sample Analytical Procedures

Plant and tailing samples were analyzed for trace metals per USEPA method 6010B by inductively coupled plasma-optical emission spectrometry (ICP-OES) techniques [26]. Plants were processed by cutting a 30-cm length of branch that was washed with 100 ml of 90°C deionized water to remove particulate dust from tailings off the leaves and branches [27-30]. The wash rinsate was acidified with 10 ml of concentrated nitric acid (HNO_3), evaporated to near-dryness on a hotplate, and cooled. The residual was weighed and treated with 10 ml of concentrated nitric acid and brought to 100 ml with deionized water to evaluate trace metals in the dust accumulated on leaves and branches.

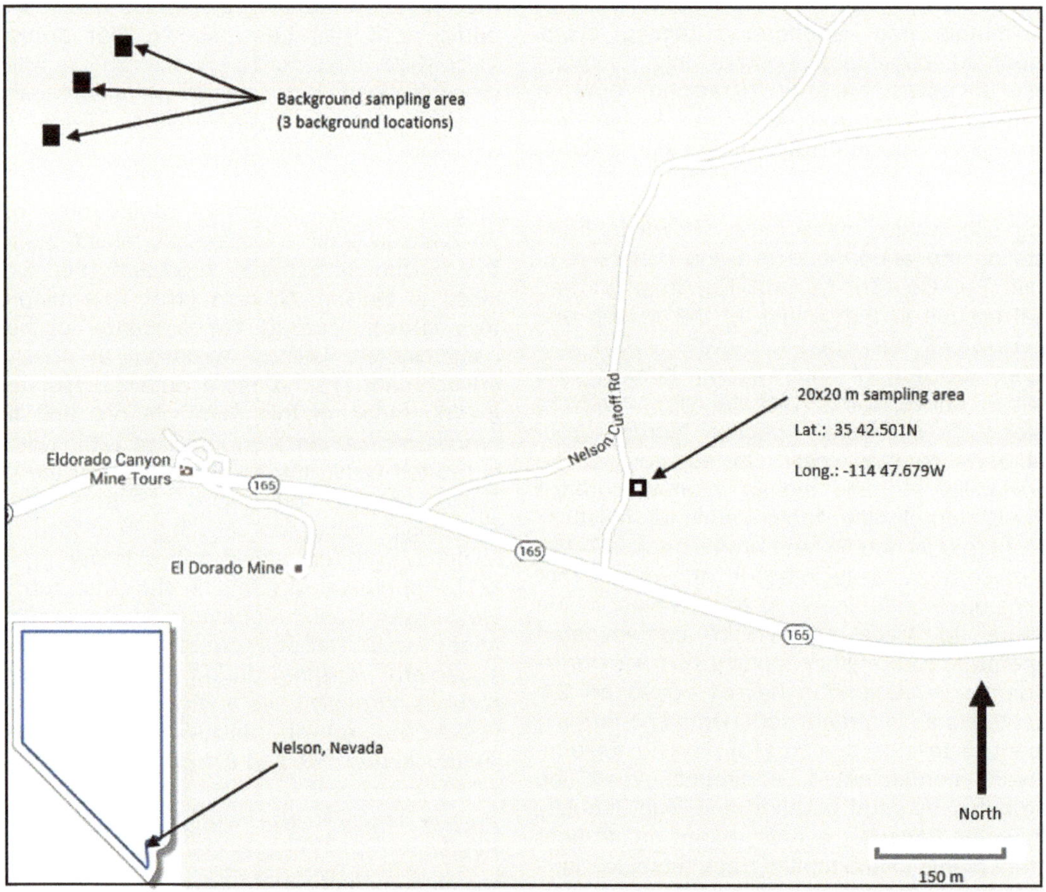

Fig. 1. Sampling locations at Techatticup Mine and Mine, Nelson, Nevada [48]

After washing the 30 cm long branches, approximately 1 gram of leaves was cut from each branch and digested per USEPA method 3050 B (hot aqua-regia digestion) and filtered through a Whatman® 541 (2 μm) to evaluate trace metals in the leaves of the *L. tridentata* [26]. The main root was cleaned with a soft bristle bush and ambient temperature deionized water to remove tailing debris. After brushing, the outer root tissue (epidermis layer and cortex) was removed with a clean plastic knife. The outer and inner root (stele) tissues were then digested (1 gram) per USEPA method 3050B and filtered through a Whatman® 541 filter (2 μm). Finally, 1 gram of mine tailings was processed per USEPA method 3050B and analyzed by USEPA method 6010B; ICP-OES [26].

2.3 Principal Components Analysis

Principal Components Analysis (PCA) was utilized to describe clustering of trace metals between sediments and plant components after log transformation, since data were not homoscedastic [31].

2.4 Quality Assurance/Quality Control

All samples analyzed for trace metals were processed per USEPA method 6010B for inductively coupled plasma-optical emission spectrometer [26]. Reagents were of ultra-pure quality, NIST traceable. Analytical integrity was confirmed with a USEPA certified reference sample (RTC Corporation: CRM022-020, Sample 5, lot D522). Quality control was further maintained by performing all analyses in triplicate.

3. RESULTS AND DISCUSSION

Metal contents varied widely within each part of the plant analyzed, and data show that there were significant differences between plant regions (Table 1). Typically, while both the sediment and dust showed high levels of metal, there was a significant reduction of metal concentrations in the biomass of the plants, particularly in the inner root and leaf tissues. Data show Cu, Pb and Zn reductions greater that 50% between sediment and outer root tissue. Cobalt (35%), Fe (35%) and Li (44%) also showed large decreases in concentration between sediment and outer root tissue (Table 1). As shown by other authors, trace metals are blocked at the outer root, possibly due to their toxicity or dehydrating nature [15,32].

It has been shown that the *L. tridentata* and analogous plants have similar patterns of trace metal uptake [32], and that similar plants block potentially toxic trace metals at the outer root [15,16]. For example, Prasad [33] demonstrated that some plants will block trace metal cations (e.g. Cd^{2+}, Cr^{3+}, Pb^{2+}, V^{3+}) and alkali metals (e.g. Li) at the outer root by affecting the water balance, and explained that when plants grow in metal cation laden sediments, Cd tended to bind to the cell membrane reducing its intake of water. Studies further suggest that the *L. tridentata* blocks trace metals in the outer root as a self-defense mechanism from certain trace metals [34,35]. For example, while B is not a metal cation, it is well documented to inhibit a plant's ability to intake water [36], and Table 1 shows that it was blocked at the outer root and absent in the inner root, although it was present at low levels in leaf tissues. Other authors suggest that in a plant's vascular system some trace metals (Cu, Cr, Mo, Zn) are not blocked and move through the tissue of the plant, most likely due to the benefit of the trace metal to the plant [16].

While the outer root tissue shows reductions in many of the metals analyzed, nearly all of the metal concentrations are significantly lower in the inner root and leaf tissues (Fig. 2). Higher concentrations of metals in the outer root compared to inner root tissue (except Mo, and leaf tissues compared to dust; except for Al and Ti) could be the result of sequestering of metal ions in the cells of the outer root, therefore preventing transport of metal ions by the xylem and phloem to other parts of the plant [37,38].

Dust on the leaves showed significantly higher levels of trace metals (Table 1). This is not unexpected, since *L. tridentata* has a resin coating on the leaves that can easily retain dust and other contaminants present in the tailing area [39]. While trace metals were present in the dust, studies show that surface dust on plant leaves, does not impact the levels of trace metals within the tissues of the plant as metals are absorbed by the roots *not* through the leaves [6,10,32,40]. Due to a non-transdermal nature of trace metals from the dust to the *L. tridentata*, its presence on the leaves has little impact on the overall health of the plant. The presence of contaminated dust on the leaves however, can impact the food-web if consumed by an animal or human.

Scatter plots and Principle Components Analysis (PCA) were utilized to identify trends and relationships between trace metals, sediments

Table 1. Average concentration of trace metals in tailings, outer and inner root, and leaf tissue and dust on the leaf of the *L. tridentata*

	Mine tailings	Outer root	Inner root	Leaf tissue	Leaf dust
Al	1416	1345	83.7	110	89
As	3.2	6.08	0.70	0.53	7.57
Ba	11.1	17.1	7.40	2.58	13.7
B	-	64.8	-	20.0	219
Cd	5.55	7.28	-	-	-
Cr	3.70	2.64	-	-	0.50
Co	2.53	1.65	0.30	0.19	0.56
Cu	74	33.7	6.43	4.84	23.6
Fe	3000	1937	129	159	1388
K	2466	13582	2759	4518	42610
Pb	264	42.9	1.05	1.29	2.15
Li	3.95	2.21	0.04	0.48	3.89
Mn	296	248	23.1	29.5	84.8
Mo	-	4.55	4.97	0.22	0.23
Ni	4.98	6.50	0.70	1.03	18.6
Ag	0.43	15.6	1.42	1.14	37.4
Ti	12.7	39.0	2.54	4.06	3.90
V	2.83	4.49	0.12	-	-
Zn	675	114	31.5	24.2	79.5

(-) indicates concentrations below detection levels, sample size was n=7, standard deviations (SD) are shown in Fig. 2, units are in mg kg^{-1}

and selected plant components *L. tridentata*. Results indicate a relationship between trace metals in the sediment, outer and inner root, but there is a diminutive relationship between the plant and dust present on the leaves (Figs. 3-4). Fig. 3 shows a significant relationship between the outer root and sediments. Fig. 4, also shows a similar relationship between sediment and inner root. There was no relationship between sediments and leaf tissue or dust on the leaf and is therefore not presented.

Figs. 3 and 4 show distinct relationships between mine tailings and the outer and inner root components; however, PCA (Figs. 5a through d) displays clustering of trace metals in plant components and mine tailings. For example, Fig. 5a through 5d shows clustering (relationships) between trace metals in the outer and inner roots and leaf tissues when compared to tailings. Data from the rinsate (dust); however, displayed no common distribution between outer and inner roots and leaf tissues. PCA clustering further exhibits a similar distribution with trace metals in plant components when compared to tailings, suggesting possible relationships with the movement through the plant's system.

Figs. 5a, b and d show that the outer root contains trace metals known to be water inhibiters (i.e. Cd), findings similar to Franzle et al. [10] and Haque et al. [16]. There is a dissimilar distribution among clusters A shown in Fig. 5a and b. Data show cluster A has little relationship between the outer and inner root; however, cluster B shows that there is a possible relationship between the inner and outer root with Al, Co, Fe, Mn, Pb and Ti (Figs. 5a and b). These finding are similar to other researchers who found that certain trace metals can move through the outer root to the inner root, but in limited concentrations due to adverse effects to the plant [10,41,42]. Fig. 5c and d display clustering with limited relationships between trace metals (i.e. Al, K and Li). PCA plots show little association between dust, leaf tissue and trace metal concentration in the leaves, these findings are similar to those of Smolders et al. [43].

The difference between the outer and inner root components may be the result of hyper-accumulation of trace metals in the outer root due to the plant's ability to block particular trace metals as suggested by Franzle et al. [10]. Trace metals that showed commonality between all three plant components were Al, Fe, Mn, Pb and Ti (Figs. 4a, b and d). While Al, Fe and Mn are within normal levels for the *L. tridentata*, Pb was also well below suggested toxic levels as reported by Smolders et al. [43]. Cluster B illustrates descending ratios for tailing-inner root transfer (Al having 16:1 and 13:1; Fe 23:1 and 18:1; Pb 26:1 and 24:1, Mn 12:1 and 10:1, and Ti 5:1 and 3:1). Although Pb is known to be toxic to plants, it was detected in plant components at approximately 1 mg kg^{-1}, lower by known standards for *L. tridentata* [32,40]. Other trace metals present (Al, Fe, Mn and Ti) are known to be beneficial for plant growth, although at significantly lower concentrations than those detected in the tailings [6].

Further analysis indicated there were some groupings of trace metals with plant components and tailings. Data show that Al, Fe, Mn, Pb and Ti were the only trace metals clustering (cluster B) across Figs. 5a, b and d; there was no commonalities with Fig. 5c. These trace metals (Al, Fe, Mn, Pb and Ti) are both beneficial and harmful to the *L. tridentata*, depending on the level of absorption [43]. Dust on the leaves of the *L. tridentata* contained no clustering between the plant, dust or tailings.

It is possible that the *L. tridentata* absorbs specific trace metals (e.g. Al, Fe, Mn, Ti) for specific reasons, such as growth [15]. It has been demonstrated that certain trace metals at high concentrations in the inner and outer root tissues of vascular plants (e.g. Al, Fe, Ni, Co, Mn) are related to a plant's selectivity [44]. Researchers have reported that Zn present in vascular plants for example, may be a defense mechanism [15,16]. This study found similar findings to Badri et al. [44] with Zn in the leaf tissues having similar concentrations to root components. The difference between leaf tissue and root tissues may be the result of hyper-accumulation of B, Cr, and Zn, and to a lesser extent Cd and V, findings similar to those reported by Franzle et al. [10].

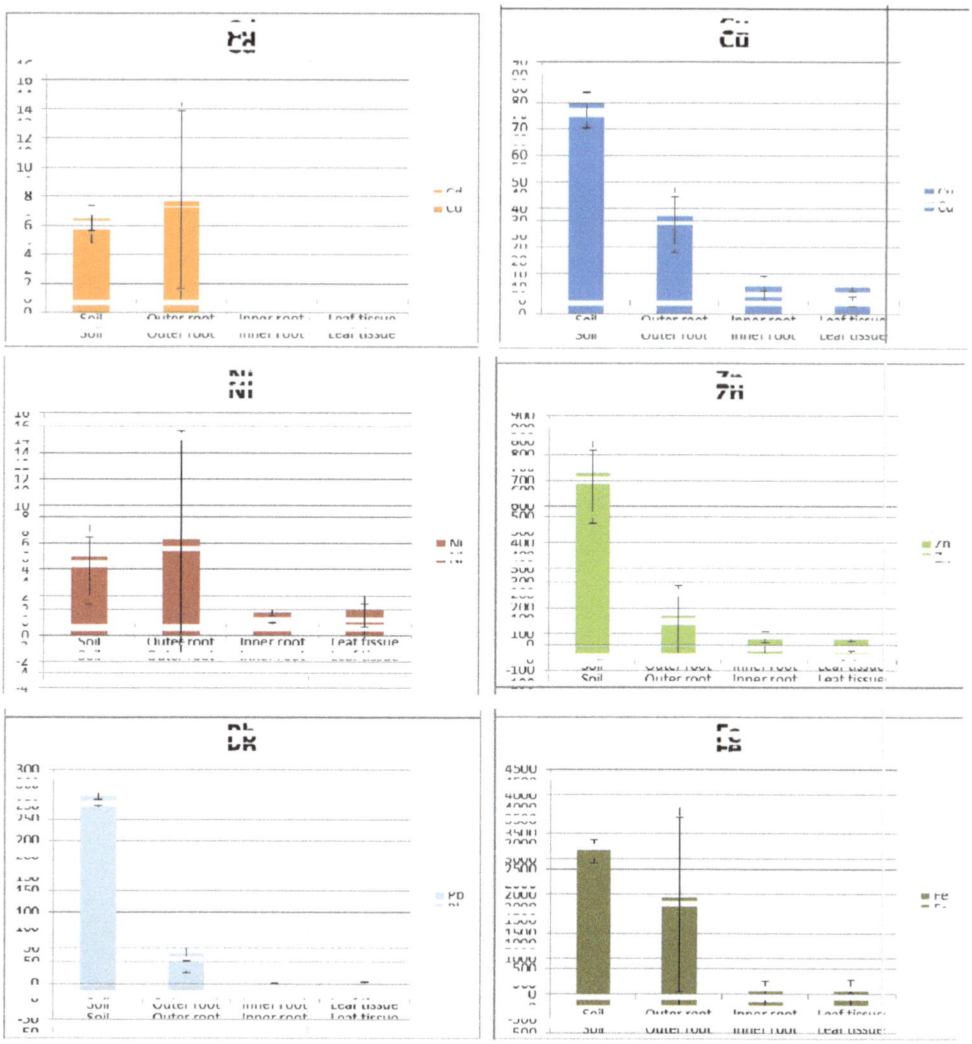

Fig. 2. Concentration of trace metals in various part of *L. tridentata*, +/- 1 standard deviation (SD), y-axis is concentration; x-axis is plant component, concentration in mg kg^{-1}

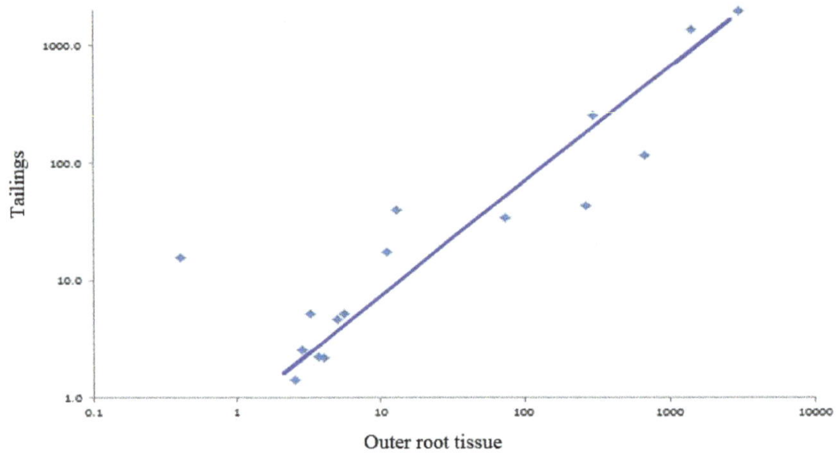

Fig. 3. Trace metal (mg kg^{-1}) relationship between tailings (y-axis) and outer root tissue (x-axis); R^2 = 0.9879

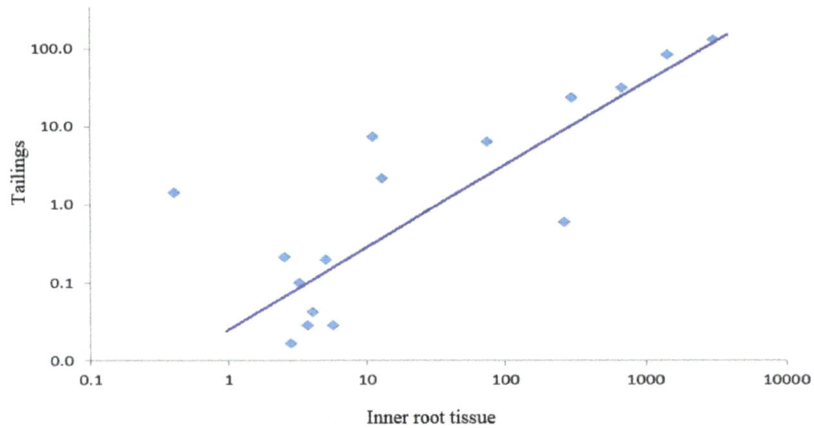

Fig. 4. Trace metal (mg kg^{-1}) relationship between tailings (y-axis) and inner root tissue (x-axis), R^2 = 0.9655

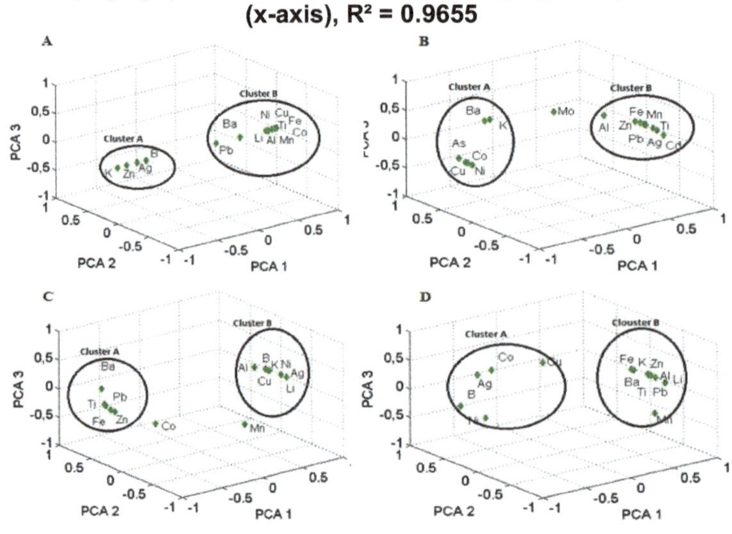

Fig. 5. PCA showing relationship of tailings to outer root (A), Inner root (B), Rinsate (C) and Leaves (D)

L. tridentata is a recent addition to the Mojave Desert and the Great Basin dating back to approximately 10,000 years B.P. [45,46]. Studies have shown that it began its migration northward at the end of the Last Glacial Maximum (Pleistocene) as the climate began to warm [45,47]. McAuliffe [45] further showed that a single *L. tridentata* plant can live more than 1000 years by cloning itself. It is possible that some of its early physiological traits, such as trace metals uptake for self-defense, persist today. *L. tridentata* may have preserved some of its physiological traits that give it no advantage in today's environment.

4. CONCLUSION

L. tridentata growing in mining regions are exposed to high levels of potentially toxic trace metals. Plants have evolved several methods of mitigating the toxic effects of these trace metals. This study shows that while some trace metals are entering the plant system through the root, levels are greatly reduced for Cu, Pb, Zn, Co, Fe, and Li when compared to levels in the contaminated tailings. Metals that enter the root and are found at high concentrations in the outer root tissue are likely blocked from entering the vascular system of the plant (inner root tissue), reducing their presence in the leaf tissue. Certain metals, while not absorbed at high concentrations, may be taken up by the plant for self-defense mechanisms or as aids in plant's growth. This lack of accumulation in tissues is of biological importance and shows the plant's ability to block intake of toxic metals. Finally, it is possible that *L. tridentata* physiological qualities remain as a result of its developmental stages, although they may not be of importance today as a result of environment change. Further research of trace metal movement through the outer root tissue and into the plant, and if possible, isolating it from mycorrhiza tissue would provide valuable information on how *L. tridentata* utilizes, or blocks trace metals in hyperarid regions.

ACKNOWLEDGEMENTS

The authors would like to thank Dr. John E. Keller, Mr. Jose Garcia and Ms. Chelsea Linares of the College of Southern Nevada for assistance in the collection and processing of samples. We would also thank the reviewers whose time and knowledge added value to the overall paper.

COMPETING INTERESTS

Authors have declared that no competing interests exist.

REFERENCES

1. Hooda PS. Assessing bioavailability of soil trace elements. In: Hooda PS. (ed.) Trace Elements in Soils. UK: Wiley. 2010;229-265.
2. Nicholson FA, Smith SR, Alloway BJ, Carlton-Smith C, Chambers BJ. An inventory of heavy metals inputs to agricultural soils in England and Wales. The Sci. of the Total Environ. 2003;311: 205-219.
3. Pancevski Z, Stafilov T, Baceva K. Distribution of heavy metals in some vegetables grown in the vicinity of leaqd and zinc smelter plants. Contributions, Sec. Nat. Math. Biotech. Sci., MASA, 2014;35(1):25-36.
4. Baceva K, Stafilov T, Matevski V. Bioaccumulation of heavy metals by endemic viola species from the soil in the vicinity of the As-Ab-Ti Mine "Allchar", Republic of Macedonia. Inter. J. of Phytoremediation. 2014;16:347-365.
5. Reglero MM, Monsalve-González L, Taggart MA, Mateo R. Transfer of metals to plants and red deer in an old lead mining area in Spain. The Sci. of the Total Environ. 2008;406:287-297.
6. Balabanova B, Stafilov T, Baceva K, Vuckovic I. Heavy metals toxicity and bioaccumulation in vegetables from potential polluted area. Scientific Works, Food Science, Engineering and Technology; 18-19 Oct., Plovdiv. 2013;60.
7. Morel F, Hering J. Principles and Applications of Aquatic Chemistry. 1st ed. 1 vols. New York, NY: John Wiley & Sons, Inc; 1993.
8. Hancock GR, Grabham MK, Martin P, Evans KG, Bollhöfer A. A methodology for the assessments of rehabilitation success of post mining landscapes-sediment and radionuclide transport at the former Nabarlek uranium mine, Northern Territory, Australia. The Sci. of the Total Environ. 2006;354:103-119.
9. Hancock GR. A catchment scale assessment of increased rainfall and storm intensity on erosion and sediment

transport for Northern Australia. Geoderma. 2009;152:350-360.
10. Franzle S, Markert B, Wu¨nschmann S. Dynamics of trace metals in organisms and ecosystems: Prediction of metal bioconcentration in different organisms and estimation of exposure risks. Environ. Pollution. 2007;150:23-33.
11. Hildebrandt U, Regvar M, Bothe H. Arbuscular mycorrhiza and heavy tolerance. Phytochemistry. 2007;68:139-146.
12. Andrade SAL, Silveira APD, Mazzafera P. Abruscular mycorrhiza alters metal uptake and the physiological response of *Coffea arbica* seedlings to increasing Zn and Cu concentrations in soil. The Sci. of the Total Environ. 2010;408:5381-5391.
13. Matula J. A Relationship between Multi-nutrient soil tests (Mehlich 3, ammonium acetate and water extraction) and bioavailability of nutrients from soils for barley. Plant Soil Environ. 2009;55(4): 173-180.
14. Peralta-Videa JR, Torresdey G, Gomez E, Tiemann KJ, Parsons JG, Carrillo G. Effect of mixed cadmium, copper, nickel, and zinc at different pHs upon alfalfa growth and heavy metal uptake. Environ. Poll. 2002;119:291-301.
15. Polette LA, Gardea-Torresdey JL, Chianelli RR, George GN, Pickering IJ, Arenas J. XAS and microscopy studies of the uptake and bio-transformation of copper in *Larrea tridentata* (Creosote Bush). Microchem J. 1998;65:277-236.
16. Haque N, Peralta-Videa JR. Jones GL. Gill TE, Gardea-Torresdey JL. Screening the Phytoremediation Potential of Desert Broom (Baccharis sarothroides Gray) Growing on Mine Tailings in Arizona, USA. Environ. Poll. 2008;153:362-368.
17. Apple ME, Thee CI, Smith-Longozo VL, Cogar CR, Wells CE, Nowak RS. Arbuscular mycorrhizal colonization of *Larrea tridentata* and Ambrosia dumosa roots varies with precipitation and season in the Mojave Desert. Symbiosis. 2005;39: 1-5.
18. Greene JM. Life in Nelson township, Eldorado Canyon, and Boulder City, Thesis MA (unpubl), History Dept, Univ of Nevada, Las Vegas, NV, USA; 1975.
19. Sims DB, Hooda PS, Gillmore GK. Sediment contamination along desert wash systems from historic mining sites in a hyperarid region of Southern Nevada, USA. Soil and Sed. Contam. 2013;22: 737–752.
20. Sims DB. Fate of contaminants at an abandoned mining site in an arid environment. Dissertation (unpubl), Geog., Geol., Env. Dept., Kingston University, London (U.K.); 2011.
21. Longwell CR, Pampeyan EH, Bowyer B, Roberts RJ. Geology and Minerals Deposits of Clark County, Nevada. A. Carlisle, Reno, NV, USA; 1965.
22. Gardea-Torresdey JL, Polette L, Arteaga S, Tiemann KJ, Bibb J, Gonzalez JH. Determination of the content of hazardous heavy metals on *Larrea tridentata* grown around a contaminated area. Proceedings of the Eleventh Annual EPA Conf. on Hazardous Waste Research, (HSRC/WERC Joint Conference on the Environment), Edited by Erickson LR, Tillison DL, Grant SC, McDonald JP. Albuquerque, NM. 1996;660-669.
23. Hartle RT, Fernandez GCJ, Nowak RS. Horizontal and vertical zones of influence for root systems of four Mojave Desert shrubs. J. of Arid Environ. 2006;64:586-603.
24. Arteaga L, Carranze T, Elitta M, Gonzalez M, Guevara C, Herrera F, Lopez B. Martinez A, Mendoza F, Narvaez A, Puentes G, Reyes R, Robles H, Sohn C. Triomphe. Paper presented at the International Workshop on Green-Manure cover crops for smallholders in tropical and subtropical regions, 6-12 Apr. 1997, Chapeco, Santa Catarina, Brazil. Rural Extension and Agricultural Research Institute of Santa Catarina, Brazil; 1997.
25. Naumburg E, Loik ME, Smith SD. Photosynthetic responses of *Larrea tridentata* to seasonal temperature extremes under elevated CO2. New Phytologist. 2004;162:323-330.
26. USEPA. SW-846, Test Methods for Evaluating Solids and Wastes – Physical/ Chemical Methods, U.S. Environmental Protection Agency. December, CD-Rom; 1986.
27. Naugle CA. Manganese transport: A preliminary study of the Three Kids mine and surrounding area. MS thesis (unpubl), University of Nevada, Las Vegas, USA; 1997.
28. Colledge BA. Preliminary study of dust transport of manganese and other selected elements at the three kids mine site.

Unpublished thesis, University of Nevada, Las Vegas; 2002.

29. Kruskal WH, Wallis WA. Use of ranks in one-criterion variance analysis. J. Am. Stat. Assoc. 1952;47:583-621.

30. Levene H. Robust tests for equality of variances. In: Olkin I (ed) Contributions to probability and statistics. Stanford Univ Press, Palo Alto. 1960;278-292.

31. Shapiro SS, Wilk MB. An analysis of variance test for normality (complete samples), Biometrika. 1965;52:591-611.

32. Gardea-Torresdey JL, Arteage S, Tiemann KJ, Chianelli R, Pingitore N, Mackay W. Absorption of Copper(II) by creosote bush (*L. Tridentata*): Use of atomic and X-Ray absorption spectroscopy. Environ. Toxicol. Chem. 2001;20(11):2572-2579.

33. Prasad MNV. Cadmium Toxicity and Tolerance in Vascular Plants. Environ. Exper. Bot. 1995;35(4):525-545.

34. Conesa HM, Robinson BH, Schulin R, Nowack B. Growth of Lygeum Sprtum in Acid Mine Tailings: Response of Plants developed from seedlings, rhizomes and at field conditions. Environ. Poll. 2007;145: 700-707.

35. Gupta DK, Nicoloso FT, Schetinger M, Rossato LV, Pereira L. Castro GY, Srivastava S, Tripathi RD. Antioxidant defense mechanism in hydroponically grown Zea mays seedlings under moderate lead stress. J. Hazard. Mater. 2009;172:479-484.

36. Obanda DN, Shupe TF, Barnes HM. Reducing leaching of Boron-based wood preservatives – A review of research. Bioresource Tech., 2008;99(15):7312-7322.

37. Ullah A, Heng S, Munis MFH, Fahad S, Yang X. Phytoremediation of heavy metals assisted by Plant Growth Promoting (PGP) bacteria: A review. Environ. Exper. Botany. 2015;117:28-40.

38. Visioli G, Vamerali T, Mattarozzi M, Dramis L, Sanangelantoni AM. Combined endophytic inoculants enhance nickel phytoextraction from serpentine soil in the hyperaccumulator *Noccaea caerulescens*. Frontiers in Plant Sci. 2015;6(638). DOI: 10.3389/fpls.2015.00638.

39. Kim SJ, Vassão DG, Moinuddin SGA, Bedgar DL, Davin LB, Lewis NG. Allyl/propenyl phenol synthases from the creosote bush and engineering production of specialty/commodity chemicals, eugenol/isoeugenol, in Escherichia coli. Archives of Biochemistry and Biophysics. 2014;541:37-46.

40. Gardea-Torresdey JL, Peralta-Videa JR, Montes M, de la Rosa G, Corral-Diaz B. Bioaccumulation of cadmium, chromium and copper by *Convolvulus arvensis* L.: impact on plant growth and uptake of nutritional elements. Bioresource Tech. 2004;92:229-235.

41. Behmer ST, Elias DO, Bernays EA. Post-ingestive feedbacks and associative learning regulate the intake of unsuitable sterols in a generalist grasshopper. J Experim. Bio. 1999;202;739-748.

42. Behmer ST, Lloyd CM, Raubenheimer D, Stewart- Clark J, Knight J, Leighton RS, Harper FA, Smith JAC. Metal hyperaccumulation in plants: Mechanisms of defence against insect herbivores. Functional Ecology. 2005;19:56-66.

43. Smolders E, Oorts K, Peeters S, Lanno R, Cheyns K. Toxicity in lead salt spiked soils to plants, invertebrates and microbial processes: Unraveling effects of acidification, salt stress and ageing reactions. The Sci. of the Total Environ. 2015;536:223-231.

44. Badri MA, Pulford ID, Springuel I. Supply and accumulation of metals in two Egyptian desert plant species growing on wadi-fill deposits. J. Arid Environ. 1996;32:421-429.

45. McAuliffe JR, Hamerlynck EP, Eppes MC. Landscape dynamics fostering the development and persistence of long-lived creosote bush (*Larrea tridentata*) clones in the Mojave Desert. J. Arid Environ. 2007;69:96-126.

46. Fuentes-Ramirez AJL, Schafer EL. Mudrak M, Schat HA, Parag C, Holzapfel KA. Moloney. Spatio-temporal impacts of fire on soil nutrient availability in *Larrea tridentata* shrublands of the Mojave Desert, USA. Geoderma. 2015;(259–260):126–133.

47. Schafer JL, Mudrak EL, Haines CE, Parag HA, Moloney KA, Holzapfel C. The association of native and non-native annual plants with *Larrea tridentata* (creosote bush) in the Mojave and Sonoran Deserts. J. Arid Environ. 2012;87:129-135.

48. Google Maps. Terrain maps; 2015. Available:http://maps.google.com/maps?hl=en&tab=wl

14

Effect of Salicylic Acid, Benzoic Acid, and *p*-Coumaric Acid on Growth, Chlorophyll, Proline, and Vitamin C of Salinity-Stressed Tobacco (*Nicotiana tabacum*)

Alaaldin Idris H. Khairy[1] and Kwang Soo Roh[1*]

[1]Department of Biology, College of Life Sciences, Keimyung University, Daegu 42601, Korea.

Authors' contributions

This work was carried out in collaboration between both authors. Both authors designed the study. Author AIHK wrote the protocol, performed the experiment, managed the experimental measurements and analysis of data, managed the literature searches and wrote the first draft of the manuscript with the guidance of author KSR. Both authors read and approved the final manuscript.

Editor(s):
(1) Mirza Hasanuzzaman, Department of Agronomy, Faculty of Agriculture, Sher-e-Bangla Agricultural University, Dhaka-1207, Bangladesh.
Reviewers:
(1) Dipak Kumar Sahoo, Iowa State University, USA.
(2) Geetika Sirhindi, Punjabi University, India.
(3) Rezzoug Waffa, Université Ibn Khaldoun, Algeria.
(4) Baskaran Kannan, University of Florida, USA.

ABSTRACT

Salinity is an important growth-limiting abiotic stress factor. Salicylic acid (S.A), Benzoic Acid (B.A) and *p*-coumaric acid (*p*-CA) are substances generally thought to play a crucial role in the enhancement of salt tolerance and regulation of growth and development in plants. This research was carried out to study the influence of Salicylic Acid (SA), Benzoic Acid (BA), and *p*-Coumaric acid (*p*-CA) on growth, contents of chlorophyll, proline and vitamin C in salinity-stressed tobacco plant (*Nicotiana tabacum*) grown *In vitro*. The results showed that Salinity (NaCl) stress impairs plant growth and inflicts severe physiological disorders. While supplementation of SA, BA and *p*-CA resulted in remarkable increases of all investigated biochemical components. Compared to the control, the high concentration of NaCl (150 mM) significantly reduced total fresh weight of tobacco. Total chlorophyll contents were remarkably increased by SA and BA treatments under 100 mM

**Corresponding author: E-mail: rks@kmu.ac.kr*

NaCl condition. The highest leaf and stem proline contents were recorded from the treatment of SA supplemented with 150 mM NaCl, while the lowest value was observed in *p*-CA + 0 mM NaCl treatment. Vitamin C contents were significantly increased by SA and BA treatments under 100 mM and 150 mM NaCl concentrations. SA and BA recovered the damage caused by high concentrations of salinity and enhanced salt tolerance in tobacco. Application of *p*-CA did not show significant differences compared to the control. This study suggests that SA and its derivatives could be used to overcome damages generated by salinity stress, enhance salt-tolerance and regulate plant growth and development.

Keywords: Benzoic acid; chlorophyll; proline; p-Coumaric acid; salicylic acid; salinity; tobacco; vitamin C.

1. INTRODUCTION

The influence of salinity on growth and development of plants has been a focus of research in plant biology for several decades because salt stress is a major environmental determinant limiting crop productivity and production [1]. Many researchers have reported on changing levels of physiological and biochemical parameters caused by salinity stress. The stresses generated by a high concentration of Salt lead to disturbance in growth and photosynthetic processes by causing changes in the accumulation of Na^+, Cl^- and nutrients, and disturbance in water and osmotic potential in plants [2]. Assimilation rate of photosynthetic CO_2 is also reduced by salinity. This reduction is partly due to a reduced stomatal conductance, both physiological and biochemical disorders created by salinity reduce crop production. In plants, toxic effects of salinity stress lead to metabolic changes such as reduction of chloroplast activity accompanied by decrease in photosynthetic rate which then leads to an increased reactive oxygen species. The decrease in photosynthesis may eventually cause reduction in both growth and yield [3]. It is estimated that at least 20% of all irrigated lands are salt-affected [4]. The huge areas of salt-affected lands on one hand and world population growth on the other have led plant scientists to the concept of enhancing salinity tolerance in plants by several approaches in order to meet the needs of ever growing world population. Proline and vitamin C are important biochemical components in plants. Proline plays a crucial role in the response of plants to salinity and other abiotic stresses [5]. Proline is also involved in flowering and development of plants [6]. Vitamin C is ubiquitous in plants, and serves a host of several physiological functions. It protects cells and organelles from oxidative damage, as well as improves the plant responses to environmental stresses like salt and drought [7]. Damages generated by salt stress can be alleviated or reduced either genetically or biochemically by application of chemical signaling molecules such as plant growth regulators. It is reported that application of signaling molecules like Salicylic Acid (SA) and its derivatives can enhance salt tolerance in plants by inducing protective effects of plants under salinity stress [8]. SA is considered as a naturally occurring molecule that can promote and develop plant growth by inducing plant tolerance against biotic and abiotic stresses such as salinity [9]. It also plays a crucial role in regulation of some plant physiological processes in plants like influences on growth and development, ion uptake, ion balance and transport and membrane permeability [10]. *p*-Coumaric acid is a hydroxycinnamic acid and it can be found in legumes. Both *p*-coumaric acid (*p*-CA) and benzoic acid (BA), are signaling molecules derived from phenylalanine [11], and has various physiological activities such as antimicrobial activity [12], cancer chemoprevention [13], antioxidant activity [14], and antimelanogenesis [15]. BA is a precursor of SA and produced by β-oxidation of cinnamic acid synthesized through phenyl propanoid metabolic pathway in plants [16]. Considerable results have been pointed concerning SA, BA and *p*-CA induced enhancements in salinity tolerance of many plants. SA improved growth and yield components of wheat plant [17]. SA and BA promote the generation of reactive oxygen species during salt stress thus play an important role in stress tolerance [18]. Despite numerous reports about the importance of signaling molecules such as SA, BA and *p*-CA, the role of *p*-CA, BA and SA in salt tolerance enhancement has not been intensively studied. Therefore, in the current research our aim is to study the influence of SA, BA and *p*-CA on salinity stressed tobacco cultivated in *In vitro* by investigating the physiological and biochemical parameters; fresh weight, chlorophyll, proline and vitamin C contents.

2. MATERIALS AND METHODS

2.1 Chemicals and Apparatuses

MS medium was purchased from Duchefa Biochemie (Haarlem, Netherland), while other chemicals were obtained from Sigma Chemical Co. [St Louis, MO, USA]. Refrigerator centrifuge (Kontron T-324), fraction collector (Bio-Rad 2110), ELISA microplate reader (Bio-Rad 680), UV-visible spectrophotometer (GeneQuant 100), were used in this experiment.

2.2 Growth Media

Plant growth medium of Murashige and Skoog 1962 [19] was supplemented with 10^{-5} mM SA, BA and *p*-CA treatments Kim and Roh 2014 [20] under four different concentrations of NaCl (0 mM, 50 mM, 100 mM and 150 mM) as interactive treatments in addition to the control treatment. The pH value was adjusted to 5.8 using pH meter. NaCl powder and 20 ml of the five-fold serial dilution of each of the three signaling molecules (SA, BA and *p*-CA) were combined with 180 ml of MS medium and poured in plant tissue culture glass jars. Culture glass bottles containing growth media were sterilized at 121°C for 15 min using autoclave. Then jars were moved to the plant tissue culture room for planting.

2.3 Planting

In vitro cultivated tobacco stems were cut into 3 cm segments. Each three explants were cultivated on the culture medium. All culture jars were maintained at 28°C under light for 16 hr. (800 uM/m2/s PFD) and dark for 8 hr. in aseptic condition for 50 days. Then measurements of the observed data were taken. The experiment was replicated three times.

2.4 Growth (Fresh weight) Determination

Tobacco growth was estimated by measuring fresh weights of tobacco roots, leaves and stems, and then total fresh weight of tobacco was calculated. To measure the fresh weight of tobacco, 50 days old tobacco plants were smoothly uprooted from the artificial medium and separated into roots, leaves and stems using sterilized forceps and scissors. Each of the three parts of the plant was put in petri dishes then fresh masses were measured using a digital balance. Total fresh weight was calculated by accumulating the fresh weight of the three separated parts of the plant.

2.5 Chlorophyll Content Determination

Chlorophyll was estimated by adopting the method described by Inskeep and Bloom 1985 [21]. Three fresh fully expanded leaves were extracted overnight with *N, N*-Dimethylformamide solution at 4°C and then the extract was centrifuged at 8000x g for 5 min. The absorbance of the supernatant was spectrophotometrically read at 647 and 664.5 nm. The following equations were used to estimate the concentration of total chlorophyll from the supernatant.

$$\text{Total chlorophyll (mg/g fw)} = 17.90\ A_{647} + 8.08\ A_{664.5}$$

2.6 Tobacco Leaf and Stem Proline Contents Determination

Proline contents were determined according to the method described by Bates et al. [22]. Using liquid nitrogen tobacco leaf and stem samples were well powdered. 0.5 g of each sample was used for the measurement. Leaf and stem materials (0.5 g) were homogenized in 10 ml of 3% aqueous sulfosalicylic acid and the homogenate was centrifuged at 8,000x g for 5 min. 2 ml of the supernatant was mixed with 2 ml of acid ninhydrin and 2 ml of glacial acetic acid in test tubes. 6 ml of the mixture was placed in a water bath for 1 h at 100°C. The reaction mixture was extracted with 4 ml toluene, and then cooled to room temperature, and the absorbance was read at 520 nm with a spectrophotometer apparatus. Appropriate proline standards were included for the calculation of proline in the samples.

2.7 Tobacco Leaf and Stem Vitamin C Determination

The content of tobacco leaf and stem vitamin contents were determined according to the method of Jagota and Dani 1982 [23]. Tobacco leaf extracts (0.5 ml) were added to 0.8 ml of 10% TCA and vigorously shaken and the mixtures were kept in ice box for 5 min and then centrifuged at 3000 x g for 5 min. 0.5 ml of this extract was then diluted to 2 ml with 1.5 ml distilled water. The 2 ml sample was added to 2 ml Folin phenol reagent was then vigorously shaken. Stored at room temperature for 10 min, then using microplate the absorbance was measured at 760 nm against distilled water as a blank and the vitamin C content was estimated through the calibration curve of ascorbic acid. Unless specified otherwise, all aforementioned processes were done at 4°C.

2.8 Statistical Analysis

The experiment was replicated three times. Results are presented as the mean ± Standard Error (SE). Significant differences were determined by one-way analysis of variance (ANOVA) using SPSS 12.0 software.

3. RESULTS

3.1 Fresh Weight (Growth)

Plant growth was significantly reduced when tobacco plants were subjected to high levels of salinity stress concentrations. Specially, fresh weights of tobacco root and stem were remarkably decreased in the presence of NaCl. The maximum reduction was observed from the treatment of 150 mM NaCl. As seen in Fig. 1, the application of SA and BA increased growth of salt-stressed tobacco, while salinity reduced the growth compared to the control. Results of tobacco growth in Fig. 1 were justified and confirmed by measuring fresh weights of tobacco roots, leaves and stems. Leaf fresh weight was increased by the exogenous application of both SA and BA supplemented with 50 and 100 mM of NaCl concentrations. While *p*-CA application didn't show significant differences in tobacco fresh weight compared to the control and salinity induced plants. However, maximum leaf fresh weight was observed in the treatment of BA and SA under 150 mM NaCl concentration as compared to control (Fig. 2).

3.2 Chlorophyll Contents

The chlorophyll contents were remarkably decreased in NaCl stressed plants. The maximum inhibition was observed when plants were exposed to the highest concentration of salt (150 mM NaCl), where the content of chlorophyll was reduced by 33.4% compared to the control, while supplementation of SA increased chlorophyll content in 100 mM NaCl induced plants. Compared to the control, the highest chlorophyll content was observed in SA treated plants under non-saline condition, while the lowest chlorophyll content was observed in 150 mM NaCl supplemented with *p*-CA treatment. Nevertheless, SA and BA treatments showed significant differences in non-saline, 50 mM and 100 mM NaCl conditions, while *P*-CA did not increase tobacco chlorophyll contents under all various concentrations of NaCl (Fig. 3).

Fig. 1. Effects of SA, BA and *p*-CA on growth of salinity stressed-tobacco plants. Tobacco plants were maintained at 28°C under light for 16 hr (800 uM/m2/s PFD) and dark for 8 hr

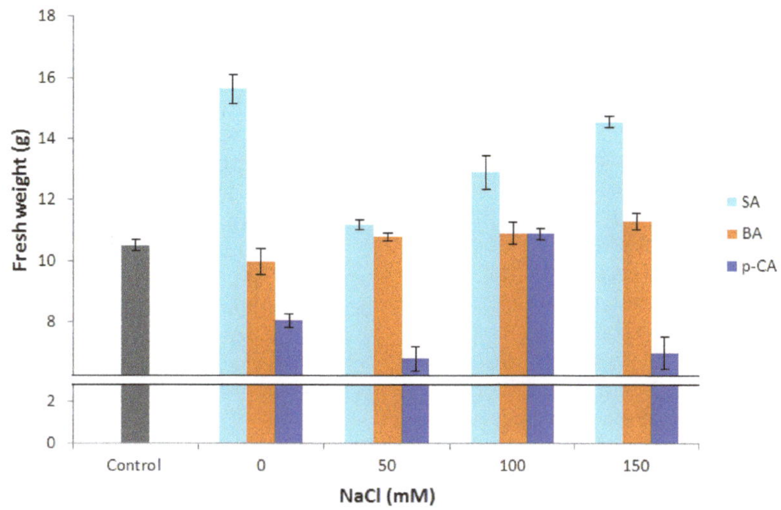

Fig. 2. Effects of SA, BA and *p*-CA on fresh weight of salinity-stressed tobacco plants. The bars indicate Standard Errors (SE)

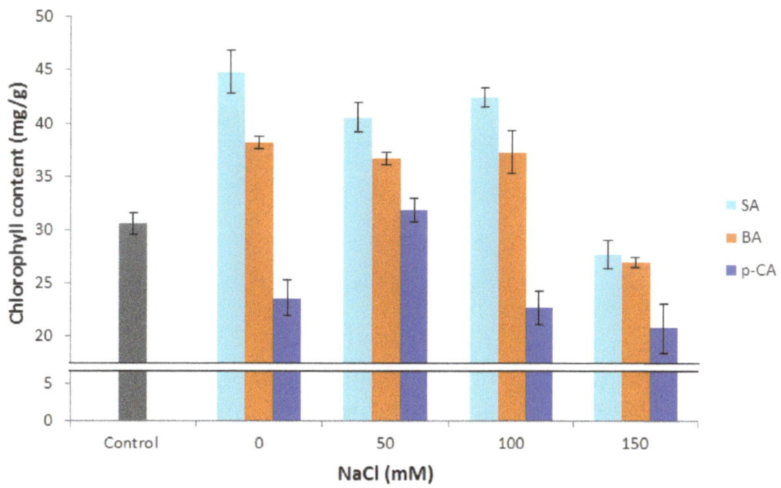

Fig. 3. Effects of SA, BA and p-CA on total chlorophyll content of salinity-stressed tobacco plants. The bars indicate Standard Errors (SE)

3.3 Proline Contents

Tobacco leaf and stem proline contents were hugely retarded by salt stress and increased by SA and BA supplementations. The stress of salinity significantly decreased tobacco leaf proline content at 0, 50 and 100 mM NaCl concentrations. Maximum reduction was observed at 100 mM NaCl concentration. Exogenous application of SA and BA increased proline content in 150 mM NaCl concentration. And the highest proline content was observed in treatment of 150 mM NaCl combined with 10^{-5} mM SA, which was 58% higher than the control. Application of *p*-CA did not increase leaf proline content of tobacco under both non-saline and saline conditions (Fig. 4). Tobacco stem proline content was significantly affected by salinity stress. Supplementation of SA remarkably increased tobacco stem proline content especially at 150 mM NaCl concentration which was higher than the control. BA treatment substantially increased proline content at 100 and 150 mM NaCl concentrations. Maximum proline content was recorded from the treatment of SA complemented with 150 mM NaCl. Both benzoic acid and salicylic acid improved tobacco salinity tolerance under elevated levels of NaCl as compared to the control. Whilst there was no significant differences observed in *P*-CA supplemented plants (Fig. 5).

3.4 Vitamin C Content

Vitamin is one of the most important components. In the present study the three signaling molecules SA, BA and *p*-CA hugely increased tobacco leaf and stem vitamin contents under all different levels of salinity. Tobacco leaf vitamin C content was not significantly affected by the stress of salt. The applications of SA and BA treatments under 50 mM and 150 mM NaCl concentrations increased leaf vitamin C contents compared to the control. The maximum leaf vitamin C contents were recorded from the treatments of SA under 50 mM and 150 mM of NaCl, while the lowest content was observed in *p*-CA treatment under non-saline condition. The exogenous application of *p*-CA slightly reduced the contents of vitamin C in NaCl stressed plants (Fig. 6). However, Salt stress did not show significant reduction of stem vitamin content. When compared to the control, SA application increased stem proline especially at high concentrations of NaCl. Maximum stem vitamin C content was observed in the treatment of SA under both 100 mM and 150 mM NaCl concentrations. Compared to NaCl treated plants, *p*-CA did not show significant differences under all various levels of salinity except 100 mM NaCl (Fig. 7).

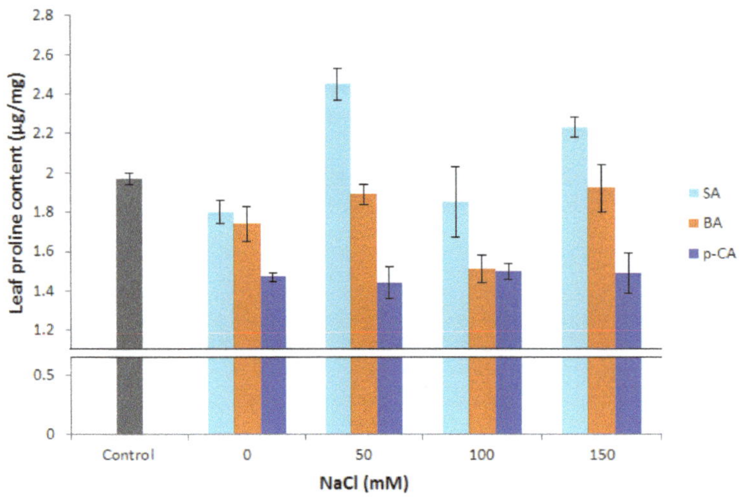

Fig. 4. Effects of SA, BA and *p*-CA on leaf proline contents in salinity-stressed tobacco plants. The bars mean Standard Errors (SE)

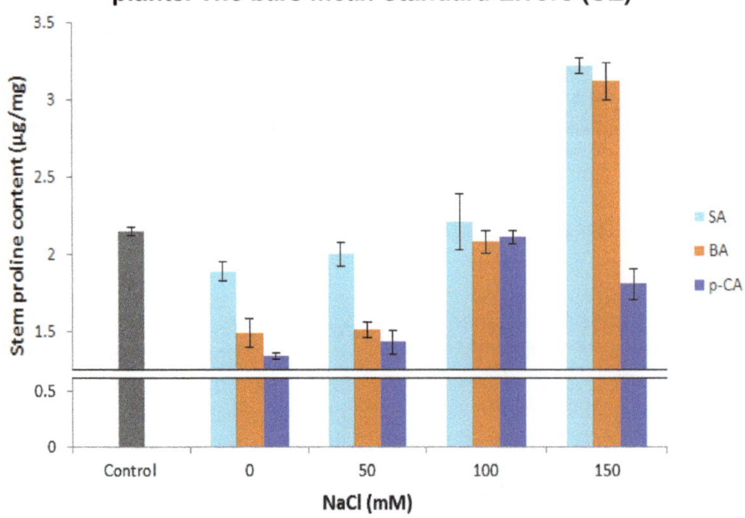

Fig. 5. Effects of SA, BA and *p*-CA on stem proline contents in salinity-stressed tobacco plants. The bars indicate Standard Errors (SE)

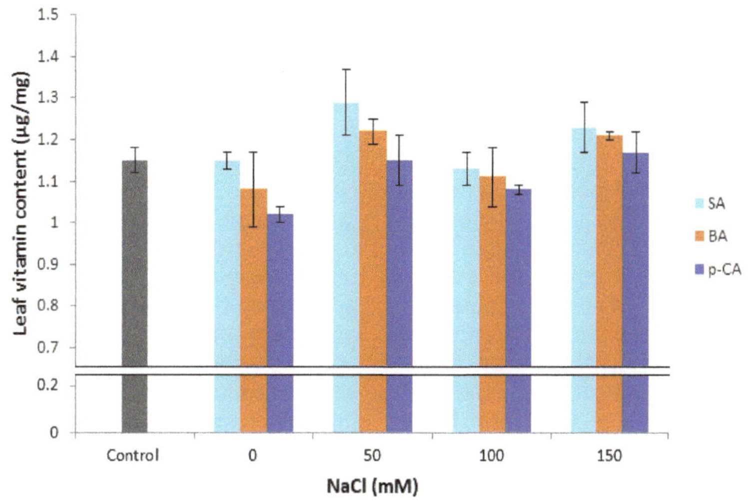

Fig. 6. Effects of SA, BA and *p*-CA on leaf vitamin C contents in salinity-stressed tobacco plants. The bars indicate standard errors (SE)

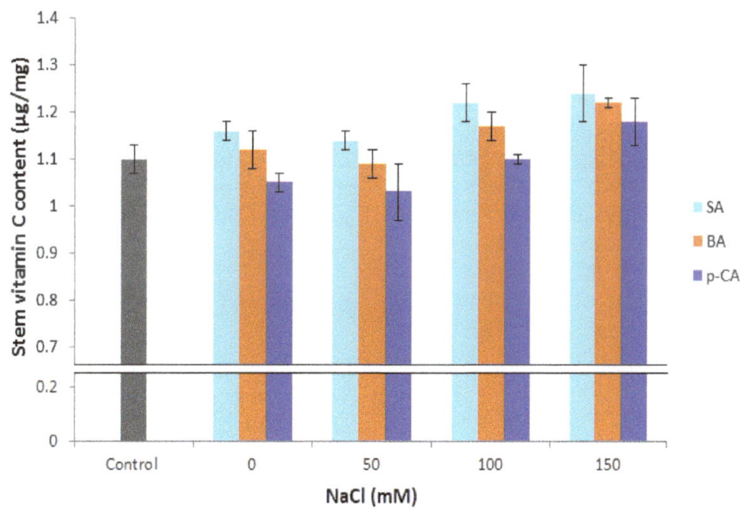

Fig. 7. Effects of SA, BA and *p*-CA on stem vitamin C contents in salinity-stressed tobacco plants. The bars indicate Standard Errors (SE)

4. DISCUSSION

Salinity is an important growth-limiting abiotic stress factor. SA, BA, and *p*-CA are substances generally thought to play a crucial role in the enhancement of salt tolerance in plants. The present research was conducted to study the influence of signaling molecules SA, BA, and *p*-CA on salt stressed tobacco physiological and biochemical characterizations. The results showed that by the increase in salinity there was decrease in all investigated parameters. Growth parameters (fresh weight of tobacco leaf, stem and root) were affected by high salt concentrations. These results are similar to those reported by Afzal et al. [24] who found that wheat fresh weight was reduced by salt stress. As well chlorophyll content, leaf and stem proline contents and vitamin contents of tobacco leaf and stem were also decreased in the presence of NaCl. Exogenous application of signaling molecules (SA, BA, and *p*-CA) had ameliorative impacts under saline conditions and enhanced tobacco salt tolerance. Tobacco plant growth, leaf and stem fresh weights were increased by SA and BA exogenous application under high concentrations of salt, 100 and 150 mM NaCl conditions. These results are in agreement with Gunes et al. [25] who observed high fresh weight by application of SA under salinity in maize plant.

Similar results have been also reported by Misra and Saxena [26] and Khodary [27] in maize plant and El-Tayeb [28] in barely plant. This may indicate that exogenous application of SA and BA exhibited a significant increase in salinity tolerance. *p*-CA didn't show significant differences in fresh weight of tobacco plants and there are no previous findings. Chlorophyll content of NaCl-treated tobacco plants was significantly decreased below that of the control. Similarly, Dela-Rosa and Maiti [29] found an inhibition in chlorophyll biosynthesis in sorghum plants due to salinity stress. Application of SA and BA increased chlorophyll content under high salinity levels 100 mM and 150 mM NaCl. Whereas, in the presence of 10^{-5} mM SA and BA treatments the damage generated by salt stress was recovered. These results are in agreement with Zhao et al. [30] and Sinha et al. [31] who pointed out that chlorophyll and carotenoid contents of maize leaves were increased upon treatment with SA. SA is supposed to enhance the functional state of the photosynthetic machinery in plants by chlorophyll biosynthesis [32]. Previous author's results support our findings. Proline is an important osmolyte and one of the components of plant defiance system to counter stresses generated by salinity, increasing proline leads to increase in salinity stress resistance. Proline is accumulated as a final product and the source of proline is total protein and total amino acid [33]. Our data showed that the accumulation of proline increased in the stems of 150 mM NaCl stressed tobacco plants by application of SA and BA. These results are in agreement with Tasgin et al. [34] who reported that the proline accumulation by SA treatment increases in wheat, oat, bean and tomato, under salt and oxidative stresses. Similar results were reported by Desnigh and Kanagaraj [35] who observed that the more tolerant plants store more proline. Leaf proline content was also increased by application of SA and BA to tobacco plants subjected to high salinity level. The highest content of leaf proline was observed in the treatment of both SA and BA under 150 mM NaCl. Similar results were observed by Shakirova and Sakhabutdinova [36], who figured out that treatment of SA under salt stress increases proline and defensive proteins. In the present study both SA and BA didn't increase proline content under low salt levels, significant levels of proline were only reported under 150 mM NaCl. While exogenous application of *p*-CA didn't increase both stem and leaf proline contents under the four different NaCl levels. Proline was not detected in tobacco roots.

Tobacco leaf and stem vitamin C contents were not significantly affected by salinity. Compared to the control, the exogenous application of SA, BA increased stem and leaf vitamin C contents under all salinity levels. Maximum stem vitamin C amounts were observed in the treatments of SA combined with 100 mM and 150 mM NaCl, while the highest contents of leaf vitamin C were recorded from the treatments of SA and BA under non-saline and 50 mM NaCl. The application of *p*-CA did not increase tobacco leaf and stem vitamin C contents under both saline and non-saline conditions. These results are similar to those reported by Mahdi Javaheri et al. [37] who reported an increase of productivity, lycopene, and vitamin C content of tomato fruit and by SA application under salinity conditions. Almost all previously documented findings support our results. Vitamin C content could not be detected in tobacco roots.

5. CONCLUSION

SA, BA and *p*-CA, can be used to alleviate salinity toxicity in plants. And enhance salt-tolerance through increasing some important physiological and biochemical characteristics such as contents of chlorophyll, proline and vitamin C.

ACKNOWLEDGEMENT

This work was supported by the Research Scholarship of the Graduate School of Keimyung University in 2015.

COMPETING INTERESTS

Authors have declared that no competing interests exist.

REFERENCES

1. Greenway H, Munns R. Mechanisms of salt tolerance in nonhalophytes. Ann. Rev. Plant Physiol. 1980;31:149-190.
2. Kirst GO. Salinity tolerance of eukaryotic marine algae. Ann. Rev. Plant Physiol. Plant Mol. Biol. 1989;40:21-53.
3. Parida AK, Das AB. Salt tolerance and salinity effects on plants. Rev. Ecotoxi. Environ. Safety. 2005;60:324-349.
4. Pitman MG, Läuchli A. Global impact of salinity and agricultural ecosystems, In: Läuchli A, Luttge U. (eds.), Salinity: Environment Plants Molecules. Kluwer

Academic Publishers: Dordrecht, the Netherlands. 2002;3-20.
5. Kaneshiro ES, Holz GG Jr, Dunham PB. Osmoregulation in a marine ciliate, *Miamiensis avidus*. II. Regulation of intracellular free amino acids. Bio. Bull. 1969;137:161-169.
6. Mauro ML, Trovato M, De Paolis A, Gallelli A, Costantino P, Altamura MM. The plant oncogene rolD stimulates flowering in transgenic tobacco plants. Devel. Bio. 1996;180:693-700.
7. Conklin PL, Barth C. Ascorbic acid, a familiar small molecule intertwined in the response of plants to ozone, pathogens, and the onset of senescence. Plant Cell Environ. 2004;27:959-970.
8. Simaei M, Khavarinejad RA, Saadatmand S, Bernard F, Fahimi H. Interactive effects of salicylic acid and nitric oxide on soybean plants under NaCl salinity. Russian J. Plant Physiol. 2011;58:783-790.
9. Horvath E, Szalai G, Janda T. Induction of abiotic stress tolerance by salicylic acid signaling. J. Plant Growth Reg. 2007;26: 290-300.
10. Raskin I. The role of salicylic acid in plants. Ann. Rev. Plant Physiol. Plant Mol. Biol. 1992;43:439-463.
11. Zang LY, Cosma G, Gardner H, Shi X, Castranova V, Vallyathan V. Effect of antioxidant protection by p-coumaric acid on low-density lipoprotein cholesterol oxidation. American J. Physiol. Cell Physiol. 2000;279:954-960.
12. Park SK, Park JC. Antimicrobial activity of extracts and coumaric acid isolated from Artemisia princeps var. orientalis. Korean J. Biotech. Bioeng. 1994;5:506-511.
13. Hudson EA, Dinh PA, Kokubun T, Simmonds MS, Gescher A. Characterization of potentially chemopreventive phenols in extracts of brown rice that inhibit the growth of human breast and colon cancer cells. Cancer Epidemiol Biomark. Prev. 2000;9:1163–1170.
14. Ferguson LR, Zhu ST, Harris PJ. Antioxidant and antigenotoxic effect of plant cell wall hydroxycinnamic acids in cultured HT-29 cell. Mol. Nutr. Food Res. 2005;49:585-593.
15. An SM, Lee SI, Choi SW, Moon SW, Boo YC. *p*-Coumaric acid, a constituent of Sasa quelpaertensis Nakai, inhibits cellular melanogenesis stimulated by a-melanocyte stimulating hormone. J. Dermatol. 2008;159:292-299.
16. Ribnicky DM, Shulaev V, Raskin I. Intermediates of salicylic acid biosynthesis in tobacco. Plant Physiol. 1998;118:565-572.
17. Shakirova FM, Bezrukova MV. Induction of wheat resistance against environmental salinization by salicylic acid. Biol. Bulletin. 1997;24:109-112.
18. Borsani O, Valpuesta V, Botella MA. Evidence for a role of salicylic acid in the oxidative damage generated by NaCl and osmotic stress in Arabidopsis seedlings. Plant Physiol. 2001;126:1024-1030.
19. Murashige T, Skoog F. A revised medium for rapid growth and bio assays with tobacco tissue cultures. Physiol. Plant. 1962;15:473-497.
20. Kim SA, Roh KS. Effect of p-coumaric acid, benzoic acid, and salicylic acid on the activity of glutathione reductase and catalase in in vitro grown tobacco plant. J. Life Sci. 2014;24:227-233.
21. Inskeep WP, Bloom PR. Extinction coefficients of chlorophyll a and b in N,N-dimethylformamide and 80% acetone. Plant Physiol. 1985;77:483-485.
22. Bates L, Waldren RP, Teare, ID. Rapid determination of free proline for water-stress studies. Plant Sci. 1973;39:205-207.
23. Jagota SK, Dani HM. A new calorimetric technique for the estimation of vitamin C using folin phenol reagent. Anal. Biochem. 1982;127:178-182.
24. Afzal IM, Shahzad BN, Ahmad MF. Optimization of hormonal priming techniques for alleviation of salinity stress in wheat. Biol. Res. 2005;17:95-109.
25. Gunes A, Inal M, Alpaslan F, Eraslan E, Bagci G, Cicek N. Salicylic acid induced changes on some physiological parameters symptomatic for oxidative stress and mineral nutrition in maize (*Zea mays* L.) grown under salinity. J. Plant Physiol. 2007;164:728-736.
26. Misra N, Saxena P. Effect of salicylic acid on proline metabolism in lentil grown under salinity stress. Plant Sci. 2009;177:181-189.
27. Khodary SEA. Effect of salicylic acid on the growth, photosynthesis and carbohydrate metabolism in salt stressed maize plants. Intl. J. Agri. Biol. 2004;6:5-8.
28. El-Tayeb MA, Response of barley grains to the interactive effect of salinity and salicylic acid. Plant Growth Reg. 2005;45:215–224.

29. Dela-Rosa IM, Maiti RK. Biochemical mechanism in glossy sorghum lines for resistance to salinity stress. J. Plant Physiol. 1995;146:515-9.
30. Zhao HJ, Lin XW, Shi, HZ, Chang, SM. The regulating effects of phenolic compounds on the physiological characteristics and yield of soybeans. Acta Agro. 1995;21:351–355.
31. Sinha SK, Srivastava HS, Tripathi RD. Influence of some growth regulators and cations on inhibition of chlorophyll biosynthesis by lead in maize. Bull. Env. Contamin. Toxic. 1993;51:241-246.
32. Shi Q, Bao Z, Zhu Z, Ying Q, Qian Q. Effects of different treatments of salicylic acid on heat tolerance, chlorophyll fluorescence, and antioxidant enzyme activity in seedlings of *Cucumis sativa* L. Plant Growth Regul. 2006;48:127-135.
33. Sairam RK, Deshmakh PS, Saxena DC. Role of antioxidant systems in wheat genotype tolerance to water stress. Plant Biol. 1998;41:387–394.
34. Tasgin E, Atici O, Nulbantoglu B, Popova, LP. Effects of salicylic acid and cold treatment on protein levels and on the activities of antioxidant enzymes in the apoplast of winter wheat leaves. Phytochem. 2006;67:710-715.
35. Desnigh R, Kanagaraj G. Influence of salinity stress on photosynthesis and antioxidative systems in two cotton varieties. Gen. Appl. Plant Physiol. 2007; 33:221-234.
36. Shakirova FM, Sakhabutdinova DR. Changes in the hormonal status of wheat seedlings induced by salicylic acid and salinity. Plant Sci. 2003;164:317-322.
37. Mahdi J, Kambiz M, Alireza D, Fateme ZT. Effects of salicylic acid on yield and quality characters of tomato fruit (*Lycopersicum esculentum* Mill.). Agri. Crop Sci. 2012;16:1184-118.

15

Growth Physiology and Membrane Permeability of Okra (*Abelmoschus esculentus* L.) Seedlings as Affected by Salinity

J. Jeyapraba[1], S. Mahendran[1] and N. Sujirtha[1*]

[1]*Department of Agricultural Biology, Faculty of Agriculture, Eastern University, Sri Lanka.*

Authors' contributions

This work was carried out in collaboration by all three authors. The authors of this manuscript worked together to design, conduct, analyze and interpret the findings of this experiment. All the authors read and approved the final manuscript.

Editor(s):
(1) Sławomir Borek, Faculty of Biology, Department of Plant Physiology, Adam Mickiewicz University, Poland.
Reviewers:
(1) Petigrosso, Lucas Ricardo, Universidad Nacional de Mar del Plata, Argentina.
(2) Edward Missanjo, Malawi College of Forestry and Wildlife, Malawi.

ABSTRACT

Aims: Among the abiotic stresses, salinity is one of the most important environmental constraints which provokes grave threats for agriculture and environment. It is a growing threat to agricultural crops and its impacts have already started becoming visible as expansion in the affected areas and deteriorating soil fertility and crop productivity. By considering this feature, a study was conducted to determine the salinity responses of selected okra cultivars on certain physiological and biochemical attributes.
Study Design: The experiment was laid out in a Completely Randomized Design with six treatments and four replications.
Place and Duration of Study: A study was conducted at the crop farm of the Eastern University, Sri Lanka during the yala season of the year 2014.
Methodology: The seeds of okra cultivar "Haritha" were surface sterilized with sodium hypochlorite 0.5% (v/v) and were allowed to germinate in petri dishes having filter papers saturated with distilled water as control and NaCl solutions of 25, 50, 75, 100 and 125 mM. The germinated seeds in distilled water and in saline solutions were transferred to plastic pots filled with soil/compost mixture

*Corresponding author: E-mail: sujee3@gmail.com

(2:1V/V). The pots were irrigated with distilled water or saline solution for two weeks. The response of okra plants to salinity stress with respect to physiological growth attributes, water content and membrane permeability were investigated.

Results: Salinity stress did not show significant ($P > 0.05$) effect on shoot dry weight. However, root dry weight decreased with an increase of NaCl concentration. The root water content and shoot/root ratio increased significantly ($P < 0.05$) with an increase of salt concentration (25, 50, 75, 100 and 125 mM). It was observed that there was an increase in membrane permeability with associated electrolyte leakage when okra seedlings were exposed to high levels of salt particularly 100 and 125 mM NaCl.

Conclusion: The property of salinity tolerance depends on different physiological interactions. The growth physiological attributes presented by the plant in response to salinity may not be enough to determine its effect.

Keywords: Dry weight; electrolyte leakage; okra; salinity; water content.

1. INTRODUCTION

Abiotic stresses like heat, cold, drought and salinity affect the plant growth and productivity but, the salt stress exerts more drastic effects [1]. Despite the essentiality of chloride as a micro-nutrient for all higher plants and of sodium as mineral nutrient for many halophytes, salt accumulation may reduce local biodiversity, limit growth and reproduction of plants and may lead to toxicity in non-salt-tolerant plants. Most of the cultivated plants are sensitive to salt stress with either as relatively low salt tolerant or severely inhibited growth at low salinity levels [2]. This change results in reduction of shoot and restricted rooting. Germination and seedling establishment constitute the most critical periods in the life cycle of plants. The conductivity test, based on solute leakage has been proposed as a good indicator of salt tolerance in plants [3]. Blokhina [4] pointed out that the cell membrane often suffered injury associated with the increases in permeability.

Okra plants have tap roots 2 inch thick and 4.5 feet deep. The stem is semi-woody and sometimes pigmented with green or reddish tinged colour. It is erect, variable in branching with many short branches. The woody stem bears leaves that are lobed and are generally hairy. Leaves are cordate, simple, usually palmately 3-7 lobed and veined. The flowers are borne vertically only on the orthotropic axis every two or three days. They are large, around 2 inches in diameter with five white to yellow petals with a red or purple spot at the base of each petal. The fruit is an elongated, conical capsule comprising for the most part and five cavities containing ovules. The fruit actually a long pod is generally ribbed developing in the leaf axil.

Okra (*Abelmoschus esculentus* L.) is recognized as an annual herbaceous plant and it is one of the most important warm season fruit vegetables grown in tropical and sub-tropical parts of the world and valued for its edible green pods that are popular vegetable in Sri Lanka. Okra requires a long, warm and humid growing period. It can successfully be grown in hot humid areas. It is sensitive to frost and extremely low temperatures. For normal growth and development, a temperature between 24 and 28°C is preferred. At higher temperatures, the plants grow faster. Temperatures beyond 40-42°C, flowers may desiccate and drop causing yield losses. Okra is grown on sandy to clay soils but, due to its well developed tap root system, relatively light, well – drained rich soils are ideal. As such, loose, friable, well manured loamy soils are desirable. A pH of 6-6.8 is ideally suited.

Okra serves as a source of carbohydrates, fats, vitamins and minerals [5]. In spite of having good nutritional value, its per hectare yield is very low. This decline in optimum yield may be due to the drastic effects of salts. High ratios of salts in root zone affect root density, root turgor pressure and its growth [6]. Okra plant at earlier growth stages is more sensitive to salinity, as it affects water relations of plants.

The present studies were conducted to evaluate the growth physiological attributes of okra seedlings to saline conditions. By measuring the seedling growth and membrane permeability, the degree of tolerance of okra plants to different salinity levels was investigated.

2. MATERIALS AND METHODS

Okra seeds cv. 'Haritha' were collected from the Agronomy farm of the Eastern University of Sri

Lanka. The seeds were surface sterilized with sodium hypochlorite 0.5% (V/V) for 20 min; washed repeatedly with distilled water and then sown in petri dishes having filter papers saturated with distilled water (control) or NaCl solutions (25, 50, 75, 100 and 125mM NaCl) at room temperature.

The seeds after eight days of germination were shifted to plastic pots filled with soil/compost mixture (2:1 V/V). The pots were irrigated with distilled water or saline solution for two weeks. There were six treatments and each one was replicated four times in a factorial arrangement laid out in a Completely Randomized Design.

2.1 Growth Measurements

A number of five plants were randomly collected from each replicate of the treatments, uprooted and washed with distilled water. A number of twenty seedlings per treatment were partitioned into hypocotyls, cotyledons and roots for the measurement of Fresh Weight (FW). The Dry Weight (DW) was recorded after drying the samples in an oven at 80℃ for 24hrs. The (hypocotyls + cotyledons) dry weight/root dry weight ratio was determined. In addition, cotyledons and hypocotyls (aerial part) water content and root water content were estimated based on (FW- DW)/DW ratio.

2.2 Membrane Permeability

Electrolyte leakage was used to assess the membrane permeability as described by [7]. The cotyledons were cleaned and washed with distilled water and were cut into discs of uniform size. These discs were placed in test tubes containing 10 ml of de-ionized water and kept at room temperature for 24h and the electrical conductivity of the solution (EC1) was recorded. The samples were autoclaved at 120℃ for 20 min and the final electrical conductivity (EC2) was measured. The Electrolyte Leakage (EL) was calculated by using the formulae EC1/EC2.

2.3 Statistical Analysis

The data were statistically analyzed and the differences between treatment means were compared using DMRT.

3. RESULTS AND DISCUSSION

3.1 Growth Attributes

The growth parameters such as hypocotyls + cotyledons and root dry weights were measured for different NaCl salinity stress levels (Table 1). It was observed that the shoot dry weight did not change much although in different NaCl concentrations.

The values were between 0.043 and 0.054g for 125 and 0 mM NaCl respectively. However, there was a significant ($P < 0.05$) reduction in the root dry weight with increasing salt concentration. This has resulted in an increase in the aerial part/root ratio. Hence, salt stress significantly inhibited the root growth. The shoot growth was not significantly ($P>0.05$) affected by increased NaCl concentration. As pointed out by Jamil [8] salt stress inhibited the growth of root greater than shoot in four vegetable species. According to Saha [9] the roots of mungbean seedlings were more sensitive to salinity than the aerial parts. Roots are directly in contact with growth media containing toxic salts that stop the long term root growth. Under saline condition, CO_2 assimilation of plant becomes decreased. It is the major energy source for growth and development so, root growth decreases. According to Renault [10], reduction in seedling dry weight in response to salinity is due to decline in weight of mobilized

Table 1. The values of the shoot (cotyledons + hypocotyls) dry weight (DW), root dry weight and the shoot/root ratio of okra cv. 'Haritha' exposed to different NaCl concentrations during the seedling stage

NaCl (mM)	shoot DW (g)	Root DW (g)	shoot / root ratio
0	0.054±0.016^{a}	0.015±0.013^{a}	3.60±0.376^{a}
25	0.052±0.014^{a}	0.013±0.007^{b}	4.00±0.362^{ab}
50	0.049±0.016^{a}	0.011±0.003^{c}	4.45±0.357^{ab}
75	0.046±0.015^{a}	0.009±0.019^{d}	5.11±0.581^{bc}
100	0.045±0.014^{a}	0.007±0.010^{e}	6.43±0.660^{bc}
125	0.043±0.015^{a}	0.005±0.022^{f}	8.60±0.741^{c}

Values in the same column followed by the same letter do not differ significantly ($P > 0.05$). Values are the means of 20 plants in four replications

seed reserve. The depressed growth of plants may be due to the toxic effect of Na+ and Cl- ions present in NaCl and low water potential in the rooting medium.

Okra plants treated with salt significantly ($P<0.05$) affected the shoot/root ratio. The 75 and 100 mM treatments increased this ratio to 5.11 and 6.43 respectively. The highest value of shoot/root ratio was obtained in plants treated with 125 mM NaCl and it was 8.6. As pointed out by Dkhil and Denden [11], salt stress resulted in a significant increase in cotyledons + hypocotyls/ root ratio which was considered as the direct effect of the root dry weight reduction of treated seedlings. The magnitude of root growth reduction was highly dependent on NaCl concentrations. The root dry weight was reduced by approximately 33% at the highest salt concentration (125 mM NaCl) compared to control (Table 1).

It was observed that there was a decrease in the water content of hypocotyls and cotyledons with increasing salt concentration (Table 2). The aerial part water content was limited to certain extent with the increase in salt stress. But, the root water content significantly increased with increasing NaCl concentration. Appraisal of water relations in plants grown under stress conditions including saline stress is necessary to ascertain that up to what extent cellular water content is maintained [12].

3.2 Membrane Permeability

It was observed that the quantity of electrolyte leakage increased under salt stress (Table 3). The extent of this increase was significant at 100 and 125 mM NaCl concentrations. It was found that the electrolyte leakage was slightly altered with 50 and 75mM NaCl salinity. The range was between 20.86 and 22.37% respectively. The highest % leakage found at 125 mM NaCl salinity indicated membrane damage. As pointed out by Kaya [13] increase in membrane permeability at seedling stage was lower than at vegetative stage at high salinity. This indicates a strong relationship between time of exposure to high salinity and membrane permeability.

Salt stress induced electrolyte leakage has also been observed in maize [14]. It is now generally accepted that the first deteriorating change during stress injury is an alteration in the structure and function of cell membranes [15]. Based on the results, it was suggested that salinity levels imposed did not create high membrane damage.

Table 2. The water content in aerial part (cotyledons + hypocotyls) and root of okra cv. 'Haritha' exposed to different NaCl concentrations during the seedling stage

NaCl (mM)	Aerial part water content (g/g) (FW-DW)/DW	Root water content (g/g) (FW-DW)/DW
0	15.13 ± 1.42^{a}	11.32 ± 1.36^{d}
25	12.57 ± 1.53^{ab}	13.47 ± 1.67^{cd}
50	13.44 ± 1.47^{ab}	13.43 ± 1.72^{cd}
75	10.28 ± 1.40^{b}	16.36 ± 1.93^{b}
100	9.50 ± 1.36^{b}	14.86 ± 2.10^{bc}
125	9.41 ± 1.39^{b}	21.16 ± 2.98^{a}

Values in the same column followed by the same letter do not differ significantly ($P> 0.05$). Values are the means of 20 plants in four replications

Table 3. Effects of NaCl treatments on electrolyte leakage (%) at seedling stage of okra cv. 'Haritha'

NaCl (mM)	Electrolyte leakage (%)
0	19.74 ± 0.713^{a}
25	19.32 ± 0.709^{a}
50	20.86 ± 0.658^{ab}
75	22.37 ± 0.633^{ab}
100	26.62 ± 0.615^{b}
125	27.08 ± 0.638^{b}

Values in the same column followed by the same letter do not differ significantly ($P> 0.05$). Values are the means of 20 plants in four replications

4. CONCLUSION

The property of salinity tolerance depends on different physiological interactions. The growth physiological attributes presented by the plant in response to salinity may not be enough to determine its effect. However, the okra seedlings were able to maintain the physiological growth attributes considerably well in the presence of relatively high degree of salinity stress during the seedling stage.

COMPETING INTERESTS

Authors have declared that no competing interests exist.

REFERENCES

1. Munns R. Comparative physiology of salt and water stress. Plant Cell Environment. 2002;25:239-250.
2. Moisender PH, Mc Clinton E, Paerl HW. Salinity effects on growth, photosynthetic parameters and nitrogenase activity in estuarine planktonic cyanobacteria. Microbiology and Ecology. 2002;43:432-442.
3. Ghoulam C, Foursy A, Fares K. Effects of salt stress on growth, inorganic ions and proline accumulation in relation to osmotic adjustment in five sugar beet cultivars. Environmental and Experimental Botany. 2002;47:39-50.
4. Blokhina O, Virolainen E, Fagerstedt KV. Antioxidants, oxidative damage and oxygen deprivation stress. A review. Annals of Botany. 2003;91:179-194.
5. Oyenuga VA. nigeria's foods and feeding stuffs, their chemistry and nutrients value, 5th Ed. University Press, Ibadan, Nigeria; 1998.
6. Maggio A, Pascale SD, Angelino G, Ruggiero C, Barbieri G. Physiological response of tomato to saline irrigation in long-term salinized soils. European Journal of Agronomy. 2004;21:149-159.
7. Lutts S, Kinet JM, Bouharmont J. Changes in plant response to NaCl during development of rice (*Oryza sativa* L.) varieties differing in salinity resistance. Journal of Experimental Botany.1995;46: 1843-1852.
8. Jamil ML, Deog Bae J, Kwang Yong M, Ashraf L, Chun S, Eui Shik, R. Effect of salt (NaCl) stress on germination and early seedling growth of four vegetable species. Journal of Central Eurpoean Agriculture. 2006;7:273-282.
9. Saha P, Chatterjee P, Biswas AK. NaCl pre-treatment alleviates salt stress by enhancement of antioxidant defense system and osmolyte accumulation in mung bean (*Vigna radiata* L.). Indian Journal of Experimental Biology. 2010;48: 593-600.
10. Renault S, Croser C, Franklin JA, Zwiazek JJ. Effect of NaCl and Na2SO4 on redosier dogwood (Cornus stolonifera Michx). Plant and Soil. 2001;233:261-268.
11. Dkhil BB, Denden M. Effect of salt stress on growth, anthocyanins, membran permeability and chlorophyll fluorescence of okra (*Abelmoschus esculentus* L) Seedlings. American Journal of Plant Physiology. 2012;7(4):174-183.
12. Ashraf M, Akram NA, Al-Qurainy F, Foolad MR. Drought tolerance: Roles of organic osmolytes, growth regulators and mineral nutrients. Advances in Agronomy. 2011;111:249-296.
13. Kaya C, Higgs D, Kirnak H. The effect of high salinity (NaCl) and supplementary phosphorus and potassium on physiology and nutrition development of spinach. Bulgarian Journal of Plant Physiology. 2001;27:47-59.
14. Hichem H, Mounir D, Naceur E. Differential responses of two maize (*Zea mays* L.) varieties to salt stress: Changes on polyphenols composition of foliage and oxidative damages. Indian Crop Production. 2009;30:144-151.
15. Surjus A, Durand M. Lipid changes in soybean root membranes in response to salt treatment. Journal of experimental botany. 1996;47(294):17-23.

16

Foliar Selenium Application to Improve the Tolerance of Eggplant Grown under Salt Stress Conditions

M. A. Abul-Soud[1] and Shaimaa H. Abd-Elrahman[2*]

[1]Central Laboratory for Agricultural Climate, Agricultural Research Centre, Giza, Egypt.
[2]Soils and Water Department, Faculty of Agriculture, Ain Shams University, Cairo, Egypt.

Authors' contributions

Both researchers have contributed in all parts of research equally. Both authors read and approved the final manuscript.

Editor(s):
(1) Sławomir Borek, Faculty of Biology, Department of Plant Physiology, Adam Mickiewicz University, Poland.
(2) Marco Trevisan, Institute of Agricultural Chemistry and Environmental Research Centre Biomass, Faculty of Agriculture, Catholic University of the Sacred Heart, Italy.
Reviewers:
(1) Linxi Yuan, University of Science and Technology of China, China.
(2) Anonymous, University of Lisbon, Portugal.
(3) Nadezhda Golubkina, All-Russian Scientific Research Institute of Vegetable Breeding and Seeds Production, Russia.

ABSTRACT

Sea level rise is one of the most risky climate change impacts under Egyptian conditions to increase the salinity of northern Delta. Increasing the tolerance for salinity in current and future crops is strongly desirable. The current experiment was carried out in the experimental station at Agriculture Research Centre, Egypt, during the summer seasons of 2014 and 2015, to evaluate the effect of selenium foliar applications (0, 5, 10, 20, 30 µM Na_2SeO_3) on eggplant grown on a sandy soil and irrigated with different concentrations of saline water (0, 30, 60, 120 mM NaCl). The results showed that the Se supplement with 20 µM showed the best effects on vegetative growth and yield of eggplants under different salinity levels of irrigation water. Increasing salinity resulted in increasing N and P contents in the leaves and fruits of eggplant, but K decreased as a result of some sort of antagonism with Na; in spite of that, N, P and K contents in leaves and fruits increased with increasing Se supplements up to 20 µM to be at higher concentrations then decreased. Regarding K/Na ratio in leaves, the treatment of EC_w 13.5 dS m^{-1} without Se supplements gave the lowest value (0.52); treatment of Se 30 µM under 0 mM NaCl irrigation water gave the highest one

**Corresponding author: E-mail: Shaimaa_Hassan@agr.asu.edu.eg*

(1.71). Also, chlorophyll contents in plant leaves increased with increasing salinity level of irrigation water, but decreased with increasing Se supplements. Regarding the proline contents in fresh leaves, the treatment of EC_w 13.5 dS m^{-1} without Se supplements gave the highest value (51 mg g^{-1}), but the lowest one (30.9 mg g^{-1}) was observed with control treatment.

Keywords: Sea level rise; salt stress; selenium supplements; eggplant; proline content.

1. INTRODUCTION

Egypt is very dependent on natural resources that are vulnerable to climate change. A large portion of the arable land in Nile Delta is particularly exposed to sea level rise. Nicholls et al. [1] estimated a mean value of 1 meter global sea level rise by the year of 2100 which would give rise to a 0.37 meter sea level rise at the Nile delta. This, combined with a non climate induced subsidence of the Nile Delta of 0.38 meters would result in the movement of the shoreline to the current 0.75 meter contour and a 5 percent loss of Egyptian agricultural land by the year of 2060 mainly at the coastal area of Nile Delta. El-Raey et al. [2] suggested land losses of 12 to 15 percent of Egypt's current arable land for a one meter sea level rise. Due to salinization and seawater intrusion, the agricultural activities will be difficult below an elevation of 1 meter.

Salinity is an abiotic stress to limit plant growth [3], and is becoming a serious agricultural problem, especially at arid and semi-arid zones, where 20-30% of the land is seriously damaged by salt [4]. High salt concentrations in the soil drastically reduce the yields of a variety of plants worldwide [5]. However, in Egypt, the saline and drainage water is one of the main waster resources for agriculture, which caused degradation of the soil and adversely affect on plant production [6].

Eggplant (*Solanum melongena* L.) is a traditional vegetable crop in many tropical, subtropical and Mediterranean countries. Eggplant is classified as a salt moderately sensitive vegetable [7]; Bresler et al. [8], on the other hand, classified it as salt sensitive vegetable. This difference in its tolerance classification could be related to differences in used varieties or cultivars and study environmental conditions. Unlukara et al. [9] found a threshold value of salinity lower than 1.5 dS m^{-1} and a slope value of 4.4% on eggplant. Such authors also reported a decrease in plant water consumption due to salinity with a decrease slope of 2.1%.

Proline could accumulate in many plant species under a broad range of stress such as water shortage, salinity, extreme temperatures and high light intensity. Proline is considered to be a compatible solute. It protects folded protein structures against denaturation, stabilizes cell membranes by interacting with phospholipids, functions as a hydroxyl radical scavenger and serves as an energy and nitrogen source. However, the contribution of proline to osmotic adjustment and tolerance of plants exposed to unfavourable environmental conditions is still controversial [10]. The metabolic effects of osmolyte accumulation may, however, be equal or even more important than their role in osmotic adjustment, since stress-regulated changes in proline synthesis and degradation may also affect expression of other genes, ensuring that the genetic response to stress is appropriate to the prevailing environmental stress conditions [11]. Proline could accumulate in leaves and roots to protect against the osmotic pressure under salt stress [12].

Although selenium (Se) is not considered an essential nutrient for plant growth, it is a vital element for human and animal nutrition in trace amounts [13]. However, a diet containing 1 mg Se kg^{-1} dry weight (DW) may lead to chronic Se poisoning in humans and animals, and one-time ingestion of plant material containing 1,000 mg Se kg^{-1} DW can lead to acute Se poisoning and death [14] as it is shown from the conducted experiments (becoming toxic for human and animals fed with these plants) . Selenium is a constituent of seleno-proteins, many of which have important functions, including antioxidant protection, energy metabolism and redox regulation during transcription and gene expression [15]. Selenium supplementation to plants enhance the production and quality of edible plant products, by increasing antioxidant activity of plants, as shown in tea leaves [16], and rice [17]. Foliar application of selenium was shown to be several times more efficient than application in soil fertilizers [18], but strongly dependent on spraying conditions. Also, [19] showed that foliar spray gave a high recovery. However, [20] found foliar application to be less efficient than application to soil at planting.

Thus, the main objective of this study is to evaluate the protective effect by foliar application of selenium supplements (0, 5, 10, 20, 30 µM Na_2SeO_3) on eggplant (vegetative growth, yield, proline and some elements content) grown on a sandy soil and irrigated with different concentrations of saline water (0, 30, 60, 120 mM NaCl).

2. MATERIALS AND METHODS

The current experiment was carried out in the experimental station at the Central Laboratory for Agricultural Climate (CLAC), Agriculture Research Centre (ARC), Egypt, during the summer seasons of 2014 and 2015. The climatic data at Dokki site during the studied seasons of 2014 and 2015 were shown in Fig. 1; these data collected from automated weather station allocated at the site.

2.1 Plant Materials

Eggplant (*Solanum melongena* L. cv. Baladi) seeds were sown on 20th and 18th January of 2014 and 2015, respectively, in polystyrene trays. After the fifth true leaf stage (26th and 23th February, respectively), the eggplant seedlings were transplanted into bedding system of sandy soil.

2.2 System Materials

Open system of sandy soil from Siwa oasis - Matroh governorate, *Typic Torripsamments*, was used and the physical and chemical characteristics of the soil were listed in Table 1. The system bed performed of bricks on cement base (60 cm width x 25 cm height x 7.5 m length). The final plant spacing was 50 cm in the row and 40 cm in-between. Black polyethylene (1 mm) was used to create the main gully which was filled by the soil. A layer of 2-3 cm of gravel takes a place in the bottom of gully bin for leaching the drainage water easily.

No organic matter or manure was applied to the soil to avoid the effect of organic matter on the salinity impacts under the investigated different treatments.

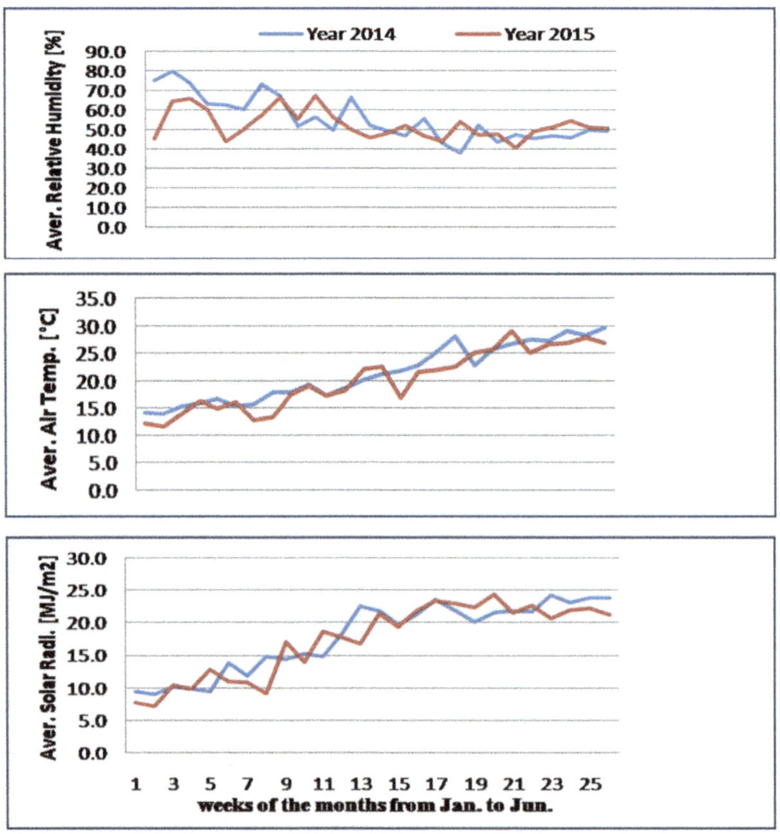

Fig. 1. Climatic data at Dokki site during the studied seasons of 2014 and 2015

Table 1. Some physical and chemical characteristics of the studied soil

Particle size distribution, %		**Soluble cations, meq L^{-1}**	
Sand	97.5	Ca^{2+}	3.40
Silt	1.50	Mg^{2+}	3.50
Clay	1.00	Na^{+}	1.50
Texture class	Sandy	K^{+}	0.60
$CaCO_3$, %	5.80	**Soluble anions, meq L^{-1}**	
OM, %	0.05	CO_3^{2-}	0.00
CEC, $cmol_c\ kg^{-1}$	9.30	HCO_3^{-}	3.20
pH (1:2.5 soil:water suspension)	8.19	Cl^{-}	4.10
EC_e, dS m^{-1}	0.90	SO_4^{2-}	1.70

Different salinity irrigation water levels were pumped via submersible pump (110 watt). A plastic tank 120 L (one per each bin system) and submersible pump (one per each tank) were used to pump the nutrient solution and different salinity irrigation water levels via polyethylene pipe (16 mm) with 2 liters per hour dripper. The nutrient solution [21] was adjusted by using EC meter to the required level (2.5 dS m^{-1}) in all treatments. The fertigation was programmed to work 2 - 4 times/day and the duration of irrigation time depended upon the season.

2.3 Investigated Treatments

The application of different treatments, after 2 weeks of transferring the eggplant seedlings, was applied. The study investigated the effects of different selenium (Se) concentrations (0, 5, 10, 20, 30 µM Na_2SeO_3) as foliar application on eggplant cultivated under different saline irrigation water levels (0, 30, 60, 120 mM NaCl) on the studied soil. The EC_w concentrations and Se supplements were applied according to [22-25].

Eggplants were harvested on 26th and 17th June 2014 and 2015, respectively. The collected samples were dried at 70°C in an air forced oven for 48hrs, and then digested by H_2SO_4/H_2O_2 mixture according to the method described by [26] and kept for the elements determination.

2.4 Experiment Design

The experimental design was a split plot with 3 replicates. Each experimental plot contained 10 plants. The saline irrigation levels were assigned as main-plots and Se concentrations as sub-plots as Fig. 2 illustrates.

2.5 Measurements

The vegetative and yield characteristics beside the chemical analysis of eggplants were measured as follows:

- Plant height (cm), before starting the flowering stage
- Number of leaves per plant, before starting the flowering stage
- Fresh weight of total fruits per plant (g/plant)
- Number of fruits per plant
- Chlorophyll content in leaves, SPAD:

Total chlorophyll of the fifth mature leaf from top was measured using Minolta chlorophyll meter Spad-501.

- Proline content in 0.5 g of fresh leaves, at harvest:

Proline content can increase upon exposure of plants to drought, salinity, cold, heavy metals, or certain pathogens. Thus, determination of proline levels is a useful assay to monitor physiological status and to assess stress tolerance of higher plants. Proline content was determined according to the method of [27] modified by [28], and was expressed as mg g^{-1} fresh weight (FW).

- Total N, P, K and Na contents in leaves and fruits, at harvest:

Total nitrogen in plant was determined using Kjeldahl method according to the procedure described by [26]; total phosphorus was determined using spectrophotometer according to [29] and both total potassium and sodium in plant was determined using Flame photometer as described by [30].

Statistical analysis was performed using the analysis of variance adopting a SAS software package [31]. Significance among treatments was evaluated using Duncan's approach (P≤0.05).

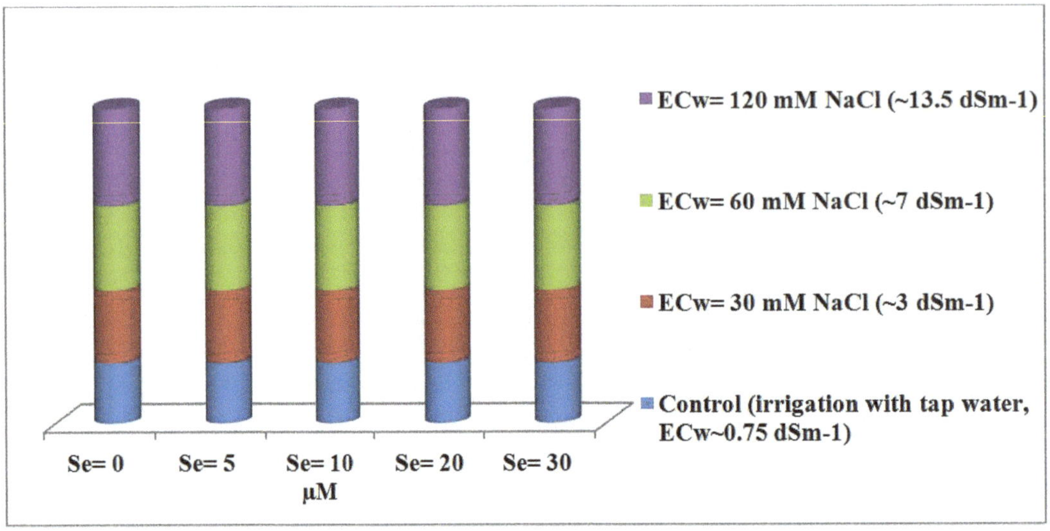

Fig. 2. The layout of experimental design

3. RESULTS AND DISCUSSION

3.1 Vegetative Growth and Yield of Eggplants

Data in Table 2 showed that the average plant height was different during the two growing seasons under the studied treatments, with decreasing along increasing irrigation water salinity. However, the plant height increased with increasing Se supplements. Regarding the interaction between irrigation water salinity and Se supplements, the irrigation with tap water and Se 20 µM were found to give the highest plant; the irrigation water with EC_w 13.5 dS m^{-1} and Se 0 µM gave the lowest one. Also, number of leaves per plant went hand by hand with the previous findings on the plant height.

The total fruit fresh weight and the number of fruits per plant (Table 2) were agreed with the findings on the plant height and number of leaves per plant during the two studied seasons. Generally, the Se supplement with 20 µM showed the best effects on vegetative growth and yield of eggplants under different irrigation water salinity treatments, with higher effect for irrigation with tap water and decreased with increasing salinity in irrigation water. These findings may be due to: (1) Se 20 µM is suitable to counter act salinity problems inside the plant. (2) The vegetative growth and yield of eggplants decreased with increasing salinity in irrigation water. Kabata-Pendias and Pendias [32] mentioned that the mean Se content in clay soils was 0.29 mg kg^{-1} and in coarse mineral soils 0.17 mg kg^{-1}, and in plant Se at 10 ppm DW is considered as phytotoxic. Yassen et al. [25] reported that Se interaction with plants depended on its concentration. At lower rates, Se stimulated growth of ryegrass seedlings, while at high doses it acted as pro-oxidant reducing yields and inducing metabolic disturbances. Terry et al. [13] found that there was a small decrease in shoot accumulation of Se with increasing salt level. Unlukara et al. [9] added that vegetative dry weight of the eggplants decreased with increasing soil salinity and with fruit yield being more sensitive.

3.2 N, P and K Contents in Leaves and Fruits of Eggplants

Data in Table 3 showed the effect of irrigation with saline water on N, P and K contents in leaves of eggplants under Se supplements, compared with the control (without any treatments) during the two studied seasons. Increasing salinity resulted in increased N and P contents in the leaves, but decreased K. Almost, N, P and K contents in leaves increased with increasing Se supplements up to 20 µM then decreased with higher concentrations. Regarding the interaction between irrigation water salinity and Se supplements, Se 20 µM with all saline water treatments generally gave the highest value of N, P and K contents in plant leaves.

Regarding the fruits of eggplant, data in Table 4 showed the effect of irrigation water salinity on N,

P and K contents under Se supplements which have similar trends with those of plant leaves. Increasing N and P contents in leaves and fruits of eggplant with increasing salinity of irrigation water may be due to increase the amino acids inside the plant with increasing the stress; amino acids also interact with phospholipids to adjust the osmotic potential according to [33]. Also, they reported that Se had a high ability to induce antioxidant and hormone balance in the plant. Yassen et al. [25] found that foliar application of Se on potato plants increased % N, P, K and protein contents in the yield of tubers. Decreasing K content with increasing salinity of irrigation water, on the other hand, may be due to the increase of NaCl concentration; Na^+ content increased in leaves and fruits indicating that the eggplant (which has a glycophytic reaction) could not control uptake of Na^+ [34]. Kong et al. [15] reported that the major influences on Se uptake by plants were soil pH and salinity; Cl^- which inhibit uptake by affecting plant metabolism. In general, increasing N, P and K contents in leaves and fruits of eggplant by increasing foliar Se supplements under irrigation with saline water may be due to the role of Se in increasing antioxidant activity of the plant to face the stress.

3.3 Some Stress Markers in Leaves of Eggplants

Data in Table 5 showed that chlorophyll content in plant leaves increased with increasing salinity of irrigation water, but decreased with increasing Se supplements. Control treatment recorded the lowest value with significance when compared to the other treatments. Increasing chlorophyll content in plant leaves indicates that plant suffered from salt stress compared to the control (irrigation with tap water). It may be due to a reduction in extension growth under salt stress, increased leaf thickness and its color became darker, which gave high readings with SPAD. Khattab [35] studied the metabolic and oxidative responses associated with exposure of rocket plants (*Eruca sativa* L.) to different levels

Table 2. Effect of irrigation water salinity and selenium supplements on average vegetative growth and yield of eggplants during the two studied seasons

Salinity levels, dS m^{-1} (A)	Selenium concentrations, µM (B)					Mean (A)
	Se 0	Se 5	Se 10	Se 20	Se 30	
			Plant height, cm			
EC_w= 0.75	55.2 g	59.7 d	62.1 c	71.7 a	65.0 b	**62.7 A**
EC_w= 3	52.3 h	57.0 f	61.2 c	64.7 b	61.3 c	**59.3 B**
EC_w= 7	45.3 j	49.7 i	58.0 e	60.3 cd	58.0 e	**54.3 C**
EC_w= 13.5	35.0 l	43.7 k	45.7 j	56.0 g	52.7 h	**46.6 D**
Mean (B)	**47.0 D**	**52.5 C**	**56.7 B**	**63.2 A**	**59.3 B**	
			No. of leaves/plant			
EC_w= 0.75	68.3 i	72.8 h	84.0 f	128 a	113 b	**93.2 A**
EC_w= 3	62.7 k	72.0 h	76.7 g	98.3 d	101 c	**82.1 B**
EC_w= 7	60.3 l	63.7 k	67.7 i	100 c	87.0 e	**75.7 C**
EC_w= 13.5	59.3 l	63.3 k	66.3 j	83.0 f	72.3 h	**68.9 D**
Mean (B)	**62.7 E**	**68.0 D**	**73.7 C**	**102 A**	**93.3 B**	
			Fruit fresh weight, g/plant			
EC_w= 0.75	1149 e	1222 d	1342 c	1686 a	1475 b	**1375 A**
EC_w= 3	748 h	880 g	1019 f	1125 e	1036 f	**962 B**
EC_w= 7	517 k	597 j	675 i	850 g	734 h	**675 C**
EC_w= 13.5	356 m	433 l	513 k	573 jk	544 k	**484 D**
Mean (B)	**693 D**	**783 C**	**887 B**	**1058 A**	**947 B**	
			No. of fruits/plant			
EC_w= 0.75	19.5 d	20.0 d	21.0 c	24.4 a	22.0 b	**21.4 A**
EC_w= 3	14.7 h	15.7 g	17.3 f	18.4 e	17.0 f	**16.6 B**
EC_w= 7	11.7 ij	11.8 i	12.3 i	14.7 h	13.8 h	**12.8 C**
EC_w= 13.5	9.15 k	9.60 k	10.8 j	10.8 j	11.2 j	**10.3 D**
Mean (B)	**13.8 C**	**14.3 C**	**15.3 B**	**17.1 A**	**16.0 B**	

 # Each value is the mean of 6 replications

This is a factorial experiment from two factors: salinity levels (A), selenium concentrations (B) in a spilt plot design, letters A B C D among the main factors, letters a b c d … among the interaction between the two factors (A×B), and different letters means significant

Table 3. Effect of irrigation water salinity and selenium supplements on average N, P and K contents of eggplant leaves during the two studied seasons

Salinity levels, dS m^{-1} (A)	Selenium concentrations, µM (B)					Mean (A)
	Se 0	Se 5	Se 10	Se 20	Se 30	
			%N			
EC_w= 0.75	2.80 l	3.36 k	3.50 j	4.34 e	3.78 h	**3.56 D**
EC_w= 3	3.78 h	3.76 h	3.64 i	4.48 d	3.99 g	**3.93 C**
EC_w= 7	4.48 d	4.85 c	4.22 f	4.90 c	4.01 g	**4.49 B**
EC_w= 13.5	5.88 a	4.94 c	4.90 c	5.32 b	4.20 f	**5.05 A**
Mean (B)	**4.24 B**	**4.23 B**	**4.07 C**	**4.76 A**	**4.00 C**	
			%P			
EC_w= 0.75	0.61 h	0.62 h	0.65 g	0.70 f	0.65 g	**0.65 C**
EC_w= 3	0.64 g	0.67 f	0.69 f	0.75 e	0.68 f	**0.69 C**
EC_w= 7	0.69 f	0.74 e	0.79 d	0.82 c	0.76 e	**0.76 B**
EC_w= 13.5	0.81 c	0.86 b	0.87 b	0.93 a	0.86 b	**0.87 A**
Mean (B)	**0.69 D**	**0.72 C**	**0.75 B**	**0.80 A**	**0.74 B**	
			%K			
EC_w= 0.75	1.62 e	1.81 c	1.94 b	1.84 c	1.70 d	**1.78 A**
EC_w= 3	1.50 g	1.77 c	1.79 c	1.62 e	1.65 e	**1.67 B**
EC_w= 7	1.26 i	1.57 f	1.58 f	2.20 a	1.58 f	**1.64 B**
EC_w= 13.5	1.09 k	1.16 j	1.23 i	1.40 h	1.53 g	**1.28 C**
Mean (B)	**1.37 D**	**1.58 C**	**1.63 B**	**1.77 A**	**1.62 B**	

 # Each value is the mean of 6 replications
This is a factorial experiment from two factors: salinity levels (A), selenium concentrations (B) in a spilt plot design, letters A B C D among the main factors, letters a b c d … among the interaction between the two factors (A×B), and different letters means significant

Table 4. Effect of irrigation water salinity and selenium supplements on average N, P and K contents of eggplant fruits during the two studied seasons

Salinity levels, dS m^{-1} (A)	Selenium concentrations, µM (B)					Mean (A)
	Se 0	Se 5	Se 10	Se 20	Se 30	
			%N			
EC_w= 0.75	2.38 k	2.41 k	2.52 j	2.55 j	2.94 g	**2.56 C**
EC_w= 3	2.40 k	2.57 i	2.68 h	3.10 f	3.02 j	**2.75 C**
EC_w= 7	3.50 d	3.08 f	3.06 f	3.22 e	3.61 c	**3.29 B**
EC_w= 13.5	3.96 a	3.94 b	3.20 e	3.50 d	3.94 b	**3.71 A**
Mean (B)	**3.06 B**	**3.00 C**	**2.87 D**	**3.09 B**	**3.38 A**	
			%P			
EC_w= 0.75	0.48 i	0.51 h	0.54 g	0.57 f	0.51 h	**0.52 C**
EC_w= 3	0.55 g	0.60 e	0.62 e	0.68 c	0.65 d	**0.62 C**
EC_w= 7	0.61 e	0.65 d	0.68 c	0.73 b	0.69 c	**0.67 B**
EC_w= 13.5	0.67 c	0.72 b	0.73 b	0.79 a	0.72 b	**0.73 A**
Mean (B)	**0.58 D**	**0.62 C**	**0.64 B**	**0.69 A**	**0.64 B**	
			%K			
EC_w= 0.75	1.16 i	1.19 i	1.24 g	1.60 b	1.57 b	**1.35 A**
EC_w= 3	1.11 i	1.14 i	1.23 hi	1.52 c	1.52 c	**1.31 B**
EC_w= 7	1.06 j	1.16 i	1.21 i	1.67 a	1.48 d	**1.32 B**
EC_w= 13.5	0.99 k	1.06 j	1.17 i	1.33 f	1.41 e	**1.19 C**
Mean (B)	**1.08 D**	**1.14 C**	**1.21 B**	**1.53 A**	**1.50 A**	

 # Each value is the mean of 6 replications
This is a factorial experiment from two factors: salinity levels (A), selenium concentrations (B) in a spilt plot design, letters A B C D among the main factors, letters a b c d … among the interaction between the two factors (A×B), and different letters means significant.

of Se (0, 5, 10, 100, 1000, 2000 and 3000 µM of sodium selenate) for 10 days. He found that Se up to 10 µM enhanced the growth and levels of chlorophylls, sugar and amino acids. However, high levels of Se (100 µM and up) exert toxic effects. Germ et al. [36] reported that Se protected chloroplasts during stress. Regarding the proline content in plant fresh leaves during the two studied seasons, data also showed that proline content increased with increasing salinity of irrigation water (indication of stress), but decreased with increasing Se supplements compared to the control. The treatment of EC_w 13.5 dS m^{-1} without Se supplements gave the highest value of proline content (51 mg g^{-1}), control treatment being the lowest one (30.9 mg g^{-1}). These findings agreed with those obtained by [33] regarding the effect of Se on proline content in cucumber seedlings grown under saline conditions. They explained the accumulation of proline in plant under the stress by increasing biosynthesis or inhibition of proline degradation. Nowak [22] reported that Se enhanced the salt tolerance of cucumber seedlings by protecting the cell membrane against lipid peroxidation. However, he explained the growth-promoting effect of low Se concentrations (5 and 10 µM) under saline conditions due to the antioxidative activity of Se, increase in proline accumulation and/or decrease in content of Cl^- ions in the shoots tissues.

Data in Table 5 showed values of K/Na ratio in leaves of eggplants as an important indicator on salt stress. K/Na ratio decreased with increasing salinity of irrigation water, but increased generally with increasing Se supplements. Regarding the interaction between irrigation water salinity and Se supplements, the treatment of EC_w 13.5 dS m^{-1} without Se supplements gave the lowest value of K/Na ratio (0.52), the treatment of Se 30 µM under irrigation with tap water being the highest one (1.71). Akinci et al. [34] reported that increasing NaCl in the solution led to a decrease in the K/Na ratio and increased Na in several eggplant varieties.

Germ et al. [36] reported that, in the senescing plants, the addition of Se strengthens the antioxidative capacity by preventing the reduction of tocopherol concentration and by enhancing superoxide dismutase (SOD) activity. Senescence processes are partly delayed due to enhanced antioxidation, which is associated with an increase of glutathione peroxidase (GPx) activity. In ryegrass (*Lolium perenne*) up to Se addition of 1.0 mg kg^{-1}, the decreased lipid peroxidation was connected with Se-induced increase in GPx activity [37]. It was shown that Se has the ability to regulate the water status of plants under conditions of drought [38], and that the protective effect of Se under drought stress conditions was achieved by increasing the water uptake capacity of the root system.

Table 5. Effect of irrigation water salinity and selenium supplements on some stress markers in eggplant leaves during the two studied seasons

Salinity levels, dS m^{-1} (A)	Selenium concentrations, µM (B)					Mean (A)
	Se 0	Se 5	Se 10	Se 20	Se 30	
	Chlorophyll content, SPAAD					
EC_w= 0.75	51.3 l	53.5 k	54.2 j	55.2 i	55.1 i	**53.9 C**
EC_w= 3	61.2 d	59.3 f	56.4 i	55.3 i	57.7 h	**58.0 C**
EC_w= 7	62.3 b	60.7 e	60.5 e	57.7 h	58.9 g	**60.0 B**
EC_w= 13.5	65.4 a	62.5 b	60.3 e	59.5 f	61.7 c	**61.9 A**
Mean (B)	**60.1 A**	**59.0 B**	**57.9 B**	**56.9 C**	**58.4 B**	
	Proline content, mg g^{-1} FW					
EC_w= 0.75	30.9 n	32.3 m	36.0 l	36.3 k	36.5 j	**34.4 D**
EC_w= 3	45.3 g	42.1 h	40.1 i	45.3 g	45.5 f	**43.7 C**
EC_w= 7	47.7 b	45.7 e	45.4 g	45.7 e	47.4 c	**46.4 B**
EC_w= 13.5	51.0 a	50.9 a	45.9 d	47.4 c	47.8 b	**48.6 A**
Mean (B)	**43.7 B**	**42.8 C**	**41.9 D**	**43.7 B**	**44.3 A**	
	K/Na ratio					
EC_w= 0.75	1.41 e	1.47 d	1.54 c	1.65 b	1.71 a	**1.56 A**
EC_w= 3	0.95 i	1.03 h	1.03 h	1.08 g	1.14 f	**1.05 B**
EC_w= 7	0.86 j	0.88 j	0.94 i	0.98 i	0.89 j	**0.91 C**
EC_w= 13.5	0.52 m	0.77 k	0.78 k	0.88 j	0.71 l	**0.73 D**
Mean (B)	**0.94 E**	**1.04 D**	**1.07 C**	**1.15 A**	**1.11 B**	

 # Each value is the mean of 6 replications

This is a factorial experiment from two factors: salinity levels (A), selenium concentrations (B) in a spilt plot design, letters A B C D among the main factors, letters a b c d ... among the interaction between the two factors (A×B), and different letters means significant

4. CONCLUSION

Under mitigation and adaption strategy of climate change impacts with the expected increase in the salinity of irrigation water especially in the Northern Egypt as a result of sea level rise, the present study recommends the applying selenium as foliar application at the concentration 10 - 20 μM to increase the tolerance of eggplants against salinity of irrigation water and to avoid salt stress impacts on the yield. These concentrations of Se supplementation present a promising potential for use in conditions of relatively high levels of NaCl in the irrigation water, due to its antioxidative activity. More work is required to investigate the effect of Se on different crops, the real role of it inside the plant in physiological stages and its content; besides studying the impact of sea level rise on the irrigation water and soil salinities.

ACKNOWLEDGEMENTS

The authors would like to thank Prof. Dr. Mohamed El-Nennah, Department of Soils and Water, Faculty of Agriculture, Ain Shams University, for his sincere guidance throughout writing the manuscript.

COMPETING INTERESTS

Authors have declared that no competing interests exist.

REFERENCES

1. Nicholls RJ, Leatherman SP. Global sea-level rise. In: Strzepek KM, Smith JB (eds.), When Climate changes: Potential impact and implications. Cambridge Univ. Press, Cambridge; 1995.
2. El-Raey M, Nasr S, Frihi S, Desouki S, Dewidar K. Potential impacts of accelerated sea level rise on Alexandria governorate, Egypt. J. Coastal Res. 1995; 14:190–204.
3. Sengupta S, Majumder AL. Insight into the salt tolerance factors of a wild halophytic rice, *Porteresia coarctata*: A physiological and proteomic approach. Planta. 2009; 229:911-929.
4. FAO. Crops and drops. Making the best use of water for agriculture. Food and Agriculture Organization of the United Nations. Roma. 2002;26.
5. Gorai M, Neffati M. Germination responses of *Reaumuria vermiculata* to salinity and temperature. Annals Appl. Biol. 2007;151:53-59.
6. Abd-Elrahman Shaimaa H. Remediation of some degraded soils using new techniques. Scholars' Press, OmniScriptum GmbH & Co. KG, Germany 2013;8.
7. Heuer B, Meiri A, Shalevet J. Salt tolerance of eggplant. Plant Soil 1986;95: 9-13.
8. Bresler E, McNeal BL, Carter DL. Saline and sodic soils. Springer-Verlag, Berlin 1982.
9. Unlukara A, Kurunc A, Kesmez GD, Yurtseven E, Suarez DL. Effects of salinity on eggplant (*Solanum melongena* L.) growth and evapotranspiration. Irrig. & Drain. 2010;59:203–214.
10. Hare PD, Cress WA. Metabolic implications of stress-induced proline accumulation in plants. Plant Growth Regulation. 1997;21:79–102.
11. Claussen W. Proline as a measure of stress in tomato plants. Plant Sci. 2005;168:241–248.
12. Chen Z, Cuin TA, Zhou M, Twomey A, Naidu BP, Shabala S. Compatible solute accumulation and stress mitigating effects in barley genotypes contrasting in their salt tolerance. J. Exp. Bot. 2007;58:4245-4255.
13. Terry N, Zayed AM, de Souza MP, Tarun AS. Selenium in higher plants. Annual Rev. Plant Physiol. Plant Mol. Biol. 2000;51:401–432.
14. Pilon-Smits, Elizabeth AH, Quinn, Colin F. Selenium metabolism in plants. In: Hell R, Mendel RR. (eds.), Cell Biology of Metals and Nutrients, Plant Cell Monographs 2010;17:225-241.
15. Kong L, Wang M, Bi D. Selenium modulates the activities of antioxidant enzymes, osmotic homeostasis and promotes the growth of sorrel seedlings under salt stress. Plant Growth Regulation. 2005;45:155-163.
16. Xu J, Yang F, Chen L, Hu Y, Hu Q. Effect of selenium on increasing the antioxidant activity of tea leaves harvested during the early spring tea producing season. J. Agric. Food Chem. 2003;51:1081-1084.
17. Xu J, Hu Q. Effect of foliar application of selenium on the antioxidant activity of aqueous and ethanolic extracts of selenium-enriched rice. J. Agric. Food Chem. 2004;52:1759-1763.
18. Aspila P. History of selenium supplemented fertilization in Finland.

Proceedings, 20 Years of Selenium Fertilization; Helsinki, Finland. 2005;8–9:8-13.

19. Curtin D, Hanson R, Lindley TN, Butler RC. Selenium concentration in wheat grain as influenced by method, rate, and timing of sodium selenate application. New Zealand J. Crop Hort. Sci. 2006;34:329-339.
20. Lyons GH, Lewis J, Lorimer MF, Holloway RE, Brace DM, Stangoulis JCR. High selenium wheat: agronomic bio-fortification strategies to improve human nutrition. Food Agric. Environ. 2004;2:171-178.
21. El-Behairy UA. The effect of levels of phosphorus and zinc in the nutrient solution on macro and micronutrients uptake and translocation in cucumber (*Cucumus sativus* L.) grown by nutrient film technique. Ph.D thesis, London University. 1994;299.
22. Nowak HB. Beneficial effects of exogenous selenium in cucumber seedlings subjected to salt stress. Biol. Trace Elem. Res. 2009;132:259-269.
23. Yao X, Chu J, Wang G. Effects of selenium on wheat seedlings under drought stress. Biol. Trace Elem. Res. 2009;130:283-290.
24. Chu J, Yao X, Zhang Z. Responses of wheat seedlings to exogenous selenium supply under cold stress. Biol. Trace Elem. Res. 2010;136:355-363.
25. Yassen AA, Safia M Adam, Sahar M Zaghloul. Impact of nitrogen fertilizer and foliar spray of selenium on growth, yield and chemical constituents of potato plants. Aust. J. Basic & Appl. Sci. 2011;5:1296-1303.
26. Chapman HD, Pratt PF. Methods of analysis for soils, plants and waters. Division of Agric. Sci. Berkeley, Univ. California, USA. 1961;150-152.
27. Troll W, Lindsley J. A photometric method for the determination of proline. J. Biol. Chem. 1955;215:655-660.
28. Petters W, Piepenbrock M, Lenz B, Schmitt JM. Cytokinine as a negative effector of phosphoenolpyruvate carboxylase induction in *Mesembryanthemum crystallinum*. J. Plant Physiol. 1997;151:362-367.
29. Watanabe FC, Olsen SR. Test of an ascorbic acid method for determining phosphorus in water and $NaHCO_3$ extracts from soils. Soil Sci. Soc. Amer. Proc. 1965;29:677-678.
30. Jackson ML. Soil chemical analysis. Prentice-Hall, Inc. Englewood Cliffs, NJ Library of Congress, USA 1958;pp 38-388.
31. SAS Institute. The SAS System for Windows; Release 6.12; Statistical Analysis System Institute Inc., Cary, NC; 1996.
32. Kabata-Pendias A, Pendias H (Eds.). Trace elements in soils and plants. 2nd Ed. CRC Press, Inc. Boca Raton, Florida; 1992.
33. Abd El-Nasser Walaa, Shatlah MA, Hossien M, Srorr HAM. Selenium induces antioxidant defensive enzymes and promotes tolerance against salinity stress in cucumber seedlings (*Cucumis sativus*). Arab Univ. J. Agric. Sci. 2010;18:65-76.
34. Akinci IE, Akinci S, Yilmaz K, Dikici H. Response of eggplant varieties (*Solanum melongena*) to salinity in germination and seedling stages. New Zealand J. Crop Hort. Sci. 2004;32:193–200.
35. Khattab HI. Metabolic and oxidative responses associated with exposure of *Eruca sativa* (Rocket) plants to different levels of selenium. Int. J. Agric. Biol. 2004;6:1101-1106.
36. Germ M, Stibilj V, Kreft I. Metabolic importance of selenium for plants. The Europ. J. Plant Sci. Biotech. 2007;1:91-97.
37. Hartikainen H, Xue T, Piironen V. Selenium as an antioxidant and prooxidant in ryegrass. Plant Soil. 2000;225:193-200.
38. Kuznetsov VV, Kholodova VP, Kuznetsov VIV, Yagodin BA. Selenium regulates the water status of plants exposed to drought. Doklady Biological Sciences. 2003;390: 266-268.

17

A Life Cycle Assessment: The Use of Sewage Sludge in Ornamental Plant Growing

TAŞ Birol[1*]

[1] *Department of Park and Horticulture, Vocational School of Technical Science, Uludag University, Bursa, Turkey.*

Author's contribution

The sole author designed, analyzed and interpreted and prepared the manuscript.

Editor(s):
(1) Sławomir Borek, Faculty of Biology, Department of Plant Physiology, Adam Mickiewicz University, Poland.
Reviewers:
(1) Abe Shegro, University of the Free State, South Africa.
(2) Melodie Claire W. Juico, Davao Doctors College, Philippines.

ABSTRACT

Today, the amount of the sewage sludge coming from domestic and industrial waste water which appear due to population increase, industrialization and urban growth has reached a level that is threatening human health and the environment. It is of crucial importance that the sewage sludge is removed with environmental-friendly methods for the protection of the environment. In this scope, the waste water sewage sludge is used in agricultural areas, forest areas, parks-gardens, recreational grounds, recreation of areas, urban landscaping as well as in arboriculture, and in growing ornamental plants as organic fertilizers, germination media and growth media. In ornamental plant growth, which has an economical value, the use of sewage sludge helps to decrease the high production costs and thus contributes to the economy of the country creating a new alternative field in removing the sewage sludge.

Keywords: Sewage sludge; reuse; ornamental plants; grassland.

1. INTRODUCTION

Although the domestic and industrial sewage sludge poses a serious threat for the future, it may be removed without damaging the environment with the developments in waste water purification sector; and the sludge is used in various forms to become useful for the environment. Especially the acceptance of these wastes in the soil-plant nutrient system as organic fertilizers ensure that they are recycled in the most economic and efficient method [1,2].

Corresponding author: E-mail: biroltas69@gmail.com

The sewage sludge whose pathogens are eliminated and which doesn't contain heavy metals being given to the soil as organic fertilizer is accepted as an efficient and environment-friendly method [3,4].

Many factors in ornamental plant growing increase the production costs and makes the production uneconomic. Among these factors that increase the production costs are the fertilizers, germination and growing media. When considered in this context, the use of sewage sludge in ornamental plant growth emerge as an alternative in decreasing the costs of the growth media and fertilizer, which are important in this field. Positive results have been obtained in recent years in the use of composted industrial and urban wastes as germination media in ornamental plant growth. The sewage sludge being rich in organic materials, macro and micro nutrients make it possible to use it alone or in combination with various media as growth media in growing ornamental plants [5-7].

However, considering that the sewage sludge also has contaminating effects, it must be used with care in areas where ornamental plants are grown, and the regulations and the audits on this field must be applied and cared for with sensitivity. Because the possibility of sewage sludge carrying toxic organic compounds that are harmful for the environment, heavy metals, eggs of pathogen microorganisms and parasitic organisms as well as organic material and nutrient elements must not be ignored.

It has been reported by many researchers that the best method to make use of the nutrients contained in sewage sludge is its use in agricultural areas; however it is also reported that before the application, the heavy metal, salt, nitrogen and pathogen amounts must be determined. The amount that may be given to the agricultural areas must also be determined before use [8-10]. The sewage sludge which is non-biodegradable and which is collected in soil may be a risk for the environment in the long-term [1,11]. The use of waste sludge in seasonal flowers growing in gardens, green areas or in flower pots decrease the possibility of using the heavy metals and toxic organic compounds in the products that are nutrients from humans and animals [12].

According to the agricultural usage criteria, when the sewage sludge with high Cd, Zn, Cu and Ni concentrations is used in some seasonal plant species like *B. semperflorens*; *O. japonicus*; *L. chindense*var. *rubrum*; *D. morifolium*; *B. radican*; *V. macrocephalum*; *O. fragrans*; *C. camphora*, they show better development when compared with the control groups; and the highest Ni, Cu, As, Pb, Cd and Cr concentration was determined in the *D. morifolium* species; and it was reported is some studies that the Cu concentration in all plant species was higher than 20 mg/kg, which is the toxic limit for plants [13].

It has been stated that in *Bougainvillea* growing, in which the sewage sludge is used in various percentages (0%, 25%, 40%, 55% and 70%) with the purpose of preparing agriculture-friendly compost, especially the *peat medium* should be preferred [14]. In a study in which the effects of heavy metals in sewage sludge on the development of *Chlorophytum comosum, Calendula officinalis* L, *Petunia hyhrida* Vilm and *Viola tricolor* were investigated, it was determined that the Cu, Zn and Pb amounts were higher in *Chlorophytum comosum* in the body; and the Cd amount was higher in *Calendula officinalis* L; and in the roots it was determined that the *Chlorophytum comosum* absorbed Cu, Zn, Pband Cd; and that the *Chlorophytum comosum* had the best phytoremediation (*phyto-remediation*) value [15].

The effect of the toxic metals in sewage sludge on some shrubs or small trees and on many outdoor plants has been investigated in various studies. For example, it has been determined that heavy metals accumulate in the plant tissues of *Terminalia arjuna*, *Prosopis juliflora*, *Populus alba*, *Eucalyptus tereticornis* and *Dendrocalamus strictus*. In other words, these plants attract the heavy metals in the medium like a magnet, and purify the soil from heavy metals. It was determined that in *E. tereticornis* species, mostly Fe, Cu, Mn and Zn metals are accumulated and in *T. arjuna* species, mostly Cd and Ni metals are accumulated. As a result, it was observed that after 1 year of the study, the toxic metals were removed from the medium in various ratios (Cr % 70.22, Ni % 59.21, Cd % 58.4, Fe % 49.75, Mn % 30.95, Zn % 22.80, Cu % 20.46 and Pb % 14.05 [16].

Normal and enriched sewage sludge was applied to *Pinus halepensis* saplings (Cu, Ni and Zn), and it was observed that the enriched sewage sludge caused an increase in the heavy metal concentrations in the plant; however, it did not reach toxic level. It was also observed that in the other application, the Cu and Zn concentrations decreased in small levels in the saplings [17].

In the study on the potential of sewage sludge being used as growth medium for *Cupressus macrocarpa* 'Gold Crest' was investigated and it was observed that the sewage sludge in increasing dosages decreased the C/N rate, and made the soil reach to the desired levels in terms of physical and chemical properties. It was also reported that the best results in terms of plant growth performance were obtained from the mixtures prepared with 30% and 50% sewage sludge [18].

In another study , various dosages (0, 7.5, 14.5 kg dry weight/plant) of sewage sludge was applied to the saplings of *Pinus halepensis* and *Quercus ilex* in forest areas and it was determined that the growth ratio was positive in *P. halepensis* species with the highest sewage sludge dosage and negative in *Q. ilex* species [19].

In a study in which the use of the mixtures of *Pinus pinea*, *Cupressus arizonica* and *Cupressus sempervirens* being used as growth medium was investigated, it was reported that *C. arizonica* and *C. sempervirens* gave very good results in the growth medium which was formed with pine bark, coir and 30% sewage sludge [20].

A study conducted on forest areas that were burnt by applying sewage sludge made into compost and green wastes (0, 20 and 40 kg/m^2) to one-year-old *Quercus ilex*, *Pinus halepensis* and *Pinus pinea* saplings for three years, and it was determined that the development in *Q. ilex* and *P. Pinea* saplings in dry period increased in all saplings; and that there was no effect on *P. halepensis*. It was also determined that generally the diameter and the length of the body and plant nutrient elements (N, P, K) increased in all saplings [21].

Many studies have been conducted on the use of sewage sludge in outdoor ornamental plants as well as cut flowers growing. Sewage sludge in various amounts was applied to *Freesia* spp. plants grown in flower pots and 60 t/ha yield was obtained in the number of the flowers per plant, 180 t/ha yield was obtained in the number of the ruddles, 90 t/ha yield was obtained in the burl diameter, and 90 t/ha yield was obtained in the number of the offspring of the burls [10].

There have been many studies conducted on the use of sewage sludge during seed germination. Sewage sludge in 4 different dosages (0, 40, 80 and 120 t/ha) were applied to grass species in which *Lolium perenne* (40%), *Festuca rubra* var. *rubra* (30%), *Poa pratensis* (15%) and *Festuca rubra* var. *comutata* (15%) were used as mixtures, and it was observed that the sewage sludge increased the Zn, Ni, Cu, Cr and Pb content when compared with the control group, and this increase was observed mostly in the Pb, Zn and Cr content; and there was a decrease in Mn content [22]

In a study in which the effects of sewage sludge on growth, development and grass area quality of *Zoysia japonica* and *Poa annua* grass species was observed and was reported that the sewage sludge applied in various dosages (0, 15, 30, 60, 120 and 150 ton/ha) increased the heavy metal content in the soil; and it was determined that the Zn, Pb and Cu did not exceed the limit values; however, Cd exceeded the limit values [9].

Industrial waste sludge in various amounts was applied to *Canna indica* in flower pot and the effects of this application on plant growth and the transport of some heavy metals in the plants were investigated. As a result, it was reported that with the increase in the amount of the sludge in the soil, the metal concentration also increased in some parts of the plant; however, *C. indica* adapted itself well to the industrial waste sludge application, and this plant is therefore recommended for green remedy works (*phytoremediation*) [23].

There have been studies conducted on which plant species might decrease the heavy metals in sewage sludge and as well as on the effects of sewage sludge applied in various dosages on plant properties. In a study in which the effect of sewage sludge and fertilizing on the fertility of *Salix viminalis* in forest areas was investigated it was reported that the sewage sludge increased the weed development in the first year because of its excessive nutrient content, and encouraged the offshoot and increased the fertility. However, it was also reported that the heavy metal content was 4-8% more than the control group, and therefore exceeded the limit values [24].

In a to another study, the tree types like *Quercus acutissimave Liriodendron tulipifera* grown in waste storage areas by applying sewage sludge were examined in terms of the effects of the sewage sludge on the soil and plants, it was determined that the moisture, organic matter, N content and respiration properties of the soil increased, and it was reported that the use of sewage sludge to purify the soil in waste storage areas was an environmental-friendly method [25].

In another study which was conducted as flower pot experiment with sewage sludge and garden compost it was reported that the mixture increased the plant development in *Ipomea aquatica* species, and that the highest fresh weight was observed in the 4% garden compost and 2% sewage sludge mixture [26].

In growth media in which nutshell and sewage sludge were mixed in various rates, the annual seasonal ornamental plants *Primula vulgaris* and *Tagates patula var. nana* were used in a study, and it was determined that with the increase of the sewage sludge in the mixture, the height of the plants also increased. It was also reported that the body diameter, canopy diameter and dry weight values of *Primula* increased in media which contained sewage sludge [5].

In a study in which the effects of the application of sewage sludge and mineral fertilizer on *Diplotaxis erucoides* species was investigated it was reported that in the plants which received sewage sludge compost, the blooming was delayed; however, the biomass increased, and the plant developed bigger root system, and also the number of the seeds increased [27].

2. CONCLUSION

In the light of all these data, it is clear that the use of the sewage sludge in growing ornamental plants as germination medium, growth medium, and organic fertilizer should be spread in a way that will not harm the environment in accordance with the regulations of each country and this will be of great use for countries both in terms of economy and ecology. The use of sewage sludge in agricultural areas as well as in growing ornamental plants will enlarge the usage area of the sewage sludge in agricultural production. In addition, the psychological negative situation that emerges due to the use of sewage sludge in plants that are consumed in food chain being not observed in its use in ornamental plant growth makes it advantageous.

The use of waste water sewage sludge for the purpose of growing various ornamental plants in dimensions that will not cause environmental problems is recommended with the opinion that it is one of the most proper methods for modern recycling concepts.

COMPETING INTERESTS

Author has declared that no competing interests exist.

REFERENCES

1. Nogueirol RC, deMelo EI, Bertoncini WJ, Alleoni LRF. Concentrations of Cu, Fe, Mn, and Zn in tropical soils amended with sewage sludge and composted sewage sludge. Environ Monit Assess. 2013; 185:2929-2938
2. Zhao S, Shang X, Duo L. Accumulation and spatial distribution of Cd, Cr and Pb in mulberry from municipal solid waste compost following application of EDTA and $(NH_4)_2SO_4$. Environ Sci Pollut Res. 2013; 20:967-975.
3. Curaqueo G, Schoebitz M, Borie F, Caravaca F, Roldán A. Inoculation with arbuscular mycorrhizal fungi and addition of composted olive-mill waste enhance plant establishment and soil properties in the regeneration of a heavy metal-polluted environment. Environ Sci.Pollut Res. 2014; 21:7403–7412.
4. Zafari A, Kianmehr MH. Factors affecting mechanical properties of biomass pellet from compost. Environ Technol. 2014; 35:478-486.
5. Dede OH, Ozdemir O, Dede G. Fındık zürufu ve arıtma çamurlarının tek yıllık süs bitkisi yetiştiriciliğinde kullanılması. II. Ulusal Arıtma Çamurları Sempozyumu, 04-06 Kasım; 2009. İzmir. (in Turkish)
6. Bozdogan E, Cetinkale G, Sogut Z. Atıksu arıtma çamurlarının yeniden kullanımları. Ulusal Katı Atık Yönetimi Kongresi –UKAY. 2009;334-342. (in Turkish)
7. Cetinkale, G ve Z. Sogut. *Cynodon dactylon* (L.) Pers. çim alanlarında kentsel su arıtım sistem çamurlarından yararlanabilme olanakları. C.Ü. Fen Bil. Ens. Fen ve Mühendislik Bilim. Dergisi. 2010;23(3):11-21. (in Turkish)
8. Ozyazici MA, Ozyazici G. Arıtma çamurunun toprağın bazı temel verimlilik parametreleri üzerine etkileri. Anadolu Tarım Bilim. Derg. 2012;27(2):101-109. (inTurkish)
9. Wang X, Chen T, Ge Y, Jia Y. Studies on land application of sewage sludge and its limiting factors. Journal of Hazardous Materials. 2008;160(2-3):554-558.
10. Unal M, Karaca A, Camci CS, Celik A. The effects of the drinking water plant sewage sludge on plant development of the *Freesia* spp. and on some soil properties. Selcuk Journal of Agriculture and Food Sciences. 2011;25(2):46-56. (in Turkish)

11. Bhattacharya T, Chakraborty S, Banerjee DK. Heavy metal uptake and its effect on macronutrients, chlorophyll, protein, and peroxidase activity of *Paspalum distichum* grown on sludge-dosed soils. Environ Monit Assess. 2010;169:15-26.
12. Silva JDC, Leal TTB, Araújo ASF, Araujo RM, Gomes RLF, Melo WJ, Singh RP. Effect of different tannery sludge compost amendment rates on growth, biomass accumulation and yield responses of *Capsicum* plants. Waste Manage. 2010;30:1976-1980.
13. Dai J, Chen L, Zhao J, Ma N. Characteristics of sewage sludge and distribution of heavy metal in plants with amendment of sewage sludge. Journal of Environmental Sciences. 2006;18(6):1094-1100.
14. Vecchietti L, Lucia B, Russo G, Rea E, Leone A. Environmental and agronomic evaluation of containerized substrates developed from sewage sludge compost for ornamental plant production. Acta Horticulturae. 2013;431-439.
15. Bai XY, Liu HH, Han BP, Su XL, Qin F. Experimental study on ornamental plants remediation of heavy metals in sewage sludge. Environmental Science & Technology China. 2010;33(10):39-44.
16. Shukla OP, Juwarkar AA, Singh SK, Khan S, Rai UN. Growth responses and metal accumulation capabilities of woody plants during the phytoremediation of tannery sludge. Waste Management. 2011;31:115-123.
17. Fuentes D, Disante BK, Valdecantos A, Cortina J, Ramo´N Vallejo V. Response of *Pinus halepensis* Mill. seedlings to biosolids enriched with Cu, Ni And Zn in three mediterranean forest soils. Environmental Pollution. 2007a;145:316-323.
18. Ozdemir S, Koseoglu G, Dede OH. Arıtma çamurlarının süs bitkisi toprağı hazırlanmasında kullanımı, Dokuz Eylül Üniversitesi, I. Ulusal Arıtma Çamurları Sempozyumu Bildiri Kitabı, İzmir, 25-26 Mart. 2005;557-564. (in Turkish)
19. Fuentes D, Valdecantos A, Cortina J,Vallejo VR. Seedling performance in sewage sludge-amended degraded mediterranean woodlands. Ecological Engineering. 2007b;31(4):281-291.
20. Hernández-Apaolaza L, Gascó AM, Gascó JM, Guerrero F. Reuse of waste materials as growing media for ornamental plants. Bioresource Technology. 2005;96(1):125-131.
21. Larchevêque M, Ballini C, Korboulewsky N, Montès N. The use of compost in afforestation of Mediterranean areas: effects on soil properties and young tree seedlings. Science of The Total Environment. 2006;369(1-3):220-230.
22. Kucukhemek M, Gur K, Berktay A. Evsel karakterli atık su arıtma çamurlarının çim bitkisi ağır metal (Mn, Zn, i Cu, Cr, Pb, Cd) içeriği üzerine etkisi. Selçuk Üniversitesi Mühendislik Mimarlık Fakültesi Dergisi. 2006;21(3-4):1-12. (in Turkish).
23. Bose S, Jain A, Rai V, Ramanathan AL. Chemical fractionation and translocation of heavy metals in *Canna indica* grown on industrial waste amended soil. Journal of Hazardous Materials. 2008;160:187–193.
24. Lazdina D, Lazdins A, Karins Z, Kaposts V. Effect of sewage sludge fertilization in short-rotation willow plantations. Journal of Environmental Engineering and Landscape Management. 2007;15(2):105-111.
25. Song U, Lee EJ. Environmental and economical assessment of sewage sludge compost application on soil and plants in a landfill. Resources, Conservation and Recycling. 2010;54:1109-1116.
26. Stabnikova O, Goh WK, Ding HB, Tay JH, Wang JY. The use of sewage sludge and horticultural waste to develop artificial soil for plant cultivation in Singapore. Bioresource Technology. 2005;96(9):1073-1080.
27. Korboulewsky N, Bonin G, Massiani C. Biological and ecophysiological reactions of white wall rocket *(Diplotaxis erucoides)* grown on sewage sludge compost. Environmental Pollution. 2002;117:365-370.

18

SDS-PAGE of *Ophioglossum vulgatum* Proteins

M. S. Chernyshenko[1], V. O. Chernyshenko[2*] and L. I. Musatenko[1]

[1]Department of Phytohormonology, MG Kholodny Institute of Botany, NAS of Ukraine, Ukraine.
[2]Department of Protein Structure and Function, Palladin Institute of Biochemistry, NAS of Ukraine, Ukraine.

Authors' contributions

All authors contributed this work equally. Author MSC collected the samples and extracted proteins. Author VOC performed SDS-PAGE and densitometry. Author LIM contributed to the work on all stages and also developed an ideology of experiments and interpreted the data obtained. All authors read and approved the final manuscript.

Editor(s):
(1) Mirza Hasanuzzaman, Department of Agronomy, Faculty of Agriculture, Sher-e-Bangla Agricultural University, Dhaka-1207, Bangladesh.
Reviewers:
(1) Ds Sheriff, Benghazi University, Benghazi, Libya.
(2) Bartosz Adamczyk, University of Helsinki, Finland.

ABSTRACT

Background: Despite that the protein composition is one of the main characteristics of plant organism, there are not many studies dedicated to the fern proteins characterisation. Protein composition as a part of biochemical characterisation of *Ophioglossum vulgatum* (an evolutionary distinct fern species related to *Psilotaceae*) is of interest from the point of view of ecology, taxonomy and plant biochemistry.
Aims: Study of the protein composition of roots and leaves of *O. vulgatum* during seasonal ontogenesis of its sporophytes.
Methodology: SDS-PAGE with following densitometry was used for the estimation of protein composition of roots and leaves of *O. vulgatum* collected from young plants, mature plants, spore-bearing and not spore-bearing plants at the peak of their seasonal development. To determine the phosphorylation of studied proteins the electrophoresis using Phos-tag™ Acrylamide Gels were applied.
Results: The simple and effective method of protein extraction from the leaves and roots of *O. vulgatum* was developed and applied. Protein fractions with the molecular weights of 112±1.3, 57±1.0, 45±1.2, 35.5±0.6, 25.8±0.5, 18±1.5 kDa were shown as the major for both leaves and

**Corresponding author: E-mail: bio.cherv@gmail.com*

roots. Huge RuBisCO accumulation during the stages of plant initial growth and spore-bearing was shown (up to 55±10% in leaves and 89±5% in roots). All large chains of detected RuBisCO were phosphorylated which suggests that the active form of *O. vulgatum* RuBisCO is phosphoprotein.
Conclusion: The dynamic change of protein content was shown for different ontogenesis stages of *O. vulgatum*. The expression of phosphorylated RuBisCO was distinctly increased during the beginning of leaf development and spore-bearing.

Keywords: SDS-PAGE; Ophioglossum vulgatum; Phos-tag; protein composition; extraction of plant proteins.

1. INTRODUCTION

Ferns are a large and diverse group of pteridophytes, commonly referred to as the lower vascular plants. More than a hundred of them are edible [1].

Despite that the protein composition is one of the main characteristics of a plant organism; there are not many studies of fern protein composition spanning evolutionary diverse branches. Evidence of the correlation between the amino acid and protein compositions and plant taxonomy was shown [2,3]. Also, some studies showed the potential nutrition value of fern-derived proteins [4-6]. It is known, for angiosperms, that growth and bloom of plants leads to distinct proteomic changes [7,8].

Ophioglossaceae are a putatively ancient lineage of ferns in which the aerial portion of the plant is composed of a single leaf [9]. *O. vulgatum* is fern species which in Europe has a wide scattered distribution [10]. Initially heterotrophic gametophytes of *O. vulgatum* form mutualistic associations with fungi of the *Glomeromycota* but it is unknown whether mature *O. vulgatum* sporophytes have such mutualism [11]. Evolutionally *Ophioglossaceae* is an early diverging fern family with reduced features that is thought to be sister group to *Psilotaceae* [12].

All mentioned above makes the biochemistry of *O. vulgatum* of interest from the points of view of evolution studies, taxonomy, phytochemistry and proteomics. However there were no studies on the change of protein composition of this remarkable plant species. That is why the aim of present work was to study the protein composition of roots and leaves of *O. vulgatum* during ontogenesis of its sporophytes.

2. MATERIALS AND METHODS

2.1 Materials

Tris (161-0716, Sigma); Glycine (161-0718, Sigma); PBS tablets (2810305, MP Biomedicals); Disposable PD 10 Desalting Columns (17-0851-01, GE Healthcare); Ready-to-Use Phos-tag™ Acrylamide Gel (195-16381, Wako); Ethylenediaminetetraacetic acid (EDTA) (03620, Sigma); Phenylmethylsulfonyl fluoride (PMSF) (329-98-6, Sigma); Amicon Ultra-0.5 mL Centrifugal Filters (UFC500324, Merck Millipore); molecular weight markers (0671, Fermentas).

2.2 Methods

2.2.1 Protein extraction

For the extraction of total protein from *Ophioglossum vulgatum* leaves and roots the methods described in [13-15] were modified as follows. Plant tissue was collected in the field during different stages of plant ontogenesis: 1 – young vegetative plant (May 7th); 2 – developing vegetative plant (June 10th); 3 – mature spore-bearing plant (June 27th); 4 – mature vegetative (not spore-bearing) plant (June 27th). Collected samples were moved on dry ice to the laboratory where they were washed, dried with paper towels, divided into leaves and roots, and then 50 mg of each was weighed and then stored in the freezer at -20℃. Samples were placed in a mortar, over them was poured liquid nitrogen and pestled until no chunks were visible. 1 ml of extraction buffer (0.05 M Tris-HCl buffer pH 7.4 with 0.3 M NaCl, 1mM PMSF and 1mM EDTA) was added to the homogenised biomass. Samples were suspended in the extraction buffer and transferred to the sample tube. Mortar and pestle were rinsed with 0.5 ml of extraction buffer to remove all traces of sample. Obtained lysates of tissues were collected to 2 ml sample tubes and spinned-down at 4 000 rpm, + 4℃ for 30 minutes. Supernatants were collected and then gel-filtered on PD-10 column to exclude non-protein components from the samples. Eluted fractions were concentrated using Amicon Ultra-0.5 mL Centrifugal Filters. The amount of total proteins in the mixture was monitored by measuring the optical density of the sample by spectrophotometry at E_{280}. The concentration of proteins in all samples were normalised at 4 mg/ml. Samples were stored at -20℃.

2.2.2 SDS-PAGE

Samples were mixed with equal volume of Laemmli sample buffer that contained 2% SDS, 5% glycerine and bromophenol blue. They were heated in a boiling-water bath and were analysed or stored no more than 3 months at -20℃. The molecular weights of proteins were determined by SDS-PAGE using 10% gels prepared accordingly to Laemmli [16]. Ready-to-Use Phos-tag™ Acrylamide Gels (10%) were used to determine the phosphorylation of studied proteins [17].

Protein samples were loaded on a gel (30 mkg/track) that was an optimum concentration for accurate detection of major proteins. All samples were normalized according to their optical density at E_{280}.

2.2.3 Densitometry

Electrophoresis gels were scanned with HP Photosmart C3100 and analysed using densitometry software TotalLab TL100 (Phoretix) [18], a tool that allows to identify molecular weights and amount of proteins separated by electrophoresis [19,20].

Molecular weight markers 0671 (Ferments) were used for the "pixel position"/"molecular weight" calibration.

2.2.4 Statistics

Statistical data analysis was performed using Microsoft Excel. All assays were performed in series of three independent experiments and the data were fitted with standard errors using "Statistica 7".

3. RESULTS AND DISCUSSION

SDS-PAGE is a simple and informative method that allows to separate proteins according to their molecular weight and detect them in comparison to molecular weight markers [21]. So it is widely used for the characterisation of biological objects providing the information of their protein compositions. The densitometry of obtained electrophoregrams allows estimating the amount of individual proteins and their ratio in the mixture [22].

Concerning plant objects SDS-PAGE was successfully used for the characterisation of seed storage proteins derived from different breeds [23], finding the common proteins in different species [24], estimation of the gall disease's effects on plant proteins [25], etc.

In our study we separated proteins of *O. vulgatum* leaves and roots samples collected at different stages of ontogenesis. According to SDS-PAGE the leaves and roots samples presented the same bands of major proteins in studied range from 10 to 110 kDa (Fig. 1). However the amounts of individual proteins differed between different stages of ontogenesis.

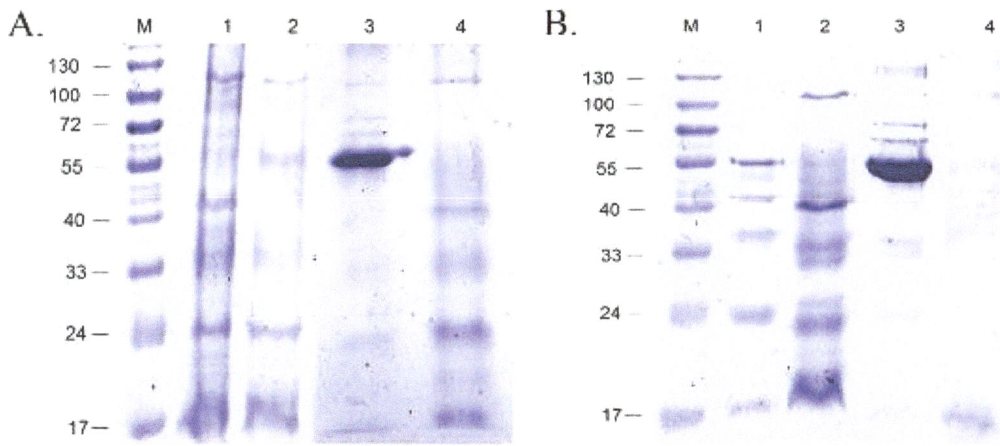

Fig. 1. SDS-PAGE of the protein extracts of *O. vulgatum* leaves (A) and roots (B); 1 – Young plant; 2 – Developing vegetative plant; 3 – Mature spore-bearing plant; 4 – Mature not spore-bearing plant. M – Molecular weight markers

For the estimation of the quantity of proteins in individual samples we used densitometry of electrophoregrams by TotalLab TL100. Densitograms corresponding to the tracks of protein samples separated by SDS-PAGE are shown on the Fig. 2. Protein fractions with the molecular weights of 112±1.3, 57±1.0, 45±1.2, 35.5±0.6, 25.8±0.5, 18±1.5 kDa (P = .05) were selected as the major protein bands and quantified using TotalLab TL100. The calculations of the series of three replicates are shown on the Fig. 3.

Here we have to point out a remarkable tendency in the protein composition of leaves and roots of *O. vulgatum*. First of all, protein band with molecular weight of 55±1.0 was shown to be the most variable fraction. It was distinctly increased on the stage of young plant (sample 1) leaf growth and highly increased when the plant started to bear spores (sample 3). From the other hand mature not spore-bearing plants (sample 4) showed the decreased amount of this protein fraction. This tendency was observed equally in the case of leaf's proteins as well as in the case of root's proteins.

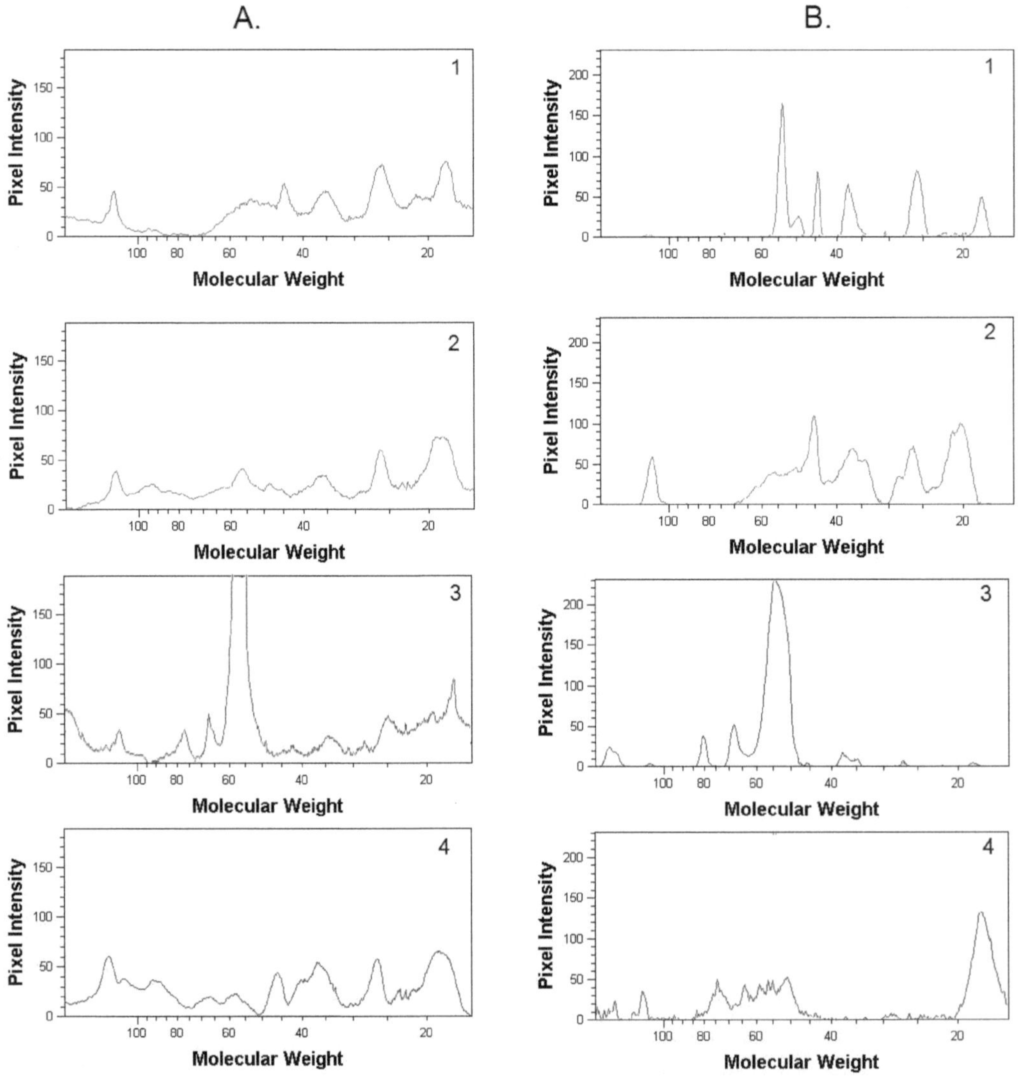

Fig. 2. Densitometry of SDS-PAGEs of the protein extracts of *O. vulgatum* leaves (A) and roots (B). 1 – Young plant; 2 – Developing vegetative plant; 3 – Mature spore-bearing plant; 4 – Mature not spore-bearing plant

According to the literature the major protein that makes up to 50% of the soluble protein of plant leaves is RuBisCO (Ribulose-1,5-bisphosphate carboxylase/oxygenase) [26]. RuBisCO has a molecular mass of 560 kDa and consists of eight small (14 kDa each) and eight large (56 kDa each) subunits arranged as eight heterodimers [27].

It is known that the expression of RuBisCO protein is activated in the growing plants and parts of plant that grow intensely [6,28,29]. So we can conclude that 55±1.0 protein fraction detected in the samples of *O. vulgatum* corresponded to the large subunit of RuBisCO. We have shown that RuBisCO had similar dynamics in leaves as well as in roots and its expression was increased during plant initial growth and spore-bearing. It is interesting that the sporogenesis in the sporangium lead to activation of synthetic processes (coincident with detected RuBisCO expression increase) in leaves and even in roots of spore-bearing plants.

Studies of RuBisCO of different plant species showed that in some species it is a phosphoprotein by its nature and its enzyme activity is highly dependent on phosphorylation, when in others it is not [30]. It was also shown that RuBisCO large-chain fragments varied significantly in protein abundance and the level of phosphorylation among ontogenetic phases [31]. To check whether *O. vulgatum* RuBisCO was phosphorylated, we used Phos-Tag SDS-PAGE. It was shown that all RuBisCO large-chains that were detected using SDS-PAGE were phosphorylated (Fig. 4).

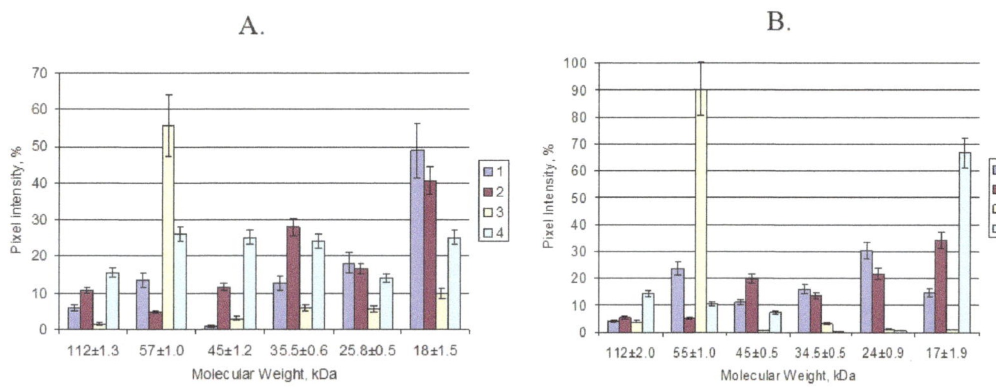

Fig. 3. The distribution of major protein fractions in the protein extracts of *O. vulgatum* leaves (A) and roots (B). 1 – Young vegetative plant; 2 – Developing vegetative plant: 3 – Spore-bearing plant; 4 – Not spore-bearing plant

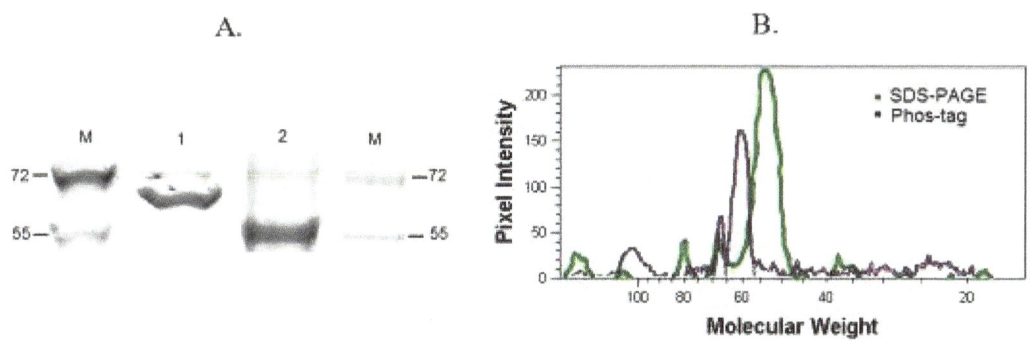

Fig. 4. SDS-PAGE (A) and corresponding densitogram (B) of RuBisCO derived from *O. vulgatum* leaves on the spore-bearing stage. 1 – Samples separated using Phos-tag electrophoresis; 2 - Samples separated using classic SDS-PAGE; M – Molecular weight markers

4. CONCLUSION

A simple and effective method of protein extraction from the leaves and roots of *O. vulgatum* was developed and applied. It allowed to obtain samples for SDS-PAGE excluding contaminants (DNA, pigments, etc) that influence proteins migration and worsen the quality of electrophoresis. SDS-PAGE analysis of *O. vulgatum* leaves and roots allowed us to reveal protein composition of this plant. Protein fractions with the molecular weights of 112±1.3, 57±1.0, 45±1.2, 35.5±0.6, 25.8±0.5, 18±1.5 kDa were shown as the major for both leaves and roots. The dynamic change of protein content was shown for different ontogenesis stages. Huge RuBisCO accumulation on the stages of plant initial growth and spore-bearing was shown. All large chains of detected RuBisCO were phosphorylated which is evidence that an active form of *O. vulgatum* RuBisCO is phosphoprotein. Presented results could be applied for further biochemical characterization of *O. vulgatum*. The comparison of data on plant hormones activation and proteins dynamics on different stages of ontogenesis could possibly help us to solve questions of the evolution of *Ophioglossaceae* and the nature of their mutualistic associations with fungi.

COMPETING INTERESTS

Authors have declared that no competing interests exist.

REFERENCES

1. Lognay G, Haubruge E, Delcarte E, Wathelet B, Mathieu F, Marlier M, Malaisse F. Ophioglossum polyphyllum A. Braun in Seub. (Ophioglossaceae, Pteridophyta), a rare potherb in south central Tibet (T.A.R, P.R. China). Geo-Eco-Trop. 2008;32:9-16.
2. Sood A, Prasanna R, Singh PK. Utilization of SDS-PAGE of whole cell proteins for characterization of Azolla. Annales Botanici Fennici. 2007;44(4):283-286.
3. Yeoh HH, Wee YC. Total amino acid composition of tropical ferns and gymnosperms. Asian Journal of Tropical Biology. 1998;3:13-18.
4. Essuman EK, Ankar-Brewoo GM, Barimah J, Ofosu IW. Functional properties of protein isolate from fern fronds. International Food Research Journal. 2014;21(5):2085-2090.
5. Fasakin EA. Nutrient quality of leaf protein concentrates produced from water fern (*Azolla africana* Desv) and duckweed (*Spirodela polyrrhiza* L. Schleiden). Bioresource Technology. 1999;69(2): 185-187.
6. Simova-Stoilova L, Stoyanova Z, Demirevska-Kepovaontogenic K. Changes in leaf pigments, total soluble protein protein and rubisco in two barley varieties in relation to yield. Bulg J Plant Physiol. 2001;27(1-2):15-24.
7. Cao X, Gao Y, Wang Y, Li CM, Zhao YB, Han ZH, Zhang XZ. Differential expression and modification of proteins during ontogenesis in Malus domestica. Proteomics. 2011;24:4688-701.
   DOI: 10.1002/pmic.201100132.
8. Garcés M, Le Provost G, Lalanne C, Claverol S, Barré A, Plomion C, Herrera R. Proteomic analysis during ontogenesis of secondary xylem in maritime pine. Tree Physiol. 2014;1-15.
   DOI: 10.1093/treephys/tpt117
9. Hauk WD, Parks CR, Chase MW. Phylogenetic studies of Ophioglossaceae: Evidence from rbcL and trnL-F plastid DNA sequences and morphology. Mol Phylogenet Evol. 2003;28(1):131-51.
10. Muller S. Assessing occurrence and habitat of *Ophioglossum vulgatum* L. and other Ophioglossaceae in European forests. Significance for nature conservation. Biodiversity & Conservation. 2000;9(5):673-681.
11. Field KJ, Leake JR, Tille S, Allinson KE, Rimington WR, Bidartondo MI, Beerling DJ, Cameron DD. From mycoheterotrophy to mutualism: Mycorrhizal specificity and functioning in *Ophioglossum vulgatum* sporophytes. New Phytol. 2015;205(4): 1492-502.
    DOI: 10.1111/nph.13263.
12. Williams E, Waller DM. Phylogenetic placement of species within the genus *Botrychium* s.s. (Ophioglossaceae) on the Basis of Plastid Sequences, amplified fragment length polymorphisms and flow cytometry. Int J Plant Sci. 2012;173(5): 516-531.
13. Wang W, Tai F, Chen S. Optimizing protein extraction from plant tissues for enhanced proteomics analysis. J Sep Sci. 2008;11:2032-2039.
    DOI: 10.1002/jssc.200800087.

14. Conlon HE, Salter MG. Plant protein extraction. Methods Mol Biol. 2007;362: 379-83.
DOI: 10.1007/978-1-59745-257-1_28.
15. Tsugama D, Liu S, Takano T. A rapid chemical method for lysing Arabidopsis cells for protein analysis. Plant Methods. 2011;7:22-29.
DOI: 10.1186/1746-4811-7-22.
16. Laemli RV. Cleavage of structural poteins during of bacteriophage T4. Nature. 1970; 227:680-685.
17. Kinoshita-Kikuta E, Kinoshita E, Koike T. A laborsaving, timesaving and more reliable strategy for separation of low-molecular-mass phosphoproteins in phos-tag affinity electrophoresis. International Journal of Chemistry. 2012;4(5):1-8.
18. Vranković J, Paunović M, Labus-Blagojević S. Expression of CYP1A protein in the freshwater clam *Corbicula fluminea* (Müller). Arch Biol Sci. 2011;63(1):37-42.
19. Young N, Chang Z, Wishart DS. Gel Scape: A web-based server for interactively annotating, manipulating, comparing and archiving 1D and 2D gel images. Bioinformatics. 2004; 20(6):976-978.
20. Wu HC, Yen CC, Tsui WH, Chen HM. A red line not to cross: Evaluating the limitation and properness of gel image tuning procedures. Anal Bio chem. 2010;396(1):42-50.
21. Shapiro AL, Viñuela E, Maizel JV Jr. Molecular weight estimation of polypeptide chains by electrophoresis in SDS-polyacrylamide gels. Biochem Biophys Res Commun. 1967;28(5):815-820.
DOI: 10.1016/0006-291X(67)90391-9.
22. Braunagel SC, Russell WK, Rosas-Acosta G, Russell DH, Summers MD. Determination of the protein composition of the occlusion-derived virus of *Autographa californica nucleopolyhedro* virus. Proc Natl Acad Sci USA. 2003;100(17): 9797-9802.
DOI: 10.1073/pnas.1733972100.
23. Ehsanpour AA, Shojaie B, Rostami F. Characterization of seed storage protein patterns of four Iranian pistachios using SDS-PAGE. Natural Science. 2010;2(7): 737-740.
24. Vishal T, Aparadh V, Amol P, Karadge BA. Comparative analysis of seed and leaf proteins by SDS PAGE gel electrophoresis within Cleome species. IJALS. 2012;3: 50-58.
25. Saini D, Sarin R. SDS-PAGE Analysis of leaf galls of *Alstonia scholaris* (L.) R. Br. J Plant Pathol Microb. 2012;3:2.
Available:http://dx.doi.org/10.4172/2157-7471.1000121
26. Andersson I, Backlund A. Structure and function of Rubisco. Plant Physiol Biochem. 2008;46(3):275-91.
DOI: 10.1016/j.plaphy.2008.01.001.
27. Ma Z, Cooper C, Kim H-J, Janick-Buckner D. A Study of Rubisco through western blotting and tissue printing techniques. CBE Life Sci Educ. 2009;8(2):140-146.
DOI: 10.1187/cbe.09-01-0003.
28. Staswick P. The occurrence and gene expression of vegetative storage proteins and a rubisco complex protein in several perennial soybean species. Journal of Experimental Botany. 1997;48(317): 2031-2036.
29. Suzuki Y, Nakabayashi K, Yoshizawa R, Mae T, Makino A. Differences in expression of the RBCS multigene family and rubisco protein content in various rice plant tissues at different growth stages. Plant Cell Physiol. 2009;50(10):1851-1855.
DOI: 10.1093/pcp/pcp120.
30. Aggarwal KK, Saluja D, Sachar RC. Phosphorylation of rubisco in Cicer rietinum: Non-phosphoprotein nature of rubisco in *Nicotiana tabacun*. Phytochemistry. 1993;34(2):329-335.
31. Wang Y, Wang Yi, Zhao YB, Chen DM, Han ZH, Zhang XZ. Rubisco large-chain fragments varied significantly in protein abundance and degree of phosphorylation among ontogenetic phases. Proteome Science. 2014;12:31.
DOI: 10.1186/1477-5956-12-31.

19

Leaf Epidermal Structures and Stomata Ontogeny in Some Members of the Family Cucurbitaceae

Sauban Musa Jibril[1] and Bello Hassan Jakada[1*]

[1]Faculty of Applied Science, Jodhpur National University, Jhanwar Road Village Narnadi, Near Boranada, Jodhpur, Rajasthan, India.

Authors' contributions

This work was carried out in collaboration between both authors. Author SMJ designed the study, and wrote the first draft of the manuscript and managed the literature searches. Author BHJ managed the experimental process and identified the species of plant. Both authors read and approved the final manuscript.

Editor(s):
(1) Krishna Kumar Shrestha, Central Department of Botany, Tribhuvan University, Kiritipur, Kathmandu, Nepal.
(2) A. Mujib, Department of Botany, Hamdard University, India.
Reviewers:
(1) Klara Kosova, Crop Research Institute, Prague, Czech Republic.
(2) Jie Zhang, Hohai University, P. R. China.
(3) Ines Cechin, Sao Paulo State University-Unesp, Brazil.
(4) Sheikh Abdul Shakoor, Guru Nanak Dev University, India.
(5) Anonymous, University of Sao Paulo, Brazil

ABSTRACT

Aim: The aim of this research work is to study the leaf epidermal structures and stomata ontogeny of some members of the family Cucurbitaceae.
Study Design: Research Article.
Place and Duration of Study: Department of Biological Science, Bayero University Kano (BUK), between November 2012 and February, 2013.
Methodology: The leaves of *Citrullus lanatus*, *Cucumis sativus* and *Cucurbita pepo* were collected from Imawa Village of Kura Local Government Kano state with a global positioning of N 11° 48' 14.6'' E 008° 48' 45.4". Epidermal peels of both abaxial (lower) and adaxial (upper) surfaces were made by placing the leaf on a clean glass slab with the surfaces to be studied facing downward. The specimens were irrigated with water holding it downward from one end, the epidermis above the desired surfaces was scrapped off carefully with a sharp razor blade, loosen cells were washed

**Corresponding author: E-mail: bellojakada@gmail.com*

away from the epidermal peel with the aid of soft camel hairbrush and water until the desired epidermis below was reached. The epidermal peels were mounted in glass slide stained with aqueous solution of safranin for 4-8 minutes, then rinsed carefully in water to remove excess stain, a drop of 50% paraffin was added and examined using light power microscope at (x40) objective magnification. For the study of stomata types, epidermal peel was made using mature leaves and for stomata ontogeny fresh immature leaves were used. Measurement of stomata was made with the aid of an ocular micrometer and stage micrometer. The data obtained of lower surface was subjected to analysis of variance (ANOVA) at 5% level of significance to determine the least significant difference (LSD).

Results: *Stomata Types:* In *Citrullus lanatus* mature stomata are anomocytic, In *Cucumis sativus* the stomata are anisocytic and present on abaxial (lower) surface only and in *Cucurbita pepo* are anomocytic stomata.

Epidermal Cells and Trichomes: In *Citrullus lanatus* epidermal cells on lower and upper surface is irregular in shape, unicellular non glandular epidermal hair (trichome) were present in upper surface, *Cucumis sativus* the epidermal cells were polygonal or irregular in shape, unicellular non-glandular trichomes were present in upper and lower surface.

Stomatal Ontogeny: In all the species studied mesogenous stomata development was observed

Conclusion: The species can be distinguished by the type of stomata and variation in stomatal index, there were similarities based on stomata development between the three species.

Keywords: Abaxial surface; adaxial surface; epidermal cell; epidermal peel; epidermal structure; systematics; taxonomy.

1. INTRODUCTION

Plant classification has been a subject of discussion among plant taxonomists and systematists over the years, plants are classified and reclassified as soon as new evidences arise and this is going to be a continuous exercise over some years to come [1]. Most plants are classified based on external morphological structures such as flowers and fruits [2]. These structures are not always available on plants because they are seasonal in production. Due to this reason, other means of classification need to be involved; one of which is anatomical studies especially of leaves [3].

The epidermis is the outer layer of cells covering the leaf [4]. It has several functions: Protecting the plants against water loss by transpiration, regulation of gaseous exchange, secretion of metabolic compounds and absorption of water [5]. Most leaves show dorsiventral anatomy: The upper (adaxial) and lower (abaxial) surfaces have somewhat different construction and may serve different functions [6]. It can also be used as taxonomic character delimiting plants, some characters or features on the epidermis which are useful taxonomically include several differentiated cell types: Epidermal cells, stomata, subsidiary cells, guard cells and epidermal hairs (i.e., trichomes) [5]. These features have been used previously to resolve some taxonomic problems or to contribute to ever increasing taxonomic data base in some genera and even families of plants [7-9]. Continuity and/or discontinuity of the characters can be used to show relationships and differences between plant taxa within a genus or between families [1].

Stomata in dicotyledon may originate by division resulting in an oblique wall in an epidermal cell, the smaller cell resulting from this division functions as the guard cell mother cell, in many monocotyledons the guard cell mother cell is formed by an asymmetric division of an epidermal cell, the process is similar to the formation of some trichoblast in the root epidermis [10]. These complexes are generated by a series of cell divisions that are stereotypical yet dynamic, during organogenesis, a protodermal cell faces a decision to undergo proliferative division to form pavement cells or undergo asymmetric division to initiate stomatal lineages [5]. A subset of protodermal cells, called meristemoid mother cells (MMC), divides asymmetrically to generate two daughter cells with distinct sizes and fates, the larger cell adopts pavement-cell identity where the smaller, triangular cell is called a meristemoid, which possesses a transient, stem-cell like property and continues a few rounds of asymmetric division [11]. In each asymmetric division of a meristemoid, the smaller daughter cell renews its property as a meristemoid, while amplifying the number of surrounding cells. The meristemoid then differentiates into a round, guard mother cell (GMC), which divides symmetrically once to give rise to a pair of guard cells [12]. The larger daughter cells produced by the amplifying asymmetric divisions of the meristemoids are termed stomatal-lineage ground cells [13]. The SLGCs resemble the pavement cells. The SLGC may divide asymmetrically to give rise to a secondary, satellite meristemoid, which differentiates into a secondary satellite stomatal complex [14]. The SLGCs of a secondary stomatal complex may divide asymmetrically to generate a tertiary meristemoid, and so on [12]. Three categories of stomatal development are known in angiosperm termed mesogenous and perigenous and a third mesoperigenous where some adjacent cells were derived and some were not derived from guard cell initials, these three main categories are further subdivided into ten categories and more recent authors have increased this total considerably [2]. It must be emphasized that there is no strict correlation between mature stomatal type and the mode of development [6]. In particular, the presence of subsidiary cells does not necessarily indicate a mesogenous development (and their absence a perigenous one) [1]. A study on stomatal development and application of these descriptive terms which are now equally applied to all vascular plants often elicits more useful taxonomic character than the use of mature characteristics alone, especially when distinctive subsidiary cells develop at a late stage in a perigenous fashion or when a mesogenous development does not give rise to a distinguishable subsidiary cells [15]. However, development studies do not give extra data for example; the anomocytic and anisocytic stomata of *Streptocarpus* develop perigenously and mesogenously respectively [16].

In most species the number and arrangement of the subsidiary cells around the stomata are relatively constant; various classifications of stomata according to these arrangements have been drawn up, and are sometimes useful to taxonomist [17,10].

Four types of stomatal complex that occur especially in certain families of dicotyledons, and to which he gave family names rather more descriptive names to these types as follows:

Anomocytic; Stoma surrounded by a limited number of cells that are indistinguishable in size, shape, or form from those of the remainder of the epidermis (irregular-celled) type.

Anisocytic; Stoma surrounded by three cells of which one is distinctly smaller or larger than the other two (unequal-celled) types.

Paracytic; Stoma accompanied on either side by one or more subsidiary cells parallel to the long axis of the pore and guard cells (parallel-celled) type.

Diacytic; Stoma enclosed by a pair of subsidiary cells whose common wall is at right angles to the guard cells the (cross-celled) type.

The leaves of family Cucurbitaceae have amphistomatic or hypostomatic and anomocytic type of stomata [18,19]. However study shows that the stomata type in Cucurbitaceae is anisocytic, paracytic and diacytic [20]. The type of stomata in *Curcurbitaceae* is anomocytic, anisocytic and paracytic, and the leaves are amphistomatic or hypostomatic [20-22,15].

1.1 Aim

The aim of this work is to show some anatomical characters of the family Cucurbitaceae.

2. MATERIALS AND METHODS

2.1 Sample Collection

The leaves of *Citrullus lanatus*, *Cucumis sativus* and *Cucurbita pepo* were collected from Imawa Village along Zaria road, Kura Local Government of Kano state. The place has a global positioning system of N 11° 48' 14.6'' E 008° 48' 45.4"

2.2 Epidermal Peel

Epidermal peels of both abaxial (lower) and adaxial (upper) surfaces were made by placing the leaf on a clean glass slab with the surfaces to be studied facing downward, the specimens were irrigated with water holding it downwards from one end, the epidermis above the desired surfaces was scrapped off carefully with a sharp razor blade loose cells were washed away from the epidermal peel with the aid of soft camel hairbrush and water until the desired epidermis below was reached, the epidermal peels were mounted in glass slide stained with aqueous solution of safranin for 4-8 minutes, then rinsed carefully in water to remove excess stain, a drop of 50% paraffin was added and examine using light power microscope at (x40) objective magnification [23].

Fig. 1.0. Photograph showing seedlings of watermelon

Fig. 1.1. Photograph showing cucumber seedlings

Fig. 1.2. Photograph showing pumpkin seedlings

2.3 Study of Stomata Types and Measurement

For the study of stomata types the epidermal peel was made using mature leaves, for stomata ontogeny a fresh immature leaves was used to study stomata ontogeny in the member of the family Cucurbitaceae, stomata measurement were made with the aid of an ocular micrometer and stage micrometer, at first stage the ocular micrometer was placed carefully in the ocular lens, the stage micrometer was then placed and focused, the point where the ocular micrometer and stage micrometer calibration met was taken and arrived at:

Ocular micrometer/stage micrometer = 20/60 = 0.3 mm.

All the measurements were later done using ocular micrometer after which each value was multiplied using above scale [24,11]. The number of stomata per field and epidermal cells per field was counted and the stomatal index was determined Using the formula: S / E+S x 100 = Stomatal Index (S.I) [19]

Where

S - Number of stomata per unit area
E - Number of epidermal cells in the same area.

2.4 Statistical Analysis

The data obtained of lower surface was subjected to analysis of variance (ANOVA) at 5% level of significance to determine the least significant difference (LSD) between the three members of the family Cucurbitaceae. In upper surface t-test was used to determine whether there was significant difference between the two species.

3. RESULTS

Three Members of the Family Cucurbitaceae were studied and the following result was obtained:

3.1 Type of Stomata

In *Citrullus lanatus* mature stomata are anomocytic, the stomata type of stomata in lower surface of *Citrullus lanatus* is shown in plate 3.0.

In *Cucumis sativus* the stomata are anisocytic and present in abaxial (lower) surface only, the anisocytic stomata in lower surface of *Cucumis sativus* is shown in plate 3.1.

Stomata in *Cucurbita pepo* are anomocytic, the anomocytic stomata on lower surface of *Cucurbita pepo* is shown in plate 3.2.

3.2 Epidermal Cells and Trichomes

In *Citrullus lanatus,* epidermal cells on lower and upper surface are irregular in shape, and unicellular non glandular epidermal hair (trichome) are present in upper surface as shown in plate 3.3.

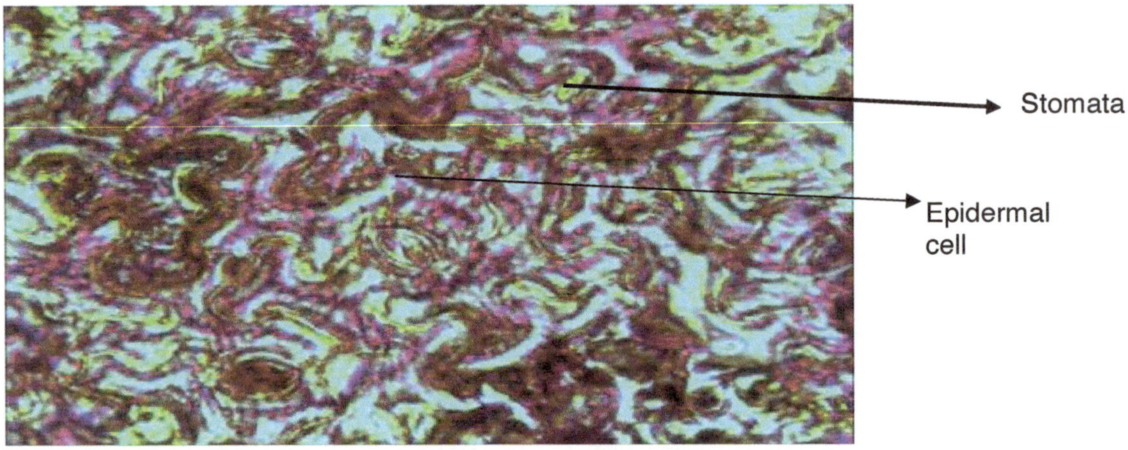

Plate 3.0. Photomicrograph showing anomocytic type of stomata in lower surface of *Citrullus lanatus* (x100)

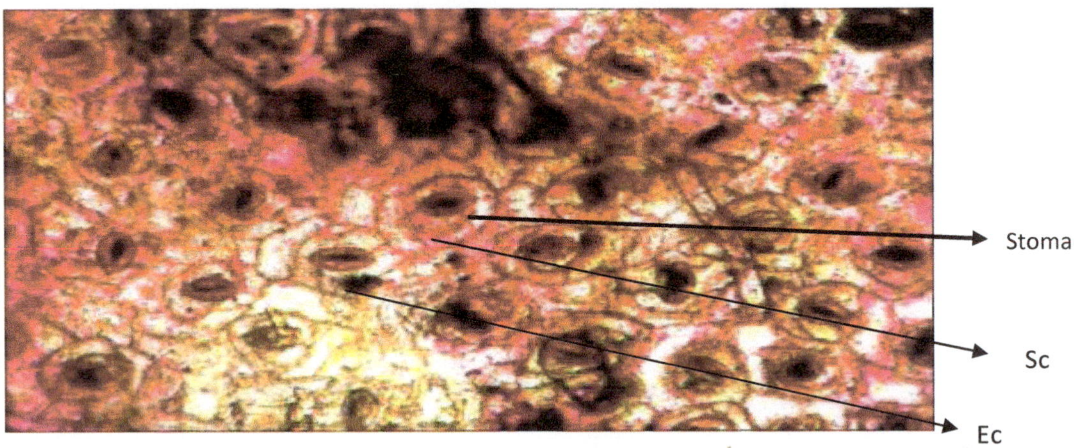

Plate 3.1. Photomicrograph showing anisocytic stomata lower surface in *Cucumis sativus* (x100)

Sc= subsidiary cell, Ec= epidermal cell

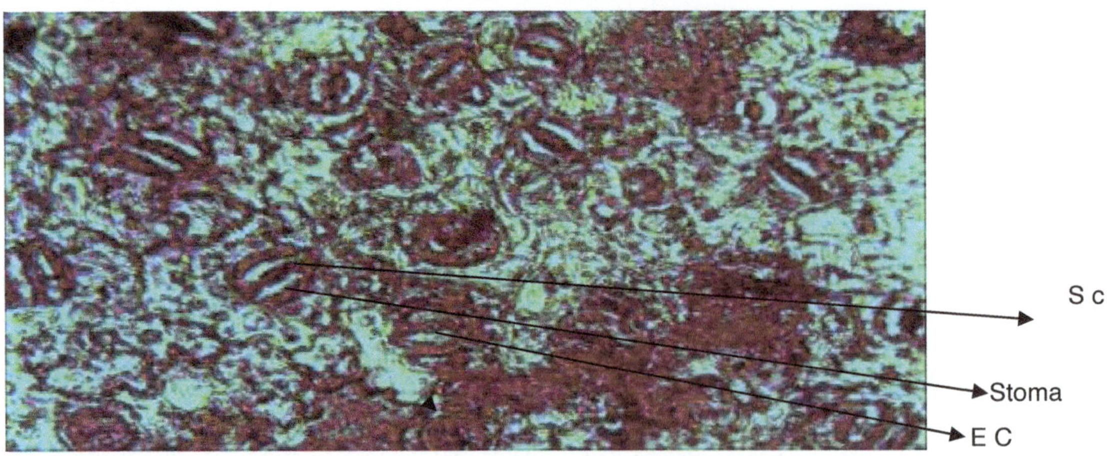

Plate 3.2. Photomicrograph showing anomocytic type of stomata in lower surface of *Cucurbita pepo* (x100)

Sc= Subsidiary cell, Ec= Epidermal cell

Cucumis sativus the epidermal cells are polygonal or irregular in shape; unicellular non-glandular trichomes are present in upper and lower surface. Trichome in the upper surface of the epidermal peel is shown in plate 3.4.

3.3 Stomatal Ontogeny

In all the species studied mesogenous stomata development were observed, the stomata develop from a meristemoid which function directly as a guard mother cell the stomata is formed by an asymmetrical division of proterdermal cells and other two division resulted in the partitioning of the precursor into the guard cell mother cell after the formation of the guard cells the growth adjustment make the subsidiary cells appear as part of stomata complex. The mesogenous stomata development is shown in plate 3.5.

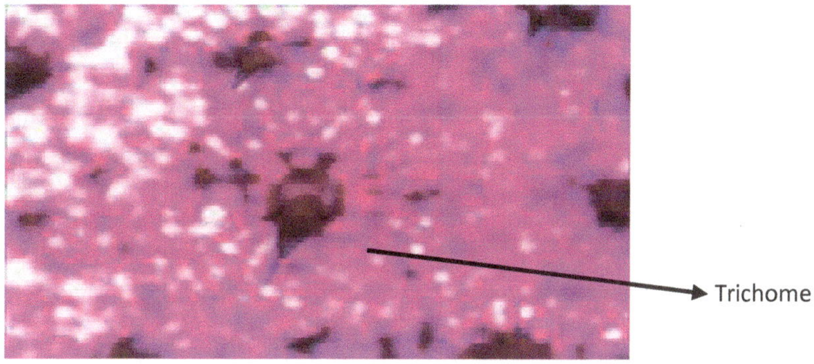

Plate 3.3. Photomicrograph showing unicellular uniseriate trichome in *Citrullus lanatus* upper surface (x100)

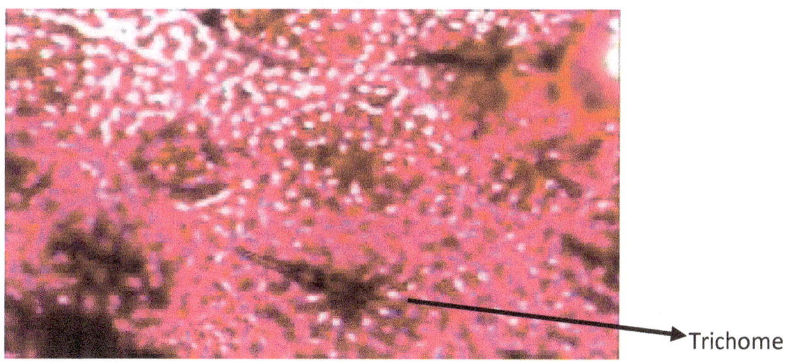

Plate 3.4 Photomicrograph showing unicellular uniseriate trichome in upper surface of *Cucumis sativus* (x100)

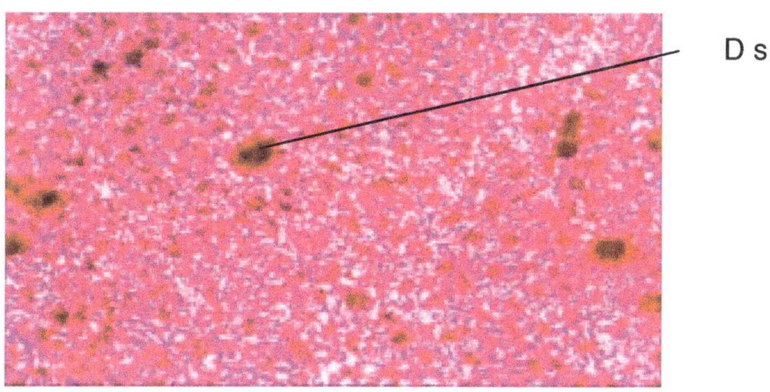

Plate 3.5. Microphotograph showing developing stomata in *Citrullus lanatus* (x100)

Key: DS= Developing stomata, for the above micrographs; On the scale of ×100 magnification; 1 mm=1000 μm

3.4 Stomata Measurement

The mean difference between stomata measurement and the least significance difference were compared at 5% level of significance to determine the least significant difference among the three species studied and the following results were obtained:

3.4.1 Lower surface

The various stomata measurement of the lower surface of the three species is shown in the Table 3.0.

3.4.2 Upper surface

For the two means of upper surface between *Citrullus lanatus and Cucurbita pepo*, t-test were used to test the differences between the two at 5% t- distribution as shown in Table 3.1.

4. DISCUSSION

4.1 Stomata

Leaf epidermal structures of the three members of the family Cucurbitaceae were studied based on the occurrence of stomata, the species can be categorized into two; those in which stomata occur on both abaxial and adaxial surface, *Citrullus lanatus* (water melon) and *Cucurbita pepo* (pumpkin). And those in which the stomata occur on only adaxial surface that is *Cucumis sativus* (cucumber). The leaves whose stomata occur on both surfaces are termed amphistomatic and those whose stomata occur on abaxial surface only are termed hypostomatic leaves. In *Citrullus lanatus* and *Cucurbita pepo* the stomata are anomocytic in which the surrounding cells are indefinite in number and do not differ from the other epidermal cells, the stomata are more numerous on abaxial surface than on adaxial surface [20,25,1,26]. In *Cucumis sativus* the stomata are anisocytic that is three subsidiary cells surround the stomata, one cell being considerably smaller or larger than the other two, the stomata occur on abaxial surface only, the leaf of *Cucumis sativus* is therefore said to be hypostomatic leaf. Similar observations were made [24,25,19]. Stomata index are higher on abaxial surfaces than on adaxial surfaces [20,27].

4.2 Epidermal Cells and Trichome

The epidermal cells are polygonal or irregular in shape in the three species studied. Trichomes are another taxonomic significance of the species studied. Unicellular non glandular, uniseriate trichome was found on both abaxial and adaxial surfaces of *Cucumis sativus* and occurs on upper surface in *Citrullus lanatus.* In *Cucubiata pepo* no trichome was observed,

Table 4.0. Mean stomata measurement in some members of Cucurbitaceae abaxial at (lower) surface of three species (x400) mm

S_no.	Species	Type of stomata	Length of stomata	Breadth of stomata	Pore length	Pore breadth	No. of stomata per field	No. of epidermal cell per field	Stomta index
1	*Citrullus lanatus*	Anomocytic	3.27	2.16	2.22	1.05	22.00	50.30	30.32
2	*Cucumis sativus*	Anisocytic	3.12	2.10	2.28	1.02	21.10	52.20	28.59
3	*Cucurbita pepo*	Anomocytic	3.24	2.13	2.28	1.08	23.20	50.10	34.14
	LSD 5%		1.837 Ns	0.522 Ns	0.244 Ns	0.142 Ns	7.095 Ns	6.896 Ns	2.120 S

Key: Ns = not significant, s = significant

Table 4.1. Mean of stomata measurement in some members of Cucurbitaceae at adaxial (upper) surface of two species (x400) mm

S/N	Species	Type of stomata	Length of stomata	Breadth of stomata	Pore length	Pore breadth	No. of stomata per field	No. of epidermal cells per field	Stomata index
1	*Citrullus lanatus*	Anomocytic	3.13	2.13	2.16	1.02	14.00	50.30	23.03
2	*Cucurbita pepo*	Anomocytic	3.18	2.06	2.06	1.05	13.50	49.30	21.47

Difference of means 0.23 1.03 1.17 0.58 0.26 0.16 0.93, table value at 5% = 2.265

Similar observation was made and many researchers have found the presence or absence and types of trichomes on the epidermal surfaces as classification tool [20,5,23,28].

4.3 Stomata Ontogeny

In all the species studied, mesogenous stomata development was observed. Stomata originate by a division resulting in an oblique wall in epidermal cell; stomata develop from a merestimoid which function directly as a guard mother cell. it is formed by asymmetrical division of protodermal cells, the first division occurs when a neutral protodermal cell becomes a meristemoid mother cell defining characteristics of the meristemoid mother cell in the asymmetrical division of the cell into one daughter cell called ameristemoid and another larger daughter cell [26]. The small meristemoid cell has two options; it may remain in the same state or undergo a second cell transition to guard mother cell which divides asymmetrically just like its mother to generate one small daughter cell. The small meristemoid daughter cell undergoes a second division to a guard mother cell. The guard meristemoid mother cell undergoes a simultaneous third cell division to become a mature guard cell, the mature guard cell does not divide, it exhibits a characteristic shape which allows it to carry out its highly specialized role in gas exchange [25,28].

4.4 Stomata Measurement

The mean of lower surface measurement was compared at 5% level of significance. Significant difference was found in stomata index among the three species studied. The variations in stomatal index observed in this study can be reasonably employed in Separation of the species [5,19,29,15], systematic relevance has been documented [24]. No significant difference was found in terms of length and breadth of stomata, pore length and breadth of the stomata. For upper surface, t-test was used to test the differences between *Citrullus lanatus and Cucurbita pepo*. After comparing the means of the two species, no significant difference was found as the table value is higher than the calculated value in the two species.

5. CONCLUSION

Based on leaf epidermal features the species studied can be distinguished from one another by stomata types. The data from this study has provided evidences for delimiting of the species to some extent based on stomata type and index. The presence of various stomata is of taxonomic interest in this study which can be used to distinguish the species; two of which are amphistomatic and another is hypostomatic. Based on stomata ontogeny, the species showed one type of development which is mesogenous. Similarities were observed based on development of stomata [25].

COMPETING INTERESTS

Authors have declared that no competing interests exist.

REFERENCES

1. Oyeleke MO, Abdulrahaman AA, Oladele FA. Stomatal anatomy and transpiration rate in some afforestation tree species. NISEB J. 2004;4:83-90.
2. Olorode O. Taxonomy of West African flowering plants. Longman Publishing Company, New York. 1984;25-32.
3. Davis PA, Heywood VH. Principle of angiosperm taxonomy. Oliver and Boyd, Edinburgh. 1963;210-230.
4. Dutta AC. Botany for degree students (26th Ed.) Oxford University Press, London. 2009;155,565-566.
5. Adedeji O, Jewoola DA. Importance of leaf epidermal characters in the asteraceae family. Notulae Botanicae. Horti, Agrobotani cluj-Napoca. 2008;36(2):7-16.
6. Hardie J. Autum leaves seen through herbivour eye. Proc. Royal Soc. Biology. Sci. 2009;276:121-127.
7. Saheed SF, Illoh HC. A taxonomic study of some species in *Cassiinae* (Leguminosae) using leaf epidermal characters. Not. Bot. hort. Agrobot. 2010;38:21-27.
8. Van Wyk AE, Robbertse PJ, Kok PDF. The genus *Eugenia L.* (Myrtaceae) in Southern Africa: The structure and taxonomic value of stomata. Botanical Journal Linnean Society. 1982;84:41-56.
9. Vishnu Swarup. Vegetable science and technology in India. Kalyani Publishers. New Delhi, India. 2006;448-483 .
10. Cuttler EG. Plant Anatomy part 1 cells and tissues contemporary biology. Published by Great Britain Williams cloves and sons limited, London. 1978;95-127
11. Essiett UA, Bala DN, Agbakali JA. Leaf epidermal studies in three Nigerian species

of aspilia (*Heliantheae* – Asteraceae) and two hybrids. Nigerian Journal of Botany. 2010;8:25-33.
12. Nadeau JA, Sack FD. Stomatal development in Arabidopsis. In Arabidopsis. Oxford University press. 2002;55-62.
13. Shpak ED, Mc Abee JM, Pillitteri LJ, Torii KU. Science. 2005;309:290-293-3.
14. Pant DD. Stomata ontogeny and other homologous structures. Clarendon press Oxford. 1965;11-24.
15. Stace CA. Cuticular studies as an aid to plant taxonomy. Bull Br. Mus. Bot. 1965;4: 1-78.
16. Clive AS. Plant taxonomy and biosystematics. Cambridge University Press. 1991;81-82.
17. Abdulrahaman AA, Oladele FA. Stomatal complex types and epidermal cells in *Jatropha species* L. (Euphorbiaceae). Nigerian Journal of Pure Applied Science. 2009;23:2160-2163.
18. Jeffrey C. Notes on Cucurbitaceae, including a proposed new classification of the family. Kew Bull. 1962;15:337-371.
19. Metcalfe CR, Chalk L. Anatomy of the dicotyledons, systematic anatomy of the leaf and stem. 1st. Ed. Clarendon Press Oxford. 1950;1:82-84.
20. Abdulrahaman AA, Oyedotun RA, Oladele FA. Diagnostic significance of leaf epidermal features in the family Cucurbitaceae. Insight Botany. 2011;1: 22-27.
DOI: 10.5567/BOTANY-IK.2011.22.27.
21. Hutchinson J, Dalziel JM. Flora of West Tropical Africa ed. 2. London. 1954; 1(1):243.
22. Kolb D, Muller M. Light, conventional and environmental scanning electron microscopy of the trichomes of *Cucurbita pepo* ssp. *pepo* var. *styriaca* and histochemistry of glandular secretory products. Annals of Botany. 2004;94: 515-526.
23. Essiett UA, Akpabio E. Aniesua & Silas; 2012. International Journal of Biotechnsol and Allied Science. 2009;4(1):424-432.
24. Abdulrahaman AA, Oladele FA. Stomatal complex types in some Nigerian species of *Ocimum* hyptis and tinnea. Nigerian Journal of Pure Applied Science. 2003;23: 2160-2163.
25. Gills LS, Karatela YY. Epidermal structure and stomatal ontogeny in some Nigerian Cucurbitaceae. Published by: Botanischer Garten and Botanisches Museum, Berlin-Dahlem. 1982;303-310.
Available:http://www.jstor.org/stable/3995942
26. Saadu RO, Abdulrahaman AA, Oladele FA. Stomatal complex types and humidification potential of some root tuber species. African Journal of Plant Science. 2009;3:107-112.
27. Watson I. The taxonomic significance of certain anatomical observation on *ericaceae.* New Phytologist. 1964;63: 274-280.
28. Inamdar JA, Gangadhara M. Structure, ontogeny, classification and organographic distribution of trichomes in some Cucurbitaceae. Feddes Repertorium. 2008;86:307-320.
29. Olatunji OA. Practical manual for plant anatomy. Obafemi Awolowo University, Ile-ife, Nigeria (Manuscript). 1983;14-19.

20

Effects of Different Sewage Sludge Concentrations on the Soil Microorganism Sensibility and Metals Accumulation during *Helianthus annuus* L. Cultivation

G. M. B. Bohm[1*], T. L. Oliveira[2], E. M. Bohm[2], E. B. Bohm[2] and P. J. Sanches[3]

[1]Departamento de Gestão Ambiental, Instituto Federal Sul-rio-grandense- IFSul, Praça 20 De Setembro 455, Pelotas, 96015-360, RS, Brazil.
[2]Graduando no Curso Superior de Gestão ambiental, Instituto Federal Sul-rio-grandense- IFSul, Praça 20 De Setembro 455, Pelotas, 96015-360, RS, Brazil.
[3]Departamento de Química, Instituto Federal Sul-rio-grandense- IFSul, Praça 20 De Setembro 455, Pelotas, 96015-360, RS, Brazil.

Authors' contributions

This work was carried out in collaboration between all authors. Authors GMBB and TLO designed the study, wrote the protocol and wrote the first draft of the manuscript. Authors GMBB and EBB managed the analyses of the study. Authors EMB and PJS performed the statistical analysis. Authors GMBB and EBB managed the literature searches, analyses of the study performed and discuss the conclusion. All authors read and approved the final manuscript.

Editor(s):
(1) Marco Trevisan, Institute of Agricultural Chemistry and Environmental Research Centre Biomass, Faculty of Agriculture, Catholic University of the Sacred Heart, Italy.
Reviewers:
(1) A. Manikandan, ICAR-Central Institute for Cotton Research, Nagpur, India.
(2) Chris O. Nwoko, Federal University of Technology, Nigeria.
(3) Petigrosso, Lucas Ricardo, Universidad Nacional de Mar del Plata, Argentina.

ABSTRACT

The agricultural use of sewage sludge have been presented as an alternative both to reduce the volume of this environmental liability as to obtain a product to be used in agricultural soils as fertilizer. However, when the household sewage is also composed of water from the industrial area, the sludge is richer in potentially toxic elements. Hence, the present study aimed to evaluating the effects of different concentrations of sewage sludge on soil microbial activity and contents of Cd, Cu, Cr, Pb and Zn in soil, plant and seed. The experiment was conducted at the greenhouse during

**Corresponding author: E-mail: bohmgiani@gmail.com*

the crop season of 2012 to 2013, with evaluations of soil total organic carbon, microbial biomass carbon, metabolic quotient and heavy metals in soil, plant and seed. According to the results, application of 20 Mg ha^{-1} of sewage sludge promoted higher microbial activity, lower metabolic quotient and in increased production of dry plant matter. Resulted in 480.77 $\mu g\ g^{-1}$ of microbial carbon and 9.75% of soil organic carbon. Heavy metal contents in seeds were 1.29, 14.26, 1.52 and 80.57 mg kg^{-1} of Cd, Cu, Pb and Zn, respectively, with the lead content above the maximum extent recommended in food by WHO (Word Health Organization).

Keywords: Sunflower; microbial activity; waste; residue.

1. INTRODUCTION

The sunflower (*Helianthus annuus L.*) is yearly crop native of North America [1]. It is one of the five oilseed crop producing edible vegetable oil in the world, It is also used as a source of protein for animal feed, grain pastry, honey production, crop rotation and cycling nutritional and production of biodiesel [2]. The area cultivated with sunflower in Brazil during the season 2012/2013, was approximately 2.37 millions of hectares [3]. With increased sunflower production also comes the increased land degradation by use of pesticides. One way to reduce this impact is the use of waste as a nutrient for plants in this context arises the potential for applying tons of solid waste generated by the treatment of industrial and domestic sewage in the world [4]. Sewage sludge has considerable percentage of organic matter and essential elements for plants and can replace, even partially, mineral fertilizers, playing an important role in agricultural production and the maintenance of soil fertility [4]. The sewage treatment is able to reduce river pollution and improve public health. However, several sewage treatment projects do not include the final destination of sludge produced and thus cancel in part the benefits of collection and treatment of wastewater [5].

Although the agricultural use of sewage sludge presents himself as one of the most attractive alternatives for final disposal of this waste, potentially toxic elements can limit your application, due to the possibility of soil contamination of aquatic systems and the atmosphere, increasing the risk of their transfer to the food chain as highlighted by [6]. The sludge from sewage treatment plants when destined for domestic sewage treatment, it is framed as waste Class II A or even B according to the Brazilian Association of Technical Standards NBR 10004 [7], allowing its use with increased safety in agriculture. This is possible since the waste is evaluated for agronomic value, and the usage limits are respected, determining the potential impacts of their application on soil properties, to meet technical and safety criteria for human and environmental health established for agricultural use of such material [8].

Interest in the production of organic compounds from these materials have been presented as an alternative both to reduce the volume of this environmental liability as to obtain a product to be used in agricultural soils [9]. However, when the household sewage are also composed of water from the industrial area, the sludge is richer in potentially toxic elements (PTE), including metals: cadmium (Cd), copper (Cu), chromium (Cr), lead (Pb) and zinc (Zn),reference prevention values are respectively 1.3, 60, 75, 72, 300 mg kg^{-1}, according to the Brazilian National Environmental Council (CONAMA) Resolution No. 420 of 2009 [10]. Repeated application of sewage sludge on agricultural soils can raise the levels of organic and inorganic compounds in the soil, requiring periodic evaluations in order to avoid environmental contamination [11].

In the face of these problems, this work was developed with the objective of evaluating the effect of different concentrations of sewage sludge on soil microbial activity and Cd, Cu, Cr, Pb and Zn contents in soil, plant and seeds. The hypothesis of this study is that the application of sewage sludge on agricultural soils enhances soil microbial activity and the concentration of heavy metals in soil and food.

2. MATERIALS AND METHODS

The experiment was conducted at the greenhouse of the experimental area of the Federal Institute of Rio Grande do Sul in the crop season of 2012 to 2013. The soil is classified as dystrophic Red-Yellow Argisol (U.S. soil taxonomy). Soil and sludge chemical and physical properties were obtained according to [12] before the experiment soil properties in the following Table 1.

The sewage sludge used was obtained from Sewage Treatment Plant - SANEP (Pelotas Sanitation stand-alone service) after treatment by anaerobic fluidized bed reactor - RALF. The plant receives around 30% of the domestic sewage from the city of Pelotas/RS. It was collected in March 2013 in a dry day with ambient temperature of 20°C and kept refrigerated. Its chemical and heavy metal properties were measured according to [12,13] (Table 1).

Table 1. Soil and sludge chemical and physical properties

Variable	Soil	Sludge
pH (H20)	5.6	5.21
clay (%)	16	-
organic matter (%)	1.1	-
N (mg dm^{-3})	-	39,960
P (mg dm^{-3})	6.5	1,560
K (mg dm^{-3})	68	1,920
Ca ($cmol_c$ dm^{-3})	1.5	22,120
Mg ($cmol_c$ dm^{-3})	0.7	1,180
Al ($cmol_c$ dm^{-3})	0.2	-
CEC ($cmol_c$ dm^{-3})	6.6	-
Cr (mg kg^{-1})	5.56	32.48
Zn (mg kg^{-1})	0.05	404.31
Cu (mg kg^{-1})	0.2	133.38
Pb (mg kg^{-1})	0.5	175.47
Fe (mg kg^{-1})	-	575.90

Chemical analysis of the sludge allowed to frame that waste as Class II A or even B according to the Brazilian Association of Technical Standards NBR 10004 [7], allowing its use with increased safety in agriculture. Current law requires that waste is disposed in special landfills denominated ARIP (Landfill of Hazardous Industrial Waste). The amounts used in this work, however, amounted to Cr additions below the maximum limits accepted by Resolution 375 of the National Environment Council Environment [10] for sludge to be applied to agricultural soils that are 1000 kg ha^{-1} Cr.

The experimental design was completely randomized with four treatments and a control, with four replications: T1 - control (no fertilizer, no sludge); T2 - NPK only; T3 - 4 Mg ha^{-1} of sludge; T4 - 12 Mg ha^{-1} of sludge; T5 - 20 Mg ha^{-1} of sludge. The experiment consisted of 20 plastic pots with 10 kg capacity, with drains and dishes. Each pot received a layer of about 3 cm of gravel. Soil and sewage sludge were air dried, homogenized and sieved (2 mm mesh size). The sludge concentration of T3 was determined to meet the N requirement of sunflower variety [13], assuming that 1/3 of N contained in the residue would be available for plants. The concentrations of T4 e T5 were 300 and 500% of the T3 concentration, respectively. For T2, the NPK fertilizers were urea, triple superphosphate and potassium chloride, and their concentrations were determined to meet the sunflower requirements according to SBCS/NRS [13], which were: N = 40 kg ha^{-1}; P_2O_5 = 180 kg ha^{-1}; and K_2O = 170 kg ha^{-1}. For pH adjustment, all pots received 3 t ha^{-1} of lime in the form of a mixture of $MgCO_3 + CaCO_3$ (2:1).

In each pot, about four sunflower seeds (Agrobel 369, 90% germination) were planted at a 2 cm depth. Fifteen days after germination, thinning was done leaving about two plants showing well-developed leaves in each pot. Soil moisture was kept near field capacity for a period of 90 days.

At 70 days after planting (DAP), four soil samples were collected from each pot at 0-10 cm depth and about 10 cm from the plants, homogenized and analyzed for soil total organic carbon (TOC), microbial biomass carbon (MBC), basal respiration (BR), and heavy metals. The TOC content was determined by the Walkley-Black method [9], and the MBC according to [14]. For MBC, soil microorganisms were killed by irradiation at 2450 Mhz for four minutes instead of fumigation with chloroform. The MBC was determined by the difference of C between the irradiated (Ci) and non-irradiated (Cni) soil sample, corrected by a correction factor (Kc) of 0.33 according to [14], as follows: Equation 1) MBC = (Ci - Cni)/Kc. The MBC to TOC ratios were also assessed.

Basal respiration rate (BR) is a measure of microbial activity that consists of quantifying the CO_2 released per hour from organic carbon decomposition by microbes. It was determined according to [14], where the quantity of CO_2 released from the soil is trapped in alkali during seven days of incubation and subsequently titrated against HCl. For the calculation of CO_2 efflux the equation 2) BR = (B - S) x M x 4 was used, where B is the volume of HCl to titrate the blank flask; S is the volume of HCl to titrate the remaining NaOH from the soil sample; M is the HCl concentration; and 4 (standard value) is the equivalent gram of respired carbon by soil microorganisms. The BR to MBC ratios, or metabolic quotients (qCO2), were derived.

When the plants reached the grain formation stage, after 86 days after planting, they were removed from the pot and proceeded then to the careful cleaning of the root system. Plants were harvested at the height of arms to separate air and roots. The shoots were dried at 60℃ to constant weight to determine the dry matter, and shoot the ground in a Wiley mill.

The shoots and sunflower seeds were analyzed for heavy metals (Cd, Cr, Cu, Ni, Pb and Zn) according to [15] using atomic absorption spectrometry.

Data were subjected to analysis of variance and comparison of means by the Tukey method (0.05 probability), using the Statistix 8.0 software (Analytical Software Inc., Tallahassee, FL, USA).

3. RESULTS AND DISCUSSION

The highest microbial biomass C (MBC) was observed in the T5 treatment, which was 38% larger than in the control treatment (T1) and 45% larger than in the NPK only treatment (T2), the average for MBC was 357.8 μg g^{-1} (Table 2). The MBC increases were associated with organic matter increases from sludge applications, according [16] the application of the sewage sludge in soil have possibility of recycling valuable components such as organic matter, N and P. In contrast, Sullivan et al. [17] did not observe changes in MBC as an effect of the application of sewage sludge (between 2.5 and 30 Mg ha^{-1}) and suggested that, in this case, a rapid adaptation of soil microbes occurred. Other negative effects on biological soil properties, such as MBC, from applications of sludge contaminated with heavy metals are reported in the literature [18], and show the harmful impact of these chemicals on soil microbiology. These deleterious effects on soil microbial communities were not observed in this study.

The largest TOC contents were observed in T5 and were 59.02% larger than in T1 and 32.8% larger than in T2. These results correlate with the sludge applications and agree with the observed MBC trends. Similar results were obtained by [19], where increasing concentrations of anaerobic sludge from parboiled rice effluent treatment station resulted in larger TOC and MBC.

The MBC/TOC ratio, which indicates the percentage of TOC represented by the microbial carbon, decreased relative to the control in T1 and T2. For qCO_2, the largest values were obtained by T2 and the smallest values by control T1, T4 and T5 treatments. The qCO_2 has been used as a biological efficiency indicator of the soil, since as the microbial biomass becomes more efficient, more carbon is incorporated in the microbial biomass and less carbon is released as CO_2 due to respiration [14]. In general, sewage sludge application resulted in qCO_2 reduction compared to T2. It is recognized that stress factors (herbicides, toxic metals and pH) cause microbial inefficiency [14]. A potential effect of the application of sewage sludge in the soil is the inhibition of soil microorganisms with detriment of important functions such as mineralization and immobilization processes [20].

However, in this study the application of sludge increased microbial efficiency (smaller qCO_2), especially at large concentrations. This is confirmed by the larger MBC/TOC ratios.

The largest DM contents were observed in T4 and T5, witch was 41.55% larger than in T1 and 40.1% larger than in the NPK only treatment (T2). These results correlate with the sludge applications (Table 2). The sludge addition favored the development of plants resulting in increased amount of dry matter. Similar results to those in soils treated sewage sludge application was obtained by [21,22].

The higher plant growth was observed in treatments with sludge applications and by the NPK only treatment (T2) (Fig. 1). During the first developments phase of plants (0-20 days) T5

Table 2. Soil Microbial Biomass Carbon (MBC), Total Organic Carbon (TOC), ratio of MBC/TOC, metabolic quotient (qCO_2) and Dry Matter (DM) from the different treatments 70 days after planting. Average values followed by the standard deviation from the four replicates

Treatments	MBC (μg g^{-1})	TOC %	MBC/TOC	qCO_2 (10^{-3})	DM(g)
T1-control	267.74±26.12d	4.50±1.59cd	1.97±0.87c	17.54±2.54c	2.36±0.34c
T2-NPK only	294.97±27.98c	6.55±1.68c	2.42±0.92bc	35.72±2.34a	2.87±0.45b
T3-sludge 4 Mg ha^{-1}	346.77±31.35b	7.90±0.87bc	3.43± 0.46b	27.63±3.05b	3.59±0.71a
T4-sludge 12 $Mgha^{-1}$	398.73±29.04b	8.05±0.97b	3.69±1.29ab	19.29±3.89c	4.16±0.54a
T5-sludge 20 $Mgha^{-1}$	480.77±16.67a	9.75±1.02a	5.13±1.40a	14.71±4.02c	4.38±0.86a

Means followed by the same letter in the same column are not statistically different according to the Tukey test at a probability < 0.05 treatment resulted in increased growth compared to other treatments. After 20 days, treatment with NPK represented a significant improvement in plant growth similar to the sludge treatments, did not differ to the applied sludge dosages.

In general, the treatments that received sewage sludge did not affect BR compared with treatments with NPK only (T2) and control (T1) (Table 2). There was increase in the production of CO_2 with increasing sewage sludge concentrations (Fig. 2), even though it was expected since MBC increased. Similar results were found by [23] when evaluating the effect of sewage sludge (0, 10, 20, 40, 80 and 160 Mg ha^{-1}) on the microbial activity of the soil, they did not found negative effect from the sludge applications.

The concentrations of Cd, Cu, Cr, Pb and Zn in soil were largest in T5 (Table 3). According to the Brazilian National Environmental Council (CONAMA) Resolution No. 420 of 2009 [10], reference prevention values for Cd, Cu, Cr, Pb and Zn are respectively 1.3, 60, 75, 72, 300 mg kg^{-1}. Therefore, the Cd in treatments with sludge and Pb in treatment T5 concentration was above the prevention limit. Larger heavy metal contents in soils with sludge applications were also observed by [9]. The metals are also present in the organic fraction of the soil, and are released to the soil solution upon the decomposition of organic matter. The soil solution provides a means for chemical reactions and transfer of metals between the soil and organisms [24]. According to the same authors, metals and non-metals are involved in a series of complex biological and chemical interactions, and the most important factors affecting their mobility are the pH and presence and concentration of inorganic and organic ligands, including humic and fulvic acids, exudates and nutrients. Furthermore, redox reactions are of great importance in the control of the mobility and toxicity of various elements such as Cr, Pb, Ni and Cu.

In parts of the plant the Cd, Cu, Pb and Zn average for treatment with sludge were 3.06, 20.12, 9.5 and 126.63 mg kg^{-1} respectively.

The contents were generally proportional to the sewage sludge additions, with largest values in T5 (Table 4). In parts of the plant the Cd, Cu, Pb and Zn average for treatment with sludge were 3.06, 20.12, 9.5 and 126.63 mg kg^{-1} respectively. The contents were generally proportional to the sewage sludge additions, with largest values in T5 (Table 4). Low levels of copper were detected in parts of the plant. This result can be attributed to the strong complexation that this element may suffer from organic matter and to the antagonism between Cu and Zn in soil solution [25]. Chromium was not detected in any of the treatments in parts of the plant (Table 4). This could be due to the soil adsorption and chelation capacity, making Cr not available for the plant [26]. This metal can be strongly complexed by humic acids, thus reducing its availability in the soil [25].

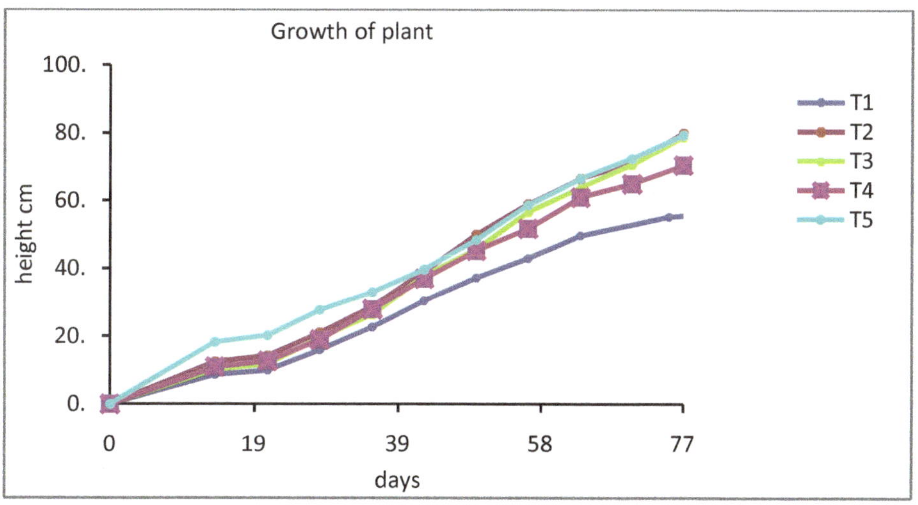

Fig. 1. Growth of plant from the different treatments 77 days during planting. T1- Control, T2- NPK only, T3- sludge 4 Mg ha^{-1}, T4- sludge 12 Mg ha^{-1} and T5- sludge 20 Mg ha^{-1}

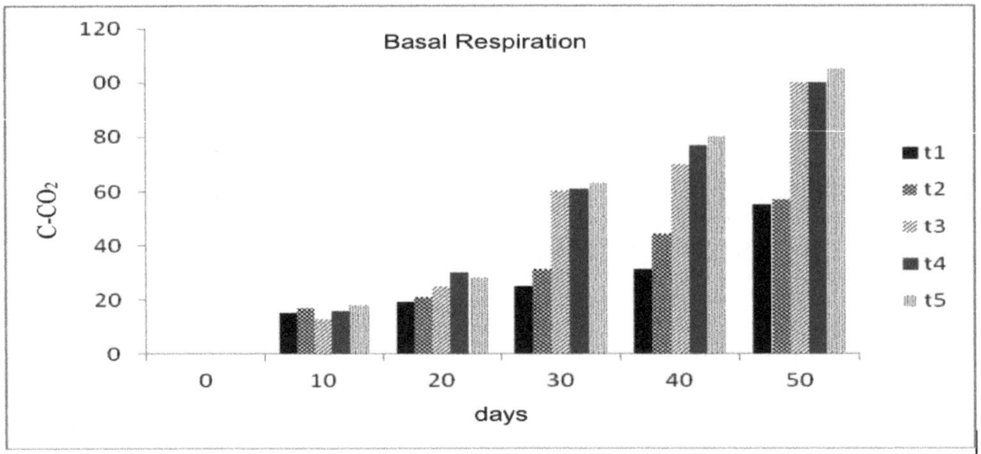

Fig. 2. Effect of sewage sludge in Argisol on the cumulative release carbon (CO_2) in the soil in 50 days. T1- Control, T2- NPK only, T3- sludge 4 Mg ha^{-1}, T4- sludge 12 Mg ha^{-1} and T5- sludge 20 Mg ha^{-1}

Table 3. Soil heavy metal (Cd, Cu, Cr, Pb and Zn) contents from the different treatments 70 days after planting. Average values followed by the standard deviation from the four replicates

Treatments	Cd	Cu	Cr	Pb	Zn
	mg kg^{-1}				
T1-control	1.02±0.76c	6.32±0.15c	2.67±0.86c	0.45± 0.12e	1.45±0.34d
T2-NPK only	1.13±1.23c	7.35±0.23c	3.44±0.98c	2.13±0.56d	55.52±10.45c
T3-sludge 4 $Mgha^{-1}$	4.36±1.43b	11.60±1.13b	6.34±1.22b	35.21±8.34c	155.53±13.60b
T4-sludge 12$Mgha^{-1}$	5.02±1.75a	13.68±1.74a	8.24±1.45a	48.35±12.35b	182.32±18.90a
T5-sludge 20 $Mgha^{-1}$	6.40±1.05a	15.45±1.87a	9.46±1.04a	75.24± 9.21a	197.76±16.43a
DL	0.051	0.02	0.014	0.06	0.03
QL	0.192	0.13	0.04	0.08	0.12

DL: detection limit; QL: quantification limit. Means followed by the same letter in the same column are not statistically different according to the Tukey test at a probability < 0.05

Lead contents in parts of the plant increased with sludge additions, with T5 showing the higher values (Table 4). T1 and T2 had statistically significant smaller Pb contents than all sludge treatments. Zn contents in parts of the plant increased with sludge additions. In [27], Zn contents in the parts of the plant except the seeds also increased with the sewage sludge additions. This suggests that a fraction of Zn contained in the sewage sludge becomes available to and is absorbed by the plant. Another study reported increases in Zn levels in parts of corn plants grown in soil treated successively with sewage sludge [28]. However, factors such as stage of plant development, exposure time and the presence of other ions in solution interfere with the distribution of metals in plants. A similar result was found by [28], evaluating the effects of application of sewage sludge on soil with improved acidity, found increases in Zn concentration in the leaves and kernels of corn, showing that other sludge constituents can alter the chemical processes occurring in the soil in such extension that liming pass to provide little protection against the absorption of this metal by plants.

In seeds the Cd, Cu, Pb and Zn the average for treatment with sludge were 1.29, 14.26, 1.52 and 80.57 mg kg^{-1}, respectively. The contents were generally proportional to the sewage sludge additions, with largest values for Cu and Zn in T5. The results obtained for Cu and Zn in the seeds are similar to those obtained by [29] where the transfer of heavy metals were those with the highest transfer rates to the plant.

Chromium was not detected in any of the treatments in seeds (Table 5).

Lead contents in seeds were increased with sludge additions, with T5 showing the largest values (Table 4). These high Pb levels found in parts of the plant are considered harmful for human consumption, according to Ordinance No. 685 of 1998 of the Ministry of Health [30], which

Table 4. Plant heavy metal (Cd, Cu, Cr, Pb and Zn) contents from the different treatments 70 days after planting. Average values followed by the standard deviation from the four replicates

Treatments	Cd	Cu	Cr	Pb	Zn
			mg kg^{-1}		
T1-control	0.59±0.19b	6.30±0.13d	Nd	0.35±0.21c	2.12±0.45e
T2-NPK only	1.03±0.42ab	9.50±1.50c	Nd	1.57±0.54c	32.61±9.34d
T3-sludge 4 $Mgha^{-1}$	2.87±1.55a	16.32±1.63b	Nd	8.42±1.23b	95.53±11.32c
T4-sludge 12$Mgha^{-1}$	2.93±1.31a	21.03±0.52a	Nd	9.57±0.84ab	120.52±20.35b
T5-sludge 20 $Mgha^{-1}$	3.40±1.03a	23.02±1.72a	Nd	10.51±0.87a	154.85±15.87a
DL	0.08	0.04	0.02	0.10	0.06
QL	0.39	0.24	0.08	0.20	0.25

DL: detection limit; QL: quantification limit. Means followed by the same letter in the same column are not statistically different according to the Tukey test at a probability < 0.05. ND- Not detected

Table 5. Seed heavy metal (Cd, Cu, Cr, Pb and Zn) contents from the different treatments 70 days after planting. Average values followed by the standard deviation from the four replicates

Treatments	Cd	Cu	Cr	Pb	Zn
			mg kg^{-1}		
T1-control	0.13±0.79a	1.10±0.12d	Nd	0.28±0.19	3.48±1.02d
T2-NPK only	0.43±0.28a	6.56±1.43c	Nd	1.08±0.94	27.15±7.14c
T3-sludge 4 $Mgha^{-1}$	1.26±0.69a	13.21±1.28b	Nd	1.29±0.59	62.80±9.80b
T4-sludge 12$Mgha^{-1}$	1.33±0.78a	12.33±0.73b	Nd	1.55±0.96	82.00±5.80a
T5-sludge 20 $Mgha^{-1}$	1.30±0.50a	17.24±1.35a	Nd	1.72±0.47	96.91±11.33a
DL	0.08	0.04	0.02	0.10	0.06
QL	0.39	0.24	0.08	0.20	0.25

DL: detection limit; QL: quantification limit. Means followed by the same letter in the same column are not statistically different according to the Tukey test at a probability < 0.05

recommend maximum allowed Pb values in food of 0.05 to 2.0 mg kg^{-1}, but there is no direct reference to vegetables. The Codex Alimentarius [31] recommends a maximum limit of 0.3 mg kg^{-1} of Pb in plants. The Pb contents found in seeds were also above this limit.

4. CONCLUSION

Sewage sludge applications, during the cultivation of sunflower, provided greater soil microbial activity, resulting in increased microbial biomass carbon, soil organic carbon decrease in the metabolic quotient and in increased production of dry plant matter. The contents of heavy metals Cd, Cu, Cr, Pb e Zn in soil were higher in treatment of sludge with higher dose. The concentrations of Cd and Pb in soil resulted in values above the limit of prevention. The lead content found in the seeds was above the maximum extent recommended by the Codex Alimentarius.

ACKNOWLEDGEMENTS

Financial support from the Federal Institute Sul-rio-grandense (IFSul) and Brazilian National Research Council (CNPq) are greatly appreciated.

COMPETING INTERESTS

Authors have declared that no competing interests exist.

REFERENCES

1. CFIA- Canadian Food Inspection Agency- The Biology of *Helianthus annuus* L. Available:http://www.inspection.gc.ca/plants/plants-with-novel-traits/applicants/directive-94-08/biology-documents/helianthus-annuus-l-/eng/1330977236841/1330977318934 (Accessed 29 out 2015)
2. Mourad LA. Main crops for obtaining vegetable oils fuels in Brazil. An. 6. Enc. Energ. Countryside; 2006. Available:http://www.proceedings.scielo.br/scielo.php?pid=MSC0000000022006000200029&script=sci_arttext Portuguese
3. Conab - National Supply Company - Monitoring Brazilian crop: grains, ninth survey, June 2013. Brasília: Conab; 2013.
4. Nascimento CWA, Fontes RLF. Correlation between oxisols characteristics and parameters of copper and of zinc adsorption equations. Journal of Soil Science. 2004;28:965-971.

5. Molina MV. Nitrogen and heavy metals in the soil and eucalyptus fifty-five months after the application of sewage sludge. 66f. Thesis (MS in Agronomy) - College of Agriculture "Luiz de Queiroz", Piracicaba; 2004.
6. Nogueira TAR, Sampaio RA, Fonseca IM, Ferreira CS, Santos SE, Ferreira LC, Gomes E, Fernandes LA. Heavy metals and pathogens in maize and cowpea intercropping system fertilized with sewage sludge. R. Bras. Eng. Agric. Amb, 2007;11:331-338. Portuguese
7. Brazilian Association of Technical Standards (ABNT) NBR 10004: Solid Waste Classification. Rio de Janeiro. 2004;71. Portuguese
8. Trannin ICB, Siqueira JO, Moreira FMS. Chemical and physical attributes of a soil treated with an industrial biosolid and cultivated with corn. R. Brazilian Environmental Agricultural Engineering, Campina Grande. 2008;12(3):223-230. Portuguese
9. Araujo JC, Nascimento CWA. Fractionation and availability of zinc in soils by different extractors incubated with sewage sludge. Soil Science Journal. 2005;29:977-985. Portuguese.
10. Brazil, National Environmental Council; Resolution No. 420 of 28/12/2009. Available:http://www.mma.gov.br/port/conama/res/res09/res42009.pdf (Accessed 15 July 2015)
11. Villanueva FC, Boaretto AE, Firme LP, Muraoka T, Nascimento Filho VF, Abreu Junior CH. Zinc bioavailability estimated by isotopic method in soil treated with sewage sludge. New Chemistry. 2012; 35(7):1348-1354.
12. Tedesco MJ, Bissani CA, Bohnen H, Volkweiss SJ. Soil analysis and other plant materials. Porto Alegre: Federal University of Rio Grande do Sul. 1995;174.
13. SBCS / NRS- Commission of Chemistry and Soil Fertility - RS / SC. Manual fertilization and liming in the states of Rio Grande do Sul and Santa Catarina. 10th ed. Porto Alegre. 2004;394. Portuguese.
14. Bohm GMB, Schneider L, Castilhos DD, Agostinetto D, Rombaldi CV. Weed control, soil microbial biomass and metabolism due to the application of glyphosate and imazethapyr in soybeans. Semina: Agricultural Sciences, Londrina. 2011;32(3):919-930.
15. Abreu MF, Furlani AMC, Abreu CA, Santos PH, Gonzalez AP. Total element concentration quantification in substrates. In: International Symposium on Soilless Culture and Hydroponics, 9th ed., Proceedings Almeria: Universidad de Almeria; 2004.
16. Martinez F, Cuevas C, Teresa M, Walter G, Iglesias CFC, Ingrid V. Urban organic wastes effects on soil chemical properties in degraded semiarid ecosystem. In: Seventeenth WCSS, Symposium No. 20, Thailand. 2002;1–9.
17. Sullivan TS, Stromberger ME, Paschke MW, Ippolito JA. Longterm impacts of infrequent biosolids applications on chemical and microbial properties of a semiarid rangeland soil. Biology and Fertility of Soils. 2006;42:258-266.
18. Kizilkaya R, Bayrakli B. Effects of N-enriched sewage sludge on soil enzyme activities. Applied Soil Ecology. 2005; 30:192-202.
19. Vieira GD, Castilhos DD, Castilhos RMV. Attributes microbial soil after addition of anaerobic sludge from rice parboiling effluent treatment station. Soil Science Journal. 2011;35:543-550. Portuguese.
20. Souza CA, Reis Júnior FB, Mendes IC, Lemainski J, Silva JE. Sewage sludge in biological soil properties and on nodulation and soybean production. Brazilian Journal of Agricultural Research. 2009;44:1319-1327. Portuguese.
21. Cunha AM, Cunha GM, Sarmento RA, Cunha GM, Amaral JTF. Effect of various substrates on seedling development sp. Revista acacia tree. 2006;30(2). Portuguese
22. Biondi CM, Nascimento CWA. Accumulation of nitrogen and dry matter production of plants in soils treated with sewage sludge. Magazine Caatinga. 2005;18:123-128.
23. Cardoso EJBN, Fortes Neto P. Applicability of biosolids in forestry plantations microbial changes in the soil. Jaguariuna: Embrapa Environment. 2000; 197-202. Portuguese.
24. Violante A, Cozzolino V, Perelomov L, Caporale AG, Pigna M. Chemistry change and mobility and bioavailability of heavy metals and metalloids in soil

environments. Jornal Soil Science Plant Nutrition. 2010;10(3):268-292.

25. Kabata-Pendias A, Pendias H. Trace elements in soils and plants. 4.ed. Boca Raton, CRC Press. 2010;413.
26. Bohm GMB, Betemps GR, Bierhals LM, Sanches Filho P, Schwans SM. Use sludge of sewage treatment plant in soybean cultivation in Argisol. Magazine Thema. 2014;11(1):12-22. Portuguese.
27. Nogueira TAR, Oliveira LR, Melo WJ, Fonseca IM, Melo MP, Melo GPM, Melo VP, Marques MO. Cadmium, chromium, lead and zinc in corn plants and oxisol after nine annual applications of sewage sludge. Journal of Soil Science. 2008; 32:2195-2207. Portuguese.
28. Rangel OJP, Silva CA, Bettiol W, Dynia JF. Effect of application of sewage sludge on heavy metal contents in corn leaves and grains. Journal of Soil Science. 2006; 30:583-594. Portuguese.
29. Chaves LHG, Mesquita EF, Araujo DL, França CP. Accumulation and distribution of copper and zinc in castor bean cultivar Paraguaçu BRS and plant growth. Magazine Environmental Engineering. 2010;7(3):263-277. Portuguese.
30. Brazil. Ordinance No. 685 / Ministry of Health of August 27, 1998. General principles for the establishment of maximum levels of chemical contaminants in food. Official Gazette [Federative Republic of Brazil], Brasilia, August 1998. Portuguese.
31. FAO - Food and Agriculture Organization. Codex Alimentarius Commission. Food additives and contaminants. Joint FAO/WHO Food Standards Programme; ALINORM 01/12A:1- 289; 2001.

21

Effect of Different Salt Concentrations on Germination and Seedling Growth of DIchondra (*Dichondra repens*)

Yaşar Özyigit[1*] and Sadik Çakmakci[2]

[1]Department of Horticulture, Akdeniz University, Korkuteli Vocational High School, Antalya, Turkey.
[2]Department of Field Crops, Faculty of Agriculture, Akdeniz University, Antalya, Turkey.

Authors' contributions

This work was carried out in collaboration between both authors. Author YÖ designed the study, performed laboratory experiments and made measurements. Author SÇ managed the statistical analyses and wrote the manuscript. Both authors read and approved the final manuscript.

Editor(s):
(1) Sławomir Borek, Faculty of Biology, Department of Plant Physiology, Adam Mickiewicz University, Poland.
Reviewers:
(1) Asghar Farajollahi, Gorgan University of Agriculture and Natural Resources, Iran.
(2) Mohamed Abdel-Ghany Aly Bekheta, National Research Centre, Giza, Egypt.

ABSTRACT

This experiment was conducted in the laboratory of the Department of Field Crops, Faculty of Agriculture, Akdeniz University in 2012. As a material the seeds of dichondra (*Dichondra repens*) which is a green area plant were used and the study was performed as randomized plots design with three replications in both plastic Petri dishes and peat media. Distilled water was used to prepare different NaCl concentrations (control, 5000, 7500, 10000, 15000, 20000 and 25000 ppm). The results showed that, germination was not observed in 20000 and 25000 ppm concentrations, so these doses were excluded from this study. While increasing salinity levels delayed the first germination day (first emergence day in peat experiment), germination rate (emergence rate in peat experiment), root length, shoot length, root weight, shoot weight and root/shoot ratio (both length and weight) characteristics decreased with increasing NaCl concentrations. This study shows that, although some plant characteristics of dichondra is effected negatively by application of salinity levels, the plant has tolerance to a certain amount of salinity (approximately 5000-7500 ppm) in germination and seedling growth stage.

**Corresponding author: E-mail: ozyigit@akdeniz.edu.tr*

Keywords: Dichondra; salinity; germination; root; shoot.

1. INTRODUCTION

Salinity is one of the main factors which has negative effects on plant production in many districts in the world. Soil salinization is an important problem especially in arid and semi-arid regions [1,2]. About 800 million hectares of land which is over 6% of the world's total land area are suffered from salinity stress, either by salinity (397 million ha) or the associated condition of sodicity (434 million ha). Also, of the current 230 million ha of irrigated land, 45 million ha are salt-affected (20%). Of the 1.5 billion ha of land farmed by dryland agriculture, 32 million (2%) are affected by secondary salinity to varying degrees [3]. However, based on salinity and sodicity criteria used in Turkish Advanced Soil Map surveys, salinity and sodicity were detected on about 1.5 million ha of the land resource [4].

However, salinity is an important problem in coastal area of the world. Because most of the coastal aquifers in the world face to the problem of saltwater intrusion due to excessive exploitation of groundwater that is primarily used for urbanization, industrialization and agricultural purposes [5]. Also seawaters inflow to rivers which riverbeds in below sea levels. So, the use of river water or groundwaters which near to riverbeds cause important salinity problem [6]. Also salinity is an important problem in turf area. For example, soil salinity is one of the most important problem in golf areas particularly in Mediterannean region [7] and salinity problem reached to important level in golf areas which established in near the sea [8]. These areas are extremly important for tourism, so increasing of turf species which can adapt to salinity conditions is very important.

Plants in germination and seedling growth stages are more sensitive to high salt concentrations [9] and prevention of water uptake is important problem especially in germination stage. So, germination and seedling growth stages are more considered to determine effect of the salinity on plants [10]. The main objective of this study was to determine the effects of different NaCl concentrations on seed germination and seedling growth in dichondra (*Dichondra repens*) which is a turf area plant.

2. MATERIALS AND METHODS

Dichondra (*Dichondra repens*) seeds were used as material in study. Dichondra is belong to *Convolvulaceae* and it can grow in warm-rainy climate. Also this plant adapt to shadow and partially to arid area. But this plant is not resistant to step on and crush [11].

This study, involved 7 different salt concentration were applied (5000, 7500, 10000, 15000, 20000 and 25000 ppm) in addition to control (distile water). Distile water and common salt (no-iodine and 99.5% purity) were used to prepare salt concentrations and 500 ml solution was prepare for each concentrations. Also electrical conductivity (EC) value of solutions were determined by using condactivity device (EC meter) in 20-21°C. Salt amount and EC value of solutions and salinity classification of this EC values according to Aydemir [12] are shown in Table 1.

Study was conducted in laboratory conditions both petri dish and peat. The experimental design was a Random-Parcel design with three replications. In petri experiments, plastic petri dishes (diameter is 9 cm) which were placed blotting paper were used. After this process, 4.5 ml solutions apply to petri dishes 3 replication for each salt concentrations and fifty seeds were placed in to petri dishes. Petri dishes were closed and wrapped with paraffin and placed in an incubator with the temperature of 20°C.

In peat experiment, the styrofor viols were filled with thin-textured peat. Its pH, salt concentration, nitrogen rate, phosphorus rate and potassium rate is 5.2-6, 0.3 $mg.lt^{-1}$, 30-80 $mg.lt^{-1}$, 20-40 and 30-90 $mg.lt^{-1}$, respectively. For each salt concentration, 3 lines of viol were splited up and each line included 8 seed bed and 9 ml solution was applied to each seed bed. After solution application, 1 seed were sown in each seed bed with 1 cm sowing depth and viols placed in an incubator with the temperature of 20°C.

In this study, measured germination properties and seedling growth are including: First germination day (day), germination percentage (%), root and shoot lenght (cm), root and shoot weight (g) and root/shoot rate (for lenght and weight). Germination was calculated as the number of seeds germinated (coleoptile emerged) as a percentage of the total number of seeds pleaced in each petri. Germination observations were made in the same time everyday and when least 5 seeds were fully germinated, this time was recorded as first germination day in petri experiment. Also, emergence day was determined instead of first germination day in peat experiment and when 3 seedlings were emerged on the peat, this time was recorded as first emergence day in peat experiment. At the end of the study, seedlings were cut at root crown parts, when they are 3 leaves stage and height and weight of roots and shoots were determined.

Data were analysed with PROC ANOVA procedure in SAS statistical programe and differences between applications were statistical tested. Duncan Multiple Range Test (DMRT) was used for comparison of means.

3. RESULTS AND DISCUSSION

In experiment, 7 different NaCl concentrations were applied as specified in material and methods divisions, but germination didn't observation or germination physiology couldn't complation at 20000 and 25000 ppm concentrations. So these concentrations were excluded in evaluation of results. Results of petri and peat experiments are shown in Tables 2 and 3, respectively.

Acorrding to results, first germination day and germination rate were significantly influenced by different NaCl concetrations. First germination day delayed depend on increasing of NaCl concentration. While the earliest germination were determined at control and 5000 ppm applications with 2 days, first germinations were in third days in other applications. Significant differences were observed for germination rate between applications. The highest germination rate was observed from control applications with 92%, while the lowest germination rate was obtained from 15000 ppm application with 54%. Also similar effects were observed on germination rate in other applications (5000, 7500 and 10000 ppm).

Similar results were obtained from peat experiment and earliest emergence was determined in control application with 3 days. Emergences were delayed depend on increasing of NaCl concentration and latest emergence were obtained from 15000 ppm application with 6 days. Also, in peat experiment, while highest germination rate were determined in control application with 100%, rates were reduced in high NaCl concentrations and lowest germination rate was observed at 15000 ppm application.

The results showed that high NaCl concentrations have negative effect on germination which is first grown stage of plants. Because water uptake prevented by osmotic pressure in high NaCl concentration. Nizam [13] recorded that water uptake rate of perennial grass (*Lolium perenne*) seeds decreased in case of NaCl concentration of solution in over 4 dS.m-1. Also in the same experiment, while highest germination rate was determined as 96.5% in control group, lowest rate was obtained as 20% in 24 dS/m. In another study, Camberato and Martin [14] reported that germination percentage of rought bluegrass (*Poa trivialis* L.) seeds decreased with increasing of NaCl concentration. The similar results were determined by Taiz and Zeiger [15] and Dai et al. [16] in different plants.

In study, better results were obtained generally in low NaCl concentrations for root and shoot features. Root lengths were determined between 2.91 cm and 0.51 cm depend on NaCl concentration in petri experiment. While the biggest root length was determined at control with 2.91 cm, lowest value was recorded at 15000 ppm application with 0.51 cm. Similarly, high shoot lengths were determined at control and 5000 ppm with 2.99 cm and 3.05 cm, respectively, shoot length was reduced to 1.23 cm at 15000 ppm.

Root length was influenced by NaCl concentrations and the highest root lengths was

Table 1. Salt amount for concentrations, EC values of concentrations and classification

Salt concentration (ppm)	Salt amount ($g.500\ ml^{-1}$)	Electrical conductivity (EC) values ($mS.cm^{-1}$)	Salinity classification
Distilled water		0.002	Non saline
5000	2.50	9.15	Highly saline
7500	3.75	13.61	Highly saline
10000	5.00	17.41	Severely saline
15000	7.50	25.4	Severely saline
20000	10.00	33.4	Severely saline
25000	12.50	39.3	Severely saline

Table 2. Effect of different NaCl concentration on some traits (Petri experiment)

	First germination (day)	Germination percentage (%)	Root length (cm)	Shoot length (cm)	Root weight (g)	Shoot weight (g)	Root/Shoot rate (for length)	Root/Shoot rate (for weight)
Distilled water	2 A	92 A	2.91 A	2.99 A	0.0053 A	0.014 BC	0.97 A	0.37 A
5000 ppm	2 A	83 B	2.53 AB	3.05 A	0.0056 A	0.018 A	0.83 A	0.31 A
7500 ppm	3 B	81 B	2.45 AB	2.76 AB	0.0049 A	0.017 AB	0.88 A	0.31 A
10000 ppm	3 B	79 B	2.06 B	2.48 B	0.0047 A	0.016 AB	0.83 A	0.31 A
15000 ppm	3 B	54 C	0.51 C	1.23 C	0.0027 B	0.012 C	0.42 B	0.23 B
F value	0.000.....	69.20**	19.31**	63.59**	10.63**	6.68**	15.52**	5.14*

*Mean differences with different letter within the same column are statistically significant *: p<0.05, **: p<0.01*

Table 3. Effect of different NaCl concentration on some traits (Peat experiment)

	First germination (day)	Germination percentage (%)	Root length (cm)	Shoot length (cm)	Root weight (g)	Shoot weight (g)	Root/Shoot rate (for length)	Root/Shoot rate (for weight)
Distilled water	3 A	100 A	5.01 A	3.41 A	0.0057 B	0.019 BC	1.48 A	0.30 A
5000 ppm	4 AB	92 AB	5.21 A	3.49 A	0.0066 A	0.023 A	1.49 A	0.21 A
7500 ppm	4 BC	84 BC	4.41 B	3.50 A	0.0055 BC	0.022 AB	1.26 B	0.25 A
10000 ppm	5 CD	75 DC	4.25 B	3.09 A	0.0048 C	0.021 AB	1.38 AB	0.23 A
15000 ppm	6 D	71 D	1.95 C	2.52 B	0.0038 D	0.017 C	0.77 C	0.22 A
F value	12.17**	14.02**	62.18**	8.20**	18.50**	6.71**	29.08**	0.83

*Mean differences with different letter within the same column are statistically significant *: p<0.05, **: p<0.01*

determined at 5000 ppm with 5.21 cm and in control with 5.01 cm in peat experiment. While the lowest root length were recorded at 15000 ppm with 1.95 cm, there was no differences between 7500 ppm and 10000 ppm in consideration of root lenght (Table 4). Shoot lengths were determined between 2.52 cm and 3.50 cm and while the lowest shoot length was obtained at 15000 ppm, there wasn't statistic difference between other NaCl concentration.

Characteristics related root and shoot are two of the most important selection criteria for determining of resistance of plants to different salinity conditions. Because roots which are directly contact with soil are uptake water and sent it to other organs of plant. Toxic effect of sodium chloride causes reduction in shoot growth through unbalanced nutrient uptake under the NaCl stress. Elongation of root and shoot can be decelerated by high salt concentrations because of prevent water and nutrient element uptake of plants [17]. This status explains results which about root and shoot length in our study. Also, similar results were reported in different plants such as wheat [18], Indian mustard [19], Kentucky bluegrass and tall fescue [20].

In study, the lowest root weigth was determined at 15000 ppm application with 0.0027 g. But a statistically significant difference was not found between other application and root weigth varied between 0.0047g and 0.0056 g. Results obtained from our study are consistent with results of study conducted by Pessarakli et al. [21] Because these researchers reported that variations in root dry weight were not statistically significant up to certain point in some turfgrass plants which are exposed to increased NaCl concentrations. We also found that root weight was affected by different NaCl concentrations. While highest root weight was determined at 5000 ppm application with 0.0066 g, root weight was affected negatively at higher concentrations and it was reduced to 0.0038 g at 15000 ppm.

While the highest shoot weight were recorded as 0.018 g at 5000 ppm application, shoot weight of seedlings were decreased in higher doses than 5000 ppm. Shoot weight values were determined as 0.022 g and 0.021 g at 7500 ppm and 10000 ppm, respectively, and finally, lowest shoot weight were obtained at 15000 ppm with 0.012 g. Similar results were observed in peat experiment and highest shoot weight value (0.023 g) was calculated at 5000 ppm application. Growth restriction can be appear in plants exposed to high salt concentrations depend on decreasing energy and reducing turgor which are required for growth [22]. In high salinity conditions, ion stability in protoplasma disrupts due to reducing water uptake of plant and competition occurs between ions and plant nutrient element and nutrient element uptake can be precluded. So high salt stress causes scrubby growth and declining root growth [23]. Both fresh and dry weights of root, shoot and leaves decrease under salt stress and this status determined in many study [24-26]. Dudeck and Peacock, [27] determined that shoot weight decreased by 50% in bermuda grass (*Cynodon dactylon*) which grown at 33 dS.m-1. Similarly, Uddin et al. [28] reported that shoot growth decreased by 50% in 8 different turfgrass plants in high salt concentrations. Also in the same study determined that significant reduction occured in root growth with increasing salt concentrations.

Generally, root/shoot ratios reduced in both petri and peat experiments with increasing NaCl concentrations. In petri experiment, while the lowest root/shoot ratio for length and weight were determined at 15000 ppm with 0.42 and 0.23 respectively, there was no statistically difference between other applications to both length and weight. According to root/shoot ratios of peat experiment, ratios for length reduced with increasing salt concentrations and lowest ratio was recorded as 0.77 at 15000 ppm. Also ratios for weight were varied between 0.30 and 0.22, but significant differences were not determined between means of different salt concentratios.

Sodium (Na+) and chlorine (Cl-) ions accumulate in higher rates in leaves compared to root. So roots are more sensitive to salinity (especially sodium ion) compared to leaves. Thus root/shoot ratio increase plants which are grown in salinity conditions [29]. But in contrast, roots are more sensitive to salinity in comparison to leaves in seedling stage and they are more affected. So in seedling stage of plants, reductions occur in root/shoot ratio in case of salt stress. This status explains reductions in root/shoot ratio obtained in our study. Kaya et al. [30] reported that root/shoot ratios of safflower (*Carthamus tinctorius* L.) seedlings were decreased with increasing salt concentrations. Similar results were recorded by Ghorashy et al. [31] in safflower, by Hussain and Rehman [32] in sunflower, by Bandeh-hagh et al. [33] in canola, Zhang et al. [34] in tall fescue and Zhang et al. [35] in buffalograss and blue grama grass.

A correlation analysis was conducted to determine the relationships among the variables. As a result of the analysis, significant relationships were recorded between characterisitcs and results are shown at Tables 4 and 5.

In petri experiment (Table 4), it was found out that germination day (GD) correlated with germination percentage (GP), root lenght (RL), shoot lenght (SL) and root weight (RW), negaitive and significant ($p<0.05$); whilst GD correlated with root/shoot ratio (R/SRW for weight), positive and significant ($p<0.05$). Also GP was positively correlated with RL, SL, RW (at 0.01) and shoot weight (SW) (at 0.05), but GP negatively correlated with root/shoot (R/SRL for length) and R/SRW at 0.01. A significantly posititive correlation ($p<0.01$) was found between RL and SL, RW and SW; R/SL was correlated negatively with R/SRW. Relationships of SL with RW and SW was determined as singnificant and positive, SL was correlated negatively with R/SRL and R/SRW. Both R/SRL and R/SW were correlated negatively with RW, positive correlation was determined between RW and SW. While there was a highly significant negative correlation between RW and R/SRL, correlation between R/SRL and R/SRW was determined as significant and positive.

Similar results to petri experiment were obtained from peat experimet for correlations and significant relationships were determined between variables (Table 5).

Table 4. Correlation coefficients between the factors among all characteristics (Petri experiment)

	FGD	GP	RL	SL	RW	SW	R/SRL
GP	-0.617*						
RL	-0.577*	0.911**					
SL	-0.622*	0.918**	0.951**				
RW	-0.574*	0.867**	0.874**	0.882**			
SW	-0.287	0.523*	0.671**	0.744**	0.685**		
R/SRL	0.418	-0.876**	-0.916**	-0.870**	-0.803**	-0.540*	
R/SRW	0.537*	-0.816**	-0.690**	-0.677**	-0.839**	-0.209	0.723**

*(FGD: First germination day, GP: Germination percentage, RL: Root length, SL: Shoot length, RW: Root weight, SW: Shoot weight, R/SRL: Root/Shoot rate (for length), R/SRW: Root/Shoot rate (for weight)); *: p<0.05, **: p<0.01*

Table 5. Correlation coefficients between the factors among all characteristics (Peat experiment)

	FGD	EP	RL	SL	RW	SW	R/SRL
GP	-0.822**						
RL	-0.731**	0.716**					
SL	-0.670**	0.642**	0.874**				
RW	-0.654**	0.762**	0.892**	0.870**			
SW	-0.400	0.306	0.697**	0.808**	0.741**		
R/SRL	0.551*	-0.530*	-0.927**	-0.712**	-0.734**	-0.541*	
R/SRW	0.607*	-0.824**	-0.754**	-0.621*	-0.831**	-0.254	0.691**

(FGD: First germination day, EP: Emergence percentage, RL: Root length, SL: Shoot length, RW: Root weight, SW: Shoot weight, R/SRL: Root/Shoot rate (for length), R/SRW: Root/Shoot rate (for weight))
**: p<0.05, **: p<0.01*

Emergence day (ED) was correlated singnificant and negatively with emergence percentage (EP), RL, SL and RW, singnificant and positively with R/SRL and R/SRW. Also while strong positive correlations were found between EP and RL, SL and RW, EP was correlated negatively with R/SRL ($p<0.05$) and R/SRW ($p<0.01$). According to another result, while RL was positively correlated with SL, RW and SW and negatively correlated with R/SRL and R/SRW. SL was significant and positively correlated with RW and SW, and negatively correlated with R/SRL and R/SRW. RW was positively correlated with SW and negatively correlated with R/SRL and R/SRW. Also while a significant negative correlations determined between SW and R/SRL, R/SRL was positively correlated with R/SRW.

4. CONCLUSION

According to the obtained results of the study, seeds and seedlings of dichondra plant are partially (approximately 5000-7500 ppm) resistant to salinity conditions, but germination percentage and growth parametres were reduce with increasing NaCl concentrations. So dichondra has a turfgrass potential in coastal areas of Aegean and Mediterranean regions which have intensive turf and golf areas and it can be use especially in salinity soils and shadow conditions.

COMPETING INTERESTS

Authors have declared that no competing interests exist.

REFERENCES

1. Akgun I, Kara B, Altındal D. Effect of salinity (NaCl) on germination, seedling growth and nutrient uptake of different triticale genotypes. Turk. J. Field Crops. 2011;16(2):225-232.
2. Turan MA, Türkmen N, Taban N. Effect of NaCl on stomatal resistance and proline, chlorophyll, Na, Cl and K concentrations of lentil plants. J. Agron. 2007;6(2):378-381.
3. Munns R. Genes and salt tolerance: Bringing them together. New Phytol. 2005;167(3):645-663.
4. Türkyılmaz B, Aktaş LY. Güven A. Salinity induced differences in growth and nutrient Accumulation in five barley cultivars. Turk J Field Crops. 2011;6(1):84-92.
5. Kilit M, Doğan A. Modeling of saltwater intrusion in coastal aquifers: Cote-orientale aquifer application in Tunusia. 6th International Advanced Technologies Symposium (IATS'11), Elazığ, Turkey, In Turkish; 2011.
6. Kaman H, Sönmez NM, Çetin M, Kurunç A, Aslan GE, Uz BY. Investigation of seawater intrusion in streams connected to the Sea: An example of acısu creek of Antalya. Res J Agric Sci. In Turkish. 2011;4(2):43-47.
7. Beltrao J, Neves A, Brito JC, Seita J. Salt removal potential of Turfgrass in golf courses in the Mediterranean Basin. WSEAS Trans Environ Dev. 2009;5(5):394-403.
8. Lee GJ, Carrow RN, Duncan RR. Salinity tolerance of selected seashore paspalums and bermudagrasses: Root and verdure responses and criteria. Hort Sci. 2004;39(5):1143-1147.
9. Bilgin O, Baser I, Korkut KZ, Balkan A, Sağlam N. The impacts on seedling root growth of water and salinity stress in maize (*Zea mays indentata* Sturt.). Bulg J Agric Sci. 2008;14(3):313-320.

10. Aydınşakir K, Erdurmuş C, Büyüktaş D, Çakmakcı S. The effect of salt (NaCl) stress on early seedling stage and germination of some silage sorghum (*Sorghum bicolor*) varieties. J Akdeniz Univ Agric Fac. In Turkish. 2012;25(1):47-52.
11. Avcıoğlu R. Turf Technique Turf Establishment & Management Ege University Press, Izmir, Turkey. In Turkish; 1997.
12. Aydemir O. Plant nutrition and soil productivity. Atatürk university publication. Erzurum. In Turkish. 1992;734.
13. Nizam İ. Effects of salinity stress on water uptake, germination and early seedling growth of perennial ryegrass. Afr J Biotechnol. 2011;10(51):10418-10424.
14. Camberato JJ, Martin SB. Salinity slows germination of rough bluegrass. Hort Sci. 2004;39:394-397.
15. Taiz L, Zeiger E. Plant physiology. Sinauer associates, Inc., Publisher; 2002. ISBN: 0-87893-823-0.
16. Dai J, Huff DR. Schlossberg MJ. Salinity effects on seed germination and vegetative growth of greens-type *Poa annua* relative to other cool-season turfgrass species. Crop Sci. 2009;49(2):696-703.
17. Janagard MS, Tobeh A, Esmailpour B. Evaluation of salinity tolerance of three canola cultivars at germination and early seedling growth stage. J Food Agric Environ. 2008;6(2):272-275.
18. Kara B, Kara (Uysal) N. Effect of different salinity (NaCl) concentrations on the first developmet stages of root and shoot organs of wheat. Anadolu J Agric Sci. In Turkish. 2010;25(1):37-43.
19. Yusuf M, Hasan SA, Ali B, Hayat S, Fariduddin Q, Ahmad A. Effect of salicylic acid on salinity-induced changes in *Brassica juncea*. J Integr Plant Biol. 2008;50(9):1096-1102.
20. Alshammary SF, Qian YL. Wallner SJ. Growth response of four turfgrass species to salinity. Agr Water Manage. 2004;66:97-111.
21. Pessarakli M, Kopec DM. Gilbert JJ. Growth responses of selected warm-season turfgrasses under salt stress. turfgrass, Landscape and Urban IPM Research Summary (P 157), Univ. of Arizona Coop. Ext., Agric. Exp. Stn., The Univ. of Arizona, Tucson. 2009;39-46. Available:http://cals.arizona.edu/pubs/crops/az1487/14872c.pdf Access time: 31.01.2014
22. Marcum KB. Use of saline and non-potable water in the turfgrass industry: Constraints and developments. Agric Water Manage. 2006;80:132-146.
23. Kacar B, Katkat AV, Öztürk S. Plant Physiology. VIPAS Press. No: 74, Bursa. In Turkish; 2002.
24. Hernandez JA, Olmos E, Corpas FJ, Sevilla F, Del Rio LA. Salt-induced oxidative stress in chloroplasts of pea plants. Plant Sci. 1995;105:151-167.
25. Ali-Dinar HM, Ebert G, Ludders P. Growth, chlorophyll content, photosynthesis and water relations in guava (*Psidium guajava* L.) under salinity and different nitrogen supply. Gartenbauwissenschaft. 1999;64: 54-59.
26. Chartzoulakis K, Klapaki G. Response of two green house pepper hybrids to NaCl salinity during different growth stages. Sci Hortic. 2000;86(3):247-260.
27. Dudeck AE, Peacock CH. Salinity effects on growth and nutrient uptake of selected warm-season turfgrasses. Int Turfgrass Soc Res J. 1993;7:680-686.
28. Uddin MK, Juraimi AS, Ismail MR, Othman R, Rahim AA. Relative salinity tolerance of warm season turfgrass species. J Environ Biol. 2011;32:309-312.
29. Tester M, Davenport R. Na+ tolerance and Na+ transport in higher plants. Ann Bot. 2003;91:503-527.
30. Kaya MD, Ipek A. Öztürk A. Effects of different soil salinity levels on germination and seedling growth of safflower (*Carthamus tinctorius* L.). Turk J Agric For. 2003;27:221-227.
31. Ghorashy SR, Sionit N, Kheradnam M. Salt tolerance of safflower varieties (*Carthamus tinctorius* L.) during emergence. Agron J. 1972;64:256-257.
32. Hussain MK, Rehman OU. Evaluation of sunflower (*Helianthus annuus* L.) germplasm for salt tolerance at the shoot stage. Helia. 1997;20(26):69-78.
33. Bandeh-Hagh A, Toorchi M, Mohammadi A, Chaparzadeh N, Salekdeh GH, Kazemnia H. Growth and osmotic adjustment of canola genotypes in response to salinity. J Food Agric Environ. 2008;6(2):201-208.

34. Zhang Q, Zuk A, Rue K. Turfgrasses responded differently to salinity, waterlogging and combined saline-waterlogging conditions. Crop Sci. 2013;53:2686-2692.

35. Zhang Q, Rue K, Wang S. Salinity effect on seed germination and growth of two warm-season native grass species. Hort Science. 2012;47:527-530.

22

Sawah Rice Farming Eco-technology Options for Enhancing Sustainable Nutrient Management and Rice Production in Degraded Inland Valleys of Southeastern Nigeria

J. C. Nwite[1*], A. O. Onunwa[2], C. A. Igwe[3] and T. Wakatsuki[4]

[1]*Department of Crop Production Technology, Federal College of Agriculture, Ishiagu, Ebonyi State, Nigeria.*
[2]*Department of Soil Science and Land Resources Management, Nnamdi Azikiwe University, Awka, Anambra State, Nigeria.*
[3]*Department of Soil Science, University of Nigeria, Nsukka, Enugu State, Nigeria.*
[4]*Faculty of Agriculture, Shimane University, Nara, Japan.*

Authors' contributions

This work was carried out in collaboration between all authors. Author JCN designed the study, wrote the protocol and wrote the first draft of the manuscript. Author AOO managed the literature searches; analyses of the study performed the spectroscopy analysis. Author CAI managed the experimental process and author TW identified the species of plant. All authors read and approved the final manuscript.

Editor(s):
(1) Yong In Kuk, Department of Development in Oriental Medicine Resources, Sunchon National University, South Korea.
(2) Scd Victoria Anatolyivna Tsygankova, Department for Chemistry of Bioactive Nitrogen-Containing Heterocyclic Compounds, Institute of Bio-organic Chemistry and Petrochemistry, Plant Physiology, NAS, Ukraine.
(3) Enrique Luis Cointry Peix, Department of Plant Breeding, Rosario National University, Argentina.
Reviewers:
(1) Filippo Sgroi, University of Palermo, Italy.
(2) Anonymous, Norway.

ABSTRACT

The decline in agricultural productivity in Nigeria is merely because the rural farmers which constitute the bulk of Nigerian crop farmers rely on the rainfall for their agricultural activities. Rice farmers in Ebonyi State, regarded as a major rice producing State in Nigeria rely on rain-fed agriculture. The water management option among the rice farmers in their lowland rice production

**Corresponding author: E-mail: johnwhite4real_2007@yahoo.com*

in the area is the use of grass materials in the demarcation of the fields into basins for water storage without any form of water diversion from one place to another as a way of controlling the field water. In an attempt to replicate the successful way of controlling water in the African agro-ecosystems, otherwise known as "Japanese Satoyama watershed management model", sawah rice cultivation technology has been introduced to West Africa in the last decades.

Sawah is generally described as a controlled water management system in the rice field which involved mainly bunding, puddling and leveling with inlets and outlets channels on the bunds for irrigation and drainage purposes. The irrigation water may be provided by rain water or underground water discharge through seepage or springs, or by rise in the level of a stream and river in an inland valley, or using modern source from well pumps, taps, canal and storage of large quantities of water in reservoirs or ponds. The study was conducted in an inland valley at Akaeze in 2010, 2011 and 2012 cropping seasons, to evaluate the effect of different water sources for *sawah* water management system and amendments on soil chemical properties and rice grain yield. A split- plot in a randomized complete block design was used to assess two factors at different levels. Three sources of water; rain-fed, spring type and pond type constituted the main plot, while the amendments, that constituted the sub- plots were replicated three times and were applied in the following manner: rice husk (RH) @ 10 t ha^{-1}, rice husk ash (RHA) @ 10 t ha^{-1}, poultry droppings (PD) @ 10 t ha^{-1}, N.P.K. @ 400 kg ha^{-1} and no amendment @ 0 t ha^{-1}. The results of the study showed that different water sources significantly ($p < 0.05$) improved the soil pH in the location. Soil organic carbon, total nitrogen and cation exchange capacity were significantly ($p < 0.05$) increased within the period of study by both the different water sources and amendments. It was observed that the exchangeable acidity was statistically reduced by different water sources and amendments within the periods. It was also recorded that available phosphorous were positively improved by different water sources and amendments in different forms in the area. The result equally gave positive improvement on the rice grain yield by the studied factors for the three years. Generally, results showed a better performance of organic amendments over mineral fertilizer in some soil chemical properties and rice grain yield improvement. The interaction of a good water source in *sawah* water management and amendment practices was observed to be a good strategy for improving some soil chemical properties in the area.

Keywords: Water sources; sawah; amendments; rice grain yield; soil properties and inland valleys.

1. INTRODUCTION

The well-established and growing demand for rice in Nigeria presently has necessitated the need for increasing rice production both to meet the country's food requirements and for the realization of rice green revolution in Nigeria. Nigeria is now one of the largest food importers in the world. In 2010 alone, Nigeria spent 356 billion naira on importation of rice. Nigeria is eating beyond its means. While we all smile as we eat rice every day, Nigerian rice farmers cry as the importations undermine domestic production [1].

Nigeria agricultural productivity fluctuates without control, mainly because the rural farmers rely on the rain for farming operations in the country. Rain-fed agriculture is a major economic activity in the developing countries and is been practiced in 80% of the total physical agricultural area with about 62 percent of the world's stable food [1-3]. According to FAO [4], 93 percent of cultivated land in sub-Saharan Africa is merely rain-fed agriculture, thus playing a crucial role in food security and water availability [5]. Rice farmers in the study area who are dependent on the rain for their rice production make straight bunds across the valley bottom to store water in the fields. The lowlands are often slightly concave; these straight bunds result in deep water in the lowest parts of the lowland, and hardly any flooding near the fringes. These traditional practices usually lead to differences in rice performance and yield from the same field, and large disparity in soil characteristics of the same field. Kadigi et al. [6] argues that land for rain-fed agriculture varies depending on the amount and distribution of rainfall in the area. Rice production in the rain-fed lowland environment being dependent on rain-fed conditions is very susceptible to climatic variability which results in low yields.

Rain-fed lowland farmers are typically challenged by poor soil quality, drought/flood conditions, and erratic yields. Study has shown that yields from rain-fed agriculture are usually low, measuring 1 t ha^{-1} in semiarid tropical agro-ecosystems [7].

Researches have revealed that the low productivity in rain-fed agriculture is majorly due to suboptimal performance related to field management aspects rather than low physical potential [8–11]. Gowing et al. [12] maintained that poor field management practices resulting to inadequate soil moisture and low soil fertility have been top challenges facing rain-fed agriculture.

The improvement of farm infrastructures like bunding, leveling of the field surface, irrigation and drainage modifications will go a long way in reducing the yield gap in rain-fed inland valley environments. The surface water could be maintained more evenly over the field's entire surface with leveling operation helping to improve soil conditions for rice production. Considering the gap yield in rain-fed agriculture and the current demand for rice in Nigeria, there is a need to sort for other water sources for supplementing the rain-fed for optimum rice production in Nigeria.

Supplementary irrigation is needed when natural precipitation is not adequate to secure grain and forage production [13].

Nigeria is blessed with enough rain and high valuable inland valleys for rice based cropping. In spite of these valuable inland valleys that abound in Nigeria especially in the Southeast for agricultural use, these areas are still facing some challenges in their exploitation.

The major limiting factors in the utilization of these inland valleys include; poor soil fertility maintenance, inadequate weed and water control [14–17]. Most soils in the West African sub-region are highly weathered and very fragile [18–22].

In order to overcome these limitations in the utilization of these inland valleys, an African adaptive *sawah* lowland farming practice with irrigation scheme for integrated watershed management have been proposed to be the most promising strategy to tackle these problems in these areas [21,23].

Sawah, an Indo-Malaysian word for padi (Malayan word for paddy) or lowland rice management system involved bunding, puddling, leveling and good water management through inlet and outlet channels for irrigation and drainage [24].

Sawah system which ensures the maintenance of water level (minimum and maximum) in the field plots during the growing period of the plant contribute to the alleviation of global warming problems through the fixation of carbon in forest and *sawah* soils in ecologically sustainable ways.

It restores/replenishes the lowland with nutrients through geological fertilization as it resists erosion. The mechanisms in *sawah* system of nutrient replenishments encourage not only rice growth, but also the breeding of various microbes, which improves biological nitrogen fixation.

Achieving high yield in most West African ecology is difficult without soil amendment, as the soils are highly leached, porous and low in essential plant nutrient. Imolehin and Wada [25] advocated a reversion to the use of organic materials in wetland rice cultivation as a more realistic option for farmers than continued reliance on inorganic fertilizers, which in addition to their deleterious effects on the soil are not readily available. Lee et al. [26] reported an improved SOC concentration and soil physical properties with continuous application of compost in a plough layer of a long-term rice paddy, relative to inorganic fertilizer application. However, the superiority of locally available organic materials over inorganic fertilizers in terms of soil properties reformation and stability after puddling of natural wetlands in our tropical environment is not yet confirmed.

The study aimed at evaluating three different water sources; spring, pond and rain-fed for *sawah* development at farmers level for sustainable nutrient management and rice production in inland valleys of Southeastern Nigeria. The study also aims at evaluating the effects of different manure sources on soil chemical properties and grain yield improvement; and to evaluate the interactions of different water sources and soil amendments on soil properties and rice grain yield.

2. MATERIALS AND METHODS

2.1 Location of the Study

This study was conducted in an inland valley at Akaeze in 2010, 2011 and 2012 cropping seasons to evaluate the effects of different sources of water for *sawah* water management system and amendments on soil properties and rice grain yield. Akaeze lies at approximately latitude 05° 56' N and longitude 07° 41' E. The annual rainfall for the area is 1,350 mm, spread from April to October with average air temperature of 29°C. The site is within the derived savanna vegetation zone with grassland and tree combinations. The soils are described as Aeric Tropoaquent [27] or Gleyic Cambisol [28]. The soils have moderate soil organic carbon (OC) content on the topsoil, low in pH and low cation exchange capacity (CEC). Soils are mainly used for rain-fed rice cultivation during the rains and vegetable production as the rain recedes.

2.2 Field Methods

The field was divided into three different main plots where the three sources of water for irrigation were located. Bulk (composite) sample was collected at 0- 20 cm soil depth in the study area for initial soil characteristics. The three main plots were demarcated into five subplots with a 0.6 m raised bunds where the soil amendments were applied (Fig. 3).

A split- plot in a randomized complete block design was used to assess the two factors at different levels. The three sources of water that constituted main plot include;

- Rain-fed *sawah* which involved plots where water supply was only from rain water and no irrigation water was allowed to flow into the plots.
- Spring type, on its own was where water source was from a spring that flows into the field and perhaps rainfall with some control, and
- Pond type involved water application to plots as supplemental irrigation with pumping machine from an artificial pond in the field.

Generally, Water was circulated in the field by manipulation of the bunds. The water flows from the spring to the plots through a constructed canal from the spring source to the field and the spring is close-by to the field, less than 100 m away (Fig. 2).

Fig. 1. Field preparation of rice field with power-tiller machine

Fig. 2. Constructed canal from the spring source and the artificial pond for supplemental irrigation

The quantity of water issued to the plots was not measured rather the depth of water was maintained at 5 cm- 10 cm throughout the growing period of the rice except in the rain-fed plots where only the water harvested by each plot during rainfall that settle in the plots. Ruled sticks with bold marks on 10 cm and 5 cm points were mounted permanently on each plot to check the water level or depth in the field. In the pumping type a pumping machine with rated power output of 2.8 kilowatts, self priming volute with 4 impeller blades and maximum discharge of 900 litres/minute, plus a total Head of 26 M, was used to pump water from an artificial pond into the field receiving pumping water as a supplemental irrigation, whenever water depth in the plots is below 5 cm (Fig. 2).

The water introduction in each case was made 2 weeks after transplanting and this was maintained till the stage of ripening of the rice grains with the help of the bunds inlets and outlets channels (Fig. 3). The water from these different sources in the field is presumed to have different qualities and as such would have different effect on the soil properties and rice yield.

The amendments that constituted the sub- plots were applied as follows:

- PD Poultry droppings @ 10 ton/ha
- F NPK fertilizer (20:10:10) @ 400 kg/ha recommended rate for rice in the zone
- RH Rice husk @ 10 t ha^{-1};
- RHA Rice husk ash @ 10 t ha^{-1}
- CT Control @ 0 t ha^{-1}

The treatments were replicated three times in each of the main-plots. The PD, RHA and RH were spread on the plots that received them and incorporated manually into the top 20 cm soil depth 2 weeks before transplanting. The nutrient contents of these organic amendments were determined (Table 2). The motivation on the selection or choice of quantities of organic amendments used was based on the soil type of study area, crop type, the affordability and availability of the amendments in the area and the recommended rates for these amendments from other researches carried in the study area.

The test crop was high-tillering rice variety *Oryza sativa var. FARO 52 (WITA 4)*. The rice seeds were first raised in the nursery and later transplanted to the main field after 3 weeks in nursery. At maturity, rice grains were harvested, dried and yield computed at 90% dry matter content. At the end of harvest, soil samples were collected from each replicate of every plot from each of the location for chemical analyses.

2.3 Laboratory Methods

Soil samples were air-dried and sieved with 2 mm sieve. Soil fractions less than 2 mm from individual samples were then analyzed using the following methods; Particle size distribution of less than 2 mm fine earth fractions was measured by the hydrometer method as described by Gee and Bauder [29]. Soil pH was measured in a 1:2.5 soil: 0.1 M KCl suspensions [30]. The soil organic carbon was determined by the wet oxidation method of Walkley and Black (1934) as modified by Nelson and Somners [31]. Total nitrogen was determined by semi-micro kjeldahl digestion method using sulphuric acid and $CuSO_4$ and Na_2SO_4 catalyst mixture [32]. Available phosphorus was measured by the Bray II method [33]. CEC was determined by the method described by Rhoades [34]. While exchangeable acidity (EA) was measured using the method of McLean [30].

Fig. 3. Construction of interceptive canals and bund making for *sawah* field development

2.4 Data Analysis

Data analysis was performed using GENSTAT 3 7.2 Edition.

Significant treatment means was separated and compared using Least Significant Difference (LSD) and all inferences were made at 1% and 5% Levels of probability.

3. RESULTS AND DISCUSSION

3.1 Soil Properties and Organic Amendments

3.1.1 Soil properties

The soil physical and chemical properties are reported in Table 1. Table 1 gave the soil of the study area as sandy loam with 100 g kg^{-1} clay and 150 g kg^{-1} silt content. The initial soil analysis showed that the soil has low pH, exchangeable bases and cation exchange capacity (Table 1).

Soil organic carbon concentration was moderate, whereas the soil total nitrogen value was 0.091%.

3.1.2 Organic amendments properties

Table 2 shows that rice husk amendment gave the highest organic carbon (33.7%), followed by rice husk ash with 23.9%, while poultry dropping had the least value. This implies that rice husk amendment has the potentials of enriching the soil with more organic carbon pools. The analysis also indicated that poultry dropping produced the highest total nitrogen percent, while the least TN was recorded in rice husk ash which could be attributed to the burning of the material. The analysis (Table 2) showed that rice husk ash had the highest values for percentage potassium and magnesium, while the highest percentage calcium was obtained from poultry dropping.

3.2 Effects of Water Sources and Amendments on the Soil pH and Organic Carbon

Tables 3 and 4 presented the effects of different sources of water and amendments on the soil pH and organic carbon for three years of study. The results (Table 3) showed that the soil pH measured in water was significantly ($p < 0.05$) higher in spring water source than other *sawah* water sources in the three years of study with pH values of 4.12, 4.64 and 4.94 in the 1st, 2nd and 3rd year of study, while the rain-fed recorded the least values (3.89, 4.31 and 4.65), 1st, 2nd and 3rd year, respectively. The result also showed that the pH-increasing trend directly followed the year of study progression. The higher pH values obtained in spring *sawah* treated plots could be linked to the fine particles and other sediments that were eroded from the adjacent uplands and moved into the spring water which are then moved to the affected plots and get accumulated.

Table 1. Some properties of the topsoil of the experimental plots (0-20 cm) before tilling and amendment

Soil property	Value
Clay (%)	10
Silt (%)	21
Total sand (%)	69
Textural class	SL
Organic matter %	2.64
Organic carbon % (OC)	1.61
Total nitrogen % (N)	0.091
pH (H_2O)	3.6
pH (KCl)	3.0
Exchangeable bases (cmolkg^{-1})	
Sodium (Na)	0.15
Potassium (K)	0.04
Calcium (Ca)	1.0
Magnesium (Mg)	0.6
Cation exchange capacity (CEC)	5.6
Exchangeable acidity (EA)	3.2
Available phosphorous (mg/kg)	4.20
Base saturation (BS)	24.70

OC= organic carbon; TN= total nitrogen; K^+= exchangeable potassium; Ca^{2+}= exchangeable calcium; Mg^{2+} = exchangeable magnesium; CEC= cation exchange capacity

This result is not in agreement with the findings of Takase et al. [35] in a research conducted in Ghana to compared river, canal, tap and well irrigation sources and observed that none of the *sawah* water types studied gave significant increase on the pH than others, but the soils irrigated with well water had the highest pH value at the end of their three months study.

Table 3 indicated that manure application within the period of study increased the soil pH measured in water significantly ($p < 0.05$) higher than plots without manure application. The soil pH was improved significantly ($p < 0.05$) higher in soils treated with rice husk ash in all the three water sources for *sawah* development in the three years of study. This was followed by plots amended with poultry dropping, while the least pH value was obtained from plots with no amendments. The values ranged from 3.44 – 4.49 in the 1st year, 3.58 – 4.84 in the 2nd year and 3.82 – 5.31 in the 3rd year of study. The results of the three years showed the pH increases as the year progresses. The significant improvement on the soil pH recorded in plots treated with RHA within the study period could be linked to the high potassium and magnesium contents in the rice husk ash material used (Table 2) which could induce a pH increase and this conforms to the submissions of Abyhammer et al. [36]; Markikainen, [37] and Nwite et al. [38]; that organic lime like ash material could induce a pH increase by as much as 0.6 – 1.0 units in humus soils. Generally, the results showed that treated soils increased pH significantly higher than untreated soils. This result agrees with the findings of Opara-Nnadi et al. [39] who reported pH increase following the application of organic wastes.

The interactions of water sources and amendments improved the soil pH significantly only in the first year of study.

Table 4 presents the effect of water source for *sawah* development and amendments on soil organic carbon. The results (Table 4) indicated that water sources and amendments increased the soil organic carbon pools (SOC) significantly ($p < 0.05$) different in the soil for the three years of study. The result shows that among the water sources, spring water source did improve the SOC pool significantly ($p < 0.05$) different from other water sources within the periods of study. It was observed that apart from the first year, pond water source did not improve the SOC significantly ($p < 0.05$) higher than the rain-fed water source. The soil organic carbon mean values varied from 1.02 – 1.36%, 1.21 – 1.47% and 1.20 – 1.49%, in the 1st, 2nd and 3rd year of study, respectively. However, the significant improvement made by spring water source over other water sources could be attributed to finer fractions or sediments that were moved into the plots by the water during flow from the spring through the canal. Follet [40] showed that organic carbon sequestration through improved soil management practices can have significant improvement on soil resources, because increasing soil C increases the functional capabilities of soils. The results (Table 4) showed that soil amendments significantly ($p < 0.05$) improved the soil organic carbon pool relatively higher than the control within the periods of study. The result also gave a higher significant improvement on the SOC pool on plots amended with rice husk dust than plots amended with other treatments. This higher improvement made by rice husk dust on the soil organic carbon could be attributed to high content/percent of carbon in the rice husk dust used as amendment (Table 2). It was also noted that all the amended plots significantly ($p < 0.05$) increased the soil organic carbon pool higher than the control. The mean values varied from 0.65 – 1.66% in the first year, 0.88 – 1.63% in the second year and 0.93 – 1.55% in the third year.

The results also showed that the interactions of water sources and amendments increased the soil organic carbon (SOC) build-up significantly ($p < 0.05$) higher than their separate performance in the second and third year of the study. This agreed with the report of Bhagat and Verma [41] that incorporation of plant residues coupled with appropriate puddling and water management build up organic carbon status of soil.

3.3 Effects of Different Water Sources and Amendments on the Soil Total Nitrogen and Exchangeable Acidity

The effects of different water sources and amendments on soil total nitrogen were presented in Table 5. The artificial application of water as supplemental irrigation was significantly ($p < 0.05$) different from the rain-fed in soil total nitrogen improvement (Table 5). The improvement could be as a result of aquatic algae activities in submerged soils that commit biological nitrogen fixation through increased photosynthesis. The result (Table 5) indicated that the supplemental irrigated plots significantly ($p < 0.05$) improved the soil total nitrogen higher than the rain-fed treated plots in the second and third year. The values varied from 0.082 – 0.095% in the second year and 0.89 – 0.104% in the third year. This implies that soil total nitrogen increase progressively as the year of the study increases. However, spring water source increased the soil total nitrogen higher than the pond and rain-fed significantly. These results implied that rain-fed agriculture does not permit proper water management systems in the field with other factors causing alternate wetting and drying of the field which do lead to loss of the element through de-nitrification process.

Table 2. Properties of the organic amendments (%)

Amendment	OC	Total N	K	Ca	Mg	P	C:N
				(%)			
PD	16.50	2.10	0.48	14.40	1.20	2.55	7.86
RH	33.70	0.70	0.11	0.36	0.38	0.49	48.14
RHA	23.90	0.06	0.65	1.00	1.40	11.94	398.33

PD= poultry droppings; RH= rice husk powder; RHA= rice husk burnt ash; OC= organic carbon

Table 3. Effects of different water source for *sawah* and amendments on soil pH

Water source for *Sawah*	Amendments					
	CT	NPK	PD	RH	RHA	Mean
		Year 1				
Rain-fed	3.37	3.93	4.07	3.83	4.23	3.89
Spring	3.57	3.70	4.23	4.33	4.77	4.12
Pond	3.40	3.90	4.03	3.93	4.47	3.95
Mean	3.44	3.84	4.11	4.03	4.49	
LSD $_{(0.05)}$ water source for *sawah*			0.1025			
LSD $_{(0.05)}$ Amendment			0.1313			
LSD $_{(0.05)}$ water source for *sawah* x Amendments			0.2157			
		Year 2				
Rain-fed	3.47	4.50	4.50	4.50	4.60	4.31
Spring	3.73	4.80	4.80	4.73	5.13	4.64
Pond	3.53	4.40	4.70	4.43	4.80	4.37
Mean	3.58	4.57	4.67	4.56	4.84	
LSD $_{(0.05)}$ water source for *sawah*			0.1105			
LSD $_{(0.05)}$ Amendment			0.1412			
LSD $_{(0.05)}$ water source for *sawah* x Amendments			NS			
		Year 3				
Rain-fed	3.60	4.77	4.90	4.97	5.03	4.65
Spring	3.97	5.03	5.13	5.03	5.53	4.94
Pond	3.90	5.00	5.03	5.00	5.37	4.86
Mean	3.82	4.93	5.02	5.00	5.31	
LSD $_{(0.05)}$ water source for *sawah*			0.0956			
LSD $_{(0.05)}$ Amendment			0.1167			
LSD $_{(0.05)}$ water source for *sawah* x Amendments			NS			

CT = control, NPK = nitrogen. phosphorous potassium, PD = poultry dropping, RH = rice husk, RHA = rice husk ash

Table 4. Effects of different water source for *sawah* and amendments on soil organic carbon (%)

Water source for *Sawah*	Amendments					
	CT	NPK	PD	RH	RHA	Mean
		Year 1				
Rain-fed	0.59	1.15	1.14	1.28	0.94	1.02
Spring	0.67	1.62	1.58	1.92	0.99	1.36
Pond	0.70	1.30	1.28	1.79	1.03	1.22
Mean	0.65	1.35	1.33	1.66	0.99	
LSD $_{(0.05)}$ water source for sawah			0.2108			
LSD $_{(0.05)}$ Amendment			0.2079			
LSD $_{(0.05)}$ water source for sawah x Amendments			NS			
		Year 2				
Rain-fed	0.85	1.35	1.24	1.36	1.26	1.21
Spring	0.99	1.81	1.46	1.89	1.20	1.47

Water source for *Sawah*	Amendments					
	CT	NPK	PD	RH	RHA	Mean
Pond	0.80	1.47	1.31	1.64	1.03	1.25
Mean	0.88	1.54	1.34	1.63	1.16	
LSD ($_{0.05}$) water source for sawah			0.1864			
LSD ($_{0.05}$) Amendment			0.1372			
LSD ($_{0.05}$) water source for sawah x Amendments			0.2540			
		Year 3				
Rain-fed	0.92	1.18	1.23	1.38	1.27	1.20
Spring	0.95	1.80	1.52	1.91	1.27	1.49
Pond	0.90	1.41	1.42	1.36	1.10	1.24
Mean	0.93	1.46	1.39	1.55	1.21	
LSD ($_{0.05}$) water source for sawah			0.1716			
LSD ($_{0.05}$) Amendment			0.1416			
LSD ($_{0.05}$) water source for sawah x Amendments			0.2530			

CT = control, NPK = nitrogen. phosphorous potassium, PD = poultry dropping, RH = rice husk, RHA = rice husk ash

Table 5. Effects of different water sources for *sawah* and amendments on soil total nitrogen (%)

Water source for *Sawah*	Amendments					
	CT	NPK	PD	RH	RHA	Mean
		Year 1				
Rain-fed	0.047	0.089	0.093	0.105	0.085	0.084
Spring	0.059	0.117	0.098	0.079	0.084	0.088
Pond	0.056	0.105	0.093	0.080	0.085	0.084
Mean	0.054	0.104	0.095	0.088	0.084	
LSD ($_{0.05}$) water source for sawah			NS			
LSD ($_{0.05}$) Amendment			0.02056			
LSD ($_{0.05}$) water source for sawah x Amendments			NS			
		Year 2				
Rain-fed	0.048	0.095	0.094	0.090	0.082	0.082
Spring	0.060	0.117	0.103	0.103	0.095	0.095
Pond	0.063	0.103	0.095	0.084	0.087	0.087
Mean	0.057	0.105	0.097	0.092	0.088	
LSD ($_{0.05}$) water source for sawah			0.006124			
LSD ($_{0.05}$) Amendment			0.006221			
LSD ($_{0.05}$) water source for sawah x Amendments			NS			
		Year 3				
Rain-fed	0.061	0.103	0.105	0.086	0.088	0.089
Spring	0.065	0.124	0.126	0.110	0.095	0.104
Pond	0.061	0.114	0.105	0.098	0.087	0.093
Mean	0.062	0.114	0.112	0.098	0.090	
LSD ($_{0.05}$) water source for sawah			0.0117			
LSD ($_{0.05}$) Amendment			0.0077			
LSD ($_{0.05}$) water source for sawah x Amendments			NS			

CT = control, NPK = nitrogen phosphorous potassium, PD = poultry dropping, RH = rice husk, RHA = rice husk ash

It has been reported that alternate wetting and drying could consequently lead to a slightly greater loss of broadcast fertilizer N and soil N by nitrification-denitrification, but this loss is expected to decrease with increasing age of the rice crop due to increased competition of rice with microorganisms for ammonium before it can be nitrified and for nitrate before it can be denitrified in uncontrolled flooded condition [42]. In a similar study by Buresh [43], it was reported that submerged soils can promotes biological nitrogen fixation (BNF) and sustain an indigenous N supply for rice as evidenced by long-term stable yields in minus-N plots in long term experiments. Buresh et al. [43] stated that uncontrolled water in lowland rice field results in alternate wetting and drying which leads to greater sequential nitrogen-denitrification than with continuous submergence.

The results (Table 5) pointed highly significant differences on the soil total nitrogen with application of amendments in all the three years of the study. Generally, all the treated plots were significantly ($p < 0.05$) different from the control in soil total nitrogen improvement. It was obtained that the soil total nitrogen was improved higher by the application of NPK fertilizer, followed by the poultry droppings in all the years of study. The soil total nitrogen values varied from 0.054 – 0.104, 0.057 – 0.105 and 0.062 – 0.114; in the 1st, 2nd and 3rd year of study, respectively. The better improvement made by NPK and poultry droppings on the soil total nitrogen higher the rice husk and rice husk ash is attributed to earlier mineralization that do occur in mineral fertilizers as against delayed or slow mineralization process that are obtained in organic amendments. This result confirms the submissions of Becker and Johnson, [44]; Sakurai, [45]; and Toure et al. [46] that *sawah* system development when used in combination with improved varieties and fertilizers can improve rice productivity in the lowlands to a great extent.

The result conforms to the submission of Kyuma and Wakatsuki, [47] and Greenland, [48] that the level of nitrogen fixation in submerged soils by microbes varies from 20 to 100 kgha^{-1}year^{-1}, and sometimes reaches up to 200 kgha^{-1}year^{-1}, depending on soil and water management as well as climatic conditions [46,47]. These natural soil fertility replenishment mechanisms are essential sustainable approach for improved productivity of lowland rice farming systems in inherently unfertile soils in West Africa and Sub-Sahara Africa [49,50].

It is important to note from the result (Table 6) that exchangeable acidity reduced significantly ($p < 0.05$) by different water sources for *sawah* development within the study period. The result (Table 6) shows that both spring and pond water sources drastically reduced the exchangeable acidity differently from the rain-fed for the three years of study. These results can be linked to higher accumulation of topsoil nutrients in the spring water source. It was recorded that even though exchangeable acidity (EA) was positively reduced within the periods, there were increasing trends in the EA as year progresses. The values ranged from 1.76 – 2.14 cmol/kg in the 1st year, 2.24 – 3.07 cmol/kg in the 2nd year and 2.57 – 3.53 cmol/kg in the 3rd year. This could be attributed to low clay and silt built in the top 0 – 20 cm as the year progresses due to downward movement of these materials.

The results also revealed that amended plots were significantly ($p < 0.05$) different from the control (non-amended plots) in decreasing the soil exchangeable acidity (EA) during the study. It was recorded that among the soil amendments, Rice husk ash (RHA) significantly ($p < 0.05$) lowered the EA more than other amendments including the control. This agrees with the findings of Errikson, [51] and Serafinelion, [52] who submitted that ashes generally have good acid-neutralizing capacity and ability to supply the soil with basic elements (Ca, K, Mg, Na) and available P; and this depends on the contents of oxides, hydroxides and carbonates of these elements. It was also obtained that there was no significant improvement due to the interactions of water sources and amendments in all the years of study.

3.4 Effects of Different Water Sources and Amendments on the Soil Available Phosphorous and Cation Exchanage Capacity (CEC)

The results (Table 7) showed that different water sources creditably increased positively ($p < 0.05$) the available phosphorous for the three years of study higher than its initial values in the soils. It was equally observed that among the three water sources, spring water source improved the soil available phosphorous significantly ($p < 0.05$) higher than other water sources in the first and third year of study, while pond water source improved the available phosphorous significantly ($p < 0.05$) higher in the second year. These results (Table 7) showed that those plots treated with supplemental irrigation significantly ($p < 0.05$) increased the available phosphorous higher than the rain-fed field in all the years. The increased available phosphorous obtained in plots treated with supplemental irrigation over rain-fed treated plots could be attributed to increased pH and reduction in ferric iron in water controlled plots as a result of neutralization of acid soils of the area, thereby liberating available phosphorous from the fixed exchange sites.

Table 6. Effects of different water sources for *sawah* and amendments on soil exchangeable acidity (EA) $cmolkg^{-1}$

Water source for *Sawah*	Amendments					
	CT	NPK	PD	RH	RHA	Mean
Year 1						
Rain-fed	3.00	2.40	2.07	1.87	1.37	2.14
Spring	2.40	1.93	1.47	2.00	1.00	1.76
Pond	2.60	2.13	1.87	2.00	0.93	1.91
Mean	2.67	2.16	1.80	1.96	1.10	
LSD ($_{0.05}$) water source for sawah			0.2317			
LSD ($_{0.05}$) Amendment			0.2056			
LSD ($_{0.05}$) water source for sawah x Amendments			NS			
Year 2						
Rain-fed	4.33	3.80	3.03	2.90	1.30	3.07
Spring	2.87	2.80	1.87	2.40	1.27	2.24
Pond	3.20	3.33	2.47	2.47	1.37	2.57
Mean	3.47	3.31	2.46	2.59	1.31	
LSD ($_{0.05}$) water source for sawah			0.166			
LSD ($_{0.05}$) Amendment			0.686			
LSD ($_{0.05}$) water source for sawah x Amendments			NS			
Year 3						
Rain-fed	5.27	4.33	3.40	3.33	1.33	3.53
Spring	3.13	3.33	2.20	2.87	1.33	2.57
Pond	3.43	4.73	2.80	2.87	1.67	3.10
Mean	3.94	4.13	2.80	3.02	1.44	
LSD ($_{0.05}$) water source for sawah			0.318			
LSD ($_{0.05}$) Amendment			1.020			
LSD ($_{0.05}$) water source for sawah x Amendments			NS			

CT = control, NPK = nitrogen. phosphorous potassium, PD = poultry dropping, RH = rice husk, RHA = rice husk ash

As a general principle, as soil drying becomes more prolonged and severe under rain-fed condition, the availability of soil available phosphorous to rice tends to decrease and the availability of zinc in acid soils tends to increase [53]. Wakatsuki et al. [54]; Hirose and Wakatsuki, [21]; Wakatsuki et al. [55]; affirmed that under flood conditions, phosphorous availability is increased through the reduction of ferric iron. Both acid and alkaline soils are neutralized or mitigated by appropriate control of flooding. Hence, micronutrient availability is also increased. These mechanisms encourage not only the growth of rice plants, but also the growth of various aquatic algae and other aerobic and anaerobic microbes, which increase nitrogen fixation through increased photosynthesis, and control oxidation and reduction potential in *sawah* systems as multifunctional wetlands.

It was also obtained (Table 7) that the applications of amendments significantly ($p < 0.05$) highly increased the availability of phosphorous differently in the studied soil within the periods. It was noted generally that all the treated plots significantly ($p < 0.05$) increased the available phosphorous higher in the studied soil than the control plots. This result is in line with the submission that achieving high yield in most West African ecology is difficult without soil amendment, as the soils are highly leached, porous and low in essential plant nutrient [56,57]. The results (Table 7) also revealed that in all the years, organic nutrient sources did significantly ($p < 0.05$) increased the available phosphorous higher than inorganic nutrient source (NPK) indicating the superiority of organic manure over inorganic in soil and crop improvement. It was observed that among the organically amended plots, rice husk ash treated plots increased the available phosphorous significantly higher than others. This was followed by poultry droppings amended plots within the period of study. This could be linked to the increased soil pH recorded in those RHA amended plots during the study which have helped to liberate soil available phosphorous in its fixed exchange site due to acidic condition. The result agrees with the findings of Imolehin and Wada [25] who advocated a reversion to the use of organic materials in wetland rice cultivation as a more realistic option for rice farmers than continued reliance on inorganic fertilizers, which in addition to their deleterious effects on the soil are not readily available.

The results (Table 8) indicated that CEC was improved differently within a short-term by use of different water sources for *sawah* development. This means that CEC of the soil gradually responds to different water sources for *sawah* development. The result (Table 8) revealed that the spring water irrigated soils in the study significantly ($p < 0.05$) increased the cation exchange capacity higher than the pond irrigated plots, while the rainfed fields gave the least CEC values throughout the period of study. The results showed the range values as; 6.05 – 8.15 cmol(+) Kg^{-1}, 7.72 – 11.37 cmol(+) Kg^{-1}, and 8.63 – 13.77 cmol(+) Kg^{-1}, in the 1st, 2nd and 3rd year of the study. The results implied that there was a progressive increase in the cation exchange capacity as the year of study progresses. The significant improvement on the CEC by spring *sawah* system attributed to edge-advantage it has for collecting eroded sediments from adjacent uplands through enhanced capacity of water harvesting. The essence of the *sawah* system is water control, not only on a field scale but also on a watershed scale [58].

Studies have shown that sawah system isnatural soil fertility replenishment mechanisms that are essential for sustainable improvement in productivity of lowland rice farming systems in inherently unfertile soils in WA and SSA [49,50].

The results (Table 8) also showed that amendments significantly ($p < 0.05$) improve the soil CEC within the period of study. Generally, all the treated plots significantly improved the CEC higher relative to the control. Poultry dropping amendment generally improved the soil CEC higher than other amendments in the 1st year, rice husk ash and rice husk dust improved the CEC higher in the 2nd and 3rd year of study, respectively. The values varied from 4.47 – 7.69 $cmolkg^{-1}$, 4.40 – 11.38 $cmolkg^{-1}$ and 5.96 – 14.91 $cmolkg^{-1}$, in the first, second and third year, respectively.

Table 7. Effects of different water source for *sawah* and amendments on soil available phosphorous ($mgkg^{-1}$)

Water source for *Sawah*	Amendments					
	CT	NPK	PD	RH	RHA	Mean
		Year 1				
Rain-fed	3.95	4.68	4.04	4.93	7.83	5.09
Spring	3.39	5.88	6.06	7.91	9.48	6.54
Pond	2.88	6.19	6.65	6.17	7.24	5.83
Mean	3.40	5.58	6.33	6.33	8.19	
LSD ($_{0.05}$) water source for sawah			1.076			
LSD ($_{0.05}$) Amendment			1.552			
LSD ($_{0.05}$) water source for sawah x Amendments			NS			
		Year 2				
Rain-fed	3.78	4.97	7.57	6.23	7.97	6.10
Spring	4.42	10.56	8.48	10.58	15.26	8.02
Pond	3.56	8.51	8.30	9.54	10.01	9.83
Mean	3.92	8.01	8.12	8.79	11.08	
LSD ($_{0.05}$) water source for sawah			2.090			
LSD ($_{0.05}$) Amendment			2.155			
LSD ($_{0.05}$) water source for sawah x Amendments			NS			
		Year 3				
Rain-fed	3.78	6.03	8.49	6.53	8.73	6.71
Spring	5.14	11.26	10.10	10.89	18.86	11.25
Pond	3.88	9.58	10.30	10.83	10.47	9.02
Mean	4.27	8.96	9.63	9.42	12.69	
LSD ($_{0.05}$) water source for sawah			1.472			
LSD ($_{0.05}$) Amendment			2.278			
LSD ($_{0.05}$) water source for sawah x Amendments			3.671			

CT = control, NPK = nitrogen. phosphorous potassium, PD = poultry dropping, RH = rice husk, RHA = rice husk ash

Table 8. Effects of different water source for *sawah* and amendments on soil cation exchange capacity CEC ($cmolkg^{-1}$)

Water source for *Sawah*	Amendments					
	CT	NPK	PD	RH	RHA	Mean
		Year 1				
Rain-fed	4.13	5.60	6.93	6.67	6.93	6.05
Spring	5.20	8.60	9.87	8.67	8.40	8.15
Pond	4.07	6.67	6.27	6.93	6.67	6.12
Mean	4.47	6.96	7.69	7.42	7.33	
LSD ($_{0.05}$) water source for sawah		1.453				
LSD ($_{0.05}$) Amendment		1.080				
LSD ($_{0.05}$) water source for sawah x Amendments		NS				
		Year 2				
Rain-fed	4.13	8.20	8.87	9.00	8.40	7.72
Spring	5.20	10.60	13.20	13.80	14.07	11.37
Pond	3.87	9.27	10.00	9.87	11.67	8.93
Mean	4.40	9.36	10.69	10.89	11.38	
LSD ($_{0.05}$) water source for sawah		2.474				
LSD ($_{0.05}$) Amendment		1.941				
LSD ($_{0.05}$) water source for sawah x Amendments		NS				
		Year 3				
Rain-fed	3.93	10.07	9.93	10.40	8.80	8.63
Spring	6.93	13.30	18.13	17.40	13.07	13.77
Pond	7.00	13.27	16.13	16.93	11.40	12.95
Mean	5.96	12.21	14.73	14.91	11.09	
LSD ($_{0.05}$) water source for sawah		1.186				
LSD ($_{0.05}$) Amendment		0.995				
LSD ($_{0.05}$) water source for sawah x Amendments		1.769				

CT = control, NPK = nitrogen. phosphorous. potassium, PD = poultry dropping, RH = rice husk, RHA = rice husk ash

3.5 Effects of Different Water Sources and Amendments on the Rice Grain Yield (t/ha)

The effects of water sources for *sawah* development and different amendments on the rice grain yield were presented on Table 9. The results (Table 9) revealed that there was significantly (P<0.05) improvement on the rice grain yield for the three years of study in the study area. The results (Figs 4 – 8 and Table 9) showed that among the three water sources, spring water source for supplemental irrigation, increased the rice grain yield significantly ($p < 0.05$) higher than other water sources within the period of study (Figs. 4 and 6). This was followed by the pond source of water, while the rain-fed type recorded the least yield performance of rice grain yield. The increased rice grain yields recorded in the spring and pond treated fields in the study as against the low yield obtained in the rain-fed treated fields could be attributed to increased water availability in those field throughout the growing period of the plant which are the desired growing environment for rice plant (a water-loving plant). The results implied that the low productivity obtained in rain-fed fields could be attributed to management aspects of the fields rather than low physical potentials. This is in line with a submission that crop yields from rain-fed agriculture are usually low, generally around 1 t ha^{-1} compared to irrigated agriculture in semiarid tropical agro-ecosystems [7], and this fact explains why rain-fed agriculture is estimated to contribute only 60% of the world crop production [4]. IRRI [59] reported that rice production in the rain-fed lowland environment being dependent on rain-fed conditions, is very susceptible to climatic variability which results in low yields.

Kadigi et al. [6] argues that land for rain-fed agriculture varies depending on the amount and distribution of rainfall in the area. Gowing et al. [12]; Barron et al. [60]; Mupangwa et al. [61]; Makurira et al. [62] maintained that inadequate soil moisture and low soil fertility have been top challenges facing rain-fed agriculture.

Table 9. Effects of different water source for *sawah* and amendments on rice grain yield (ton/ha)

Water source for *Sawah*	Amendments					
	CT	NPK	PD	RH	RHA	Mean
	Year 1					
Rain-fed	1.80	4.40	4.20	3.10	4.00	3.50
Spring	2.03	5.37	5.73	5.37	5.23	4.75
Pond	1.77	4.63	4.83	3.13	3.17	3.51
Mean	1.87	4.8	4.92	3.87	4.13	
LSD $_{(0.05)}$ water source for sawah		0.7156				
LSD $_{(0.05)}$ Amendment		0.6250				
LSD $_{(0.05)}$ water source for *sawah* x Amendments		NS				
	Year 2					
Rain-fed	2.10	4.73	4.70	4.53	4.53	4.12
Spring	1.97	5.77	5.77	5.30	4.80	4.72
Pond	1.83	5.10	5.13	4.93	4.67	4.33
Mean	1.97	5.20	5.20	4.92	4.67	
LSD $_{(0.05)}$ water source for sawah		0.2132				
LSD $_{(0.05)}$ Amendment		0.400				
LSD $_{(0.05)}$ water source for sawah x Amendments		NS				
	Year 3					
Rain-fed	2.60	6.31	6.45	4.98	6.45	5.36
Spring	4.21	7.30	8.27	7.22	7.78	6.96
Pond	3.93	7.01	6.17	6.66	6.45	6.04
Mean	3.58	6.87	6.96	6.29	6.89	
LSD $_{(0.05)}$ water source for sawah		1.081				
LSD $_{(0.05)}$ Amendment		0.809				
LSD $_{(0.05)}$ water source for sawah x Amendments		NS				

CT = control, NPK = nitrogen. phosphorous potassium, PD = poultry dropping, RH = rice husk, RHA = rice husk ash

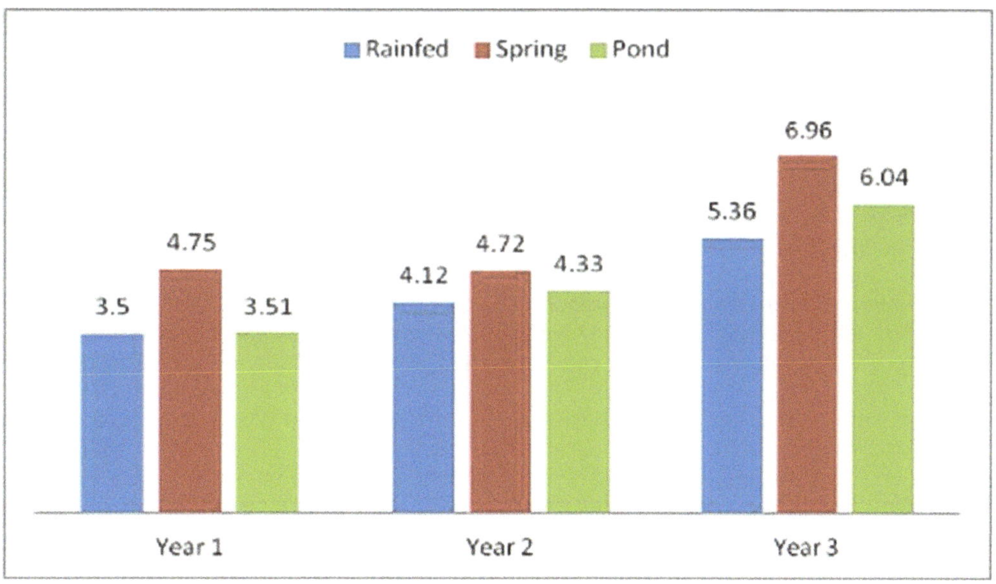

Fig. 4. Effect of different water sources on the rice grain yield (t/ha) as affected by water sources

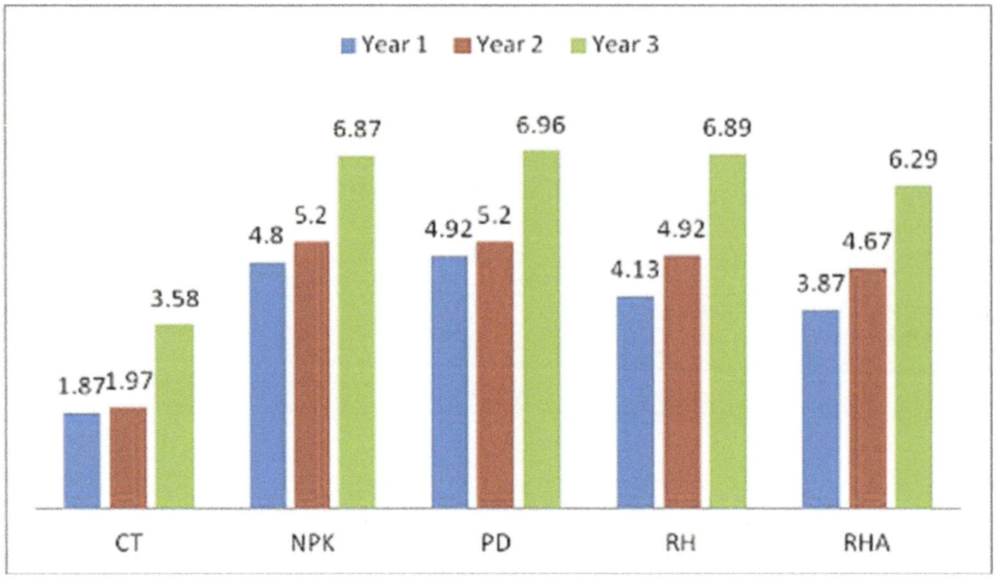

Fig. 5. Effect of soil amendments on the rice grain yield (ton/ha) as affected by amendments

Fig. 6. Yield from spring *Sawah* adopted rice field

Fig. 7. Yield from Pond *Sawah* adopted rice field

Fig. 8. Yield from Rain-fed *Sawah* adopted rice field

However, the higher yield recorded in rain-fed plots above the standard 2 t/ha yield for traditional rice production in the studied area could be attributed to high management practices such as improved water control and soil amendments adopted in this study.

The above result also agrees with the findings of Buri et al. [63] who maintained that lowlands constitute one of the largest and appropriate environments suitable for rice cultivation. They further stated that, within these environments, crop is traditionally grown without any structures to control water, minimal use of fertilizers and most often than not local varieties are used. Paddy yields are therefore normally low under the traditional system and vary sharply due to yearly variation in total rainfall and its distribution. They further reported that rice yield in the *sawah* system is usually about 2–3 t ha^{-1} without any fertilizer application, and this yield is continuously attainable at least for several decades without any fallow period. The results (Fig. 5) also revealed the short-term superiority of organic amendments over mineral (inorganic) fertilizer in a lowland rice production. It was obtained that among the amendments; rice grain yield was increased significantly ($p < 0.05$) higher in poultry dropping (PD) treated plots than NPK fertilizer amended plots in all the years studied (Fig. 5). This result is in line with the findings of Imolehin and Wada [25] who suggested that it is better to revert to the use of organic materials in wetland rice cultivation as a more realistic option for farmers than continued reliance on inorganic fertilizers, which not only affect the soil negatively, but cannot be readily available. It was also recorded that rice husk (RH) followed the PD in improving the grain yield of rice on the third year of the study. The results generally indicated that all the amended plots increased the rice grain yield significantly higher than the control.

4. CONCLUSION

The study revealed the successful improvement of spring water source on both soil chemical properties and rice grain yield over other water sources within the study period, through its mechanisms of regular geological fertilization process that do occur in inland valley *sawah* system. The study showed that supplemental irrigation gave higher significant improvement than the rain-fed water source on the soil chemical properties studied and rice grain yield on a short-term basis. Organic amendments have been observed to have superior improvement on some chemical properties of the studied soil over mineral fertilizer on a short-term basis. The integration of irrigation for *sawah* management system and amendment practices could be advocated for sustainable improvement of soil properties and rice grain yield in degraded inland valleys of Southeastern Nigeria. Therefore, *sawah* eco-technology is possibly the most promising rice production strategy for sustainable restoration of degraded inland valley soils in the Southeastern Nigeria.

COMPETING INTERESTS

Authors have declared that no competing interests exist.

REFERENCES

1. Adesina A. Keynote address on: Achieving a doubly green agricultural transformation in Nigeria, delivered by Dr. Akinwunmi Adesina, Hon. Minister of Agric. and Rural Dev., on the occasion of 36th annual conference of the Soil Science Society of Nigeria, University of Nigeria, Nsukka, Enugu State, 12th March 2012. Proceedings of the 36th Annual Conference of the Soil Science Society of Nigeria (SSSN) 12th – 16th March 2012. University of Nigeria, Nsukka. Pp 17–14.
2. FAOSTAT. Database. Food and Agriculture Organization, Rome; 2005. Available:http://faostat.fao.org/ (Accessed November 2005)
3. Bhattacharya A. Sustainable Livelihood Based Watershed Management – Watershed plus Approach, 2nd Working Group meeting of ERIA, Japan IGES; 2008.
4. FAO. Nepal country profile; 2002. Available:http://www.fao.org/countryprofiles
5. Wani SP, Sreedevi TK, Rockstrom J, Ramakrishna YS. Rainfed agriculture – Past trends and future prospects. In: Wani SP, Rockstrom J, Oweis T (eds). Rainfed Agriculture: unlocking the potential, Oxfordshire: CABI International; 2009.
6. Kadigi RMJ, Kashaigili JJ, Mdoe NS. The economics of irrigated paddy in Usangu Basin in Tanzania: Water utilization, productivity, income and livelihood implications. Phys. Chem. Earth, 29/15-18. 2004;1091-1100.
7. Rockstrom J. Green water security for the food makers of tomorrow: Windows of opportunity in drought-prone savannahs. Water Science and Technology. 2001; 43(4):71-78.
8. Agarwal A, Narain S. Dying wisdom. Rise, Fall and Potential of India's Traditional Water Harvesting System. Centre for Science and Environment, Faridabad. India: Thomson Press Ltd; 1997.
9. Benites J, Chuma E, Fowle R, Kienzle J, Molapong K, Manu J, Nyagumbo I, Steiner K, van Veenhuizen R. (eds). Conservation Tillage for Sustainable Agriculture. Proceedings from an International Workshop, Harare, 22–27 June. Part 1. Workshop Report. Deutsche Gesellschaft, GTZ, Eschborn, Germany. 1998;59.
10. Rockström J, Falkenmark M. Semiarid crop production from a hydrological perspective: Gap between potential and actual yields. Critical Reviews in Plant Sciences. 2000;19(4):319-346.
11. SIWI. Water Harvesting for Upgrading of Rain-fed Agriculture. Problem Analysis and Research Needs. SIWI Report 11, Stockholm International Water Institute (SIWI), Stockholm, Sweden. 2001;97.
12. Gowing JW, Young MDB, Hatibu N, Mahoo HF, Rwehumbiza F, Mzirai OB. Developing improved dryland cropping systems for maize in semi-arid Tanzania. Part II. Use of a Model to Extrapolate and Add Value to Experimental Results, Exp. Agric. 2003; 9(3):293-306.
13. Abu – Awwad M, Kharabshed A. Influence of supplemental irrigation and soil surface furrow on barley yield in arid areas affected by surface crust. J. Arid Environ. 2000;46: 227–237.
14. Moormann FR. Problem in characterizing and classifying wetland soils. In wetland soils. Characterization, classification,

utilization. Proceeding of a workshop 26 mar. to 5 April 1984. 1985;53-68. IRRI, Los Banos, Philippines.
15. Wakatsuki T, Koski T, Palada M. Ecological engineering for sustainable rice farming in inland valley (Ivs) in West Africa. Paper presented at the second WAFSRN symposium. Accra, Ghana; 1989.
16. Windmeijer PN, Andriesse W. Inland valleys in West Africa: An agro-ecological characteristic of rice- growing environment. ILRI. Wageningen, The Netherlands. 1993;28-37.
17. Otoo E, Asubonteng KO. Reconnaissance characterization of inland valleys in Southern Ghana. In characterization of inland valley Agron-ecosystems. A tool for their sustainable use. Proceeding of a workshop, 6 to 10 Nov. 1995;149-160. WARDA, Bouake, Ivory Coast.
18. Mbagwu JSC. The agricultural soils of Nigeria: Properties and agronomic significance for increased productivity. Beitrage für Tropical Landwirtschaften und Veterinari Medizin. 1989;27:395-409.
19. Nnabude PC, Mbagwu JSC. Soil water relations of a Nigerian typic haplustult amended with fresh and burnt rice mill wastes. Soil and Tillage Res. 1999;50:207-214.
20. Mbagwu JSC. Improving the productivity of a De-graded ultisol in Nigeria using organic and inorganic amendments. Part 2: Changes in physical properties. Bioresource Technology. 1992;42(3)167-175.
21. Hirose S, Wakatsuki T. Restoration of inland valley ecosystems in West Africa. 2002;56-86,222-2224. Association of agriculture and forestry statistics. Megro-Sumiya building, Tokyo, Japan.
22. Igwe CA. Erodibility of soils of the upper rainforest zone, Southeastern Nigeria. Land Degradation & Development. 2003; 14(3):323-334.
23. Hayashi K, Wakatsuki T. Sustainable soil fertility management by indigenous and scientific knowledge in Sahel zone of Niger, in the CD- ROM Transactions of the 17th World congress of soil science, symposium No. 15. perceptions of soil management: Matching indigenous and scientific knowledge systems, paper No. 1251; 2002.
24. Wakatsuki T, Masunaga T. Ecological engineering for sustainable food production and the restoration of degraded watersheds in Tropics of low pH soils: Focus on West Africa. Soil Sci. Plant Nutri. 2005;51:629-636.
25. Imolehin ED, Wada AC. Meeting the rice production and consumption needs of Nigeria with improved technologies. Int Rice Commiss Newsl FAO, Rome. 2000;49:33–41.
26. Lee SB, Lee CB, Jung KY, Park KD, Lee D, Kim PJ. Changes of soil organic carbon and its fractions in relation to soil physical properties in a long-term fertilized paddy. Soil Till Res. 2009;104:227–232.
27. USDA. Keys to soil taxonomy. Natural Resources Conservation Services, United States Department of Agriculture, Washington, D.C; 1998.
28. FAO. Soil Map of the World: 1:5 million (Revised Legend). World Soil Resources Report, 60. Food and Agricultural Organization (FAO), Rome; 1988.
29. Gee GW, Bauder JW. Particle size analysis. In: Klute A, (ed.): Methods of soil analysis, Part 1: Physical and mineralogical properties. Monograph No. 9. American Society of Agronomy, Madison. 1986;91–100.
30. McLean EO. Soil pH and Lime requirement. In: Page AL, et al. (eds.), Methods of soil analyses (No. 9, part 2), 1982;199-224. Amer. Soc. of Agron. Soil Sci. Soc. Am; Inc. Madison, Wisconsin, U.S.A.
31. Nelson DW, Sommers LE. Total carbon, organic carbon and organic matter. In: Page AL, et al. (eds). Methods of soil analyses (No. 9, part 2), 552-553. 1982. Amer. Soc. of Agron. In: Soil Sci. Soc. Amer., Inc, Madison, Wisconsin, U.S.A.
32. Bremner JM, Mulvancy CS. Total nitrogen. In: Page AL, et al (eds). Methods of soil analyses. No. 9; part 2, Amer. Soc. of Agron. Inc, Madison, Wisconsin, USA. 1982;595-624.
33. Bray RH, Kurtz LT. Determination of total organic carbon and available forms of phosphorous in soils. Soil Sci. J. 1945;59: 39-43.
34. Rhoades JD. Cation exchange capacity. In: Page AL, Miller RH, Keeny DR, (eds). Methods of Soil Analysis, Part 2. Am. Soc. Agron., Madison. 1982;149-157.
35. Takase M, Sam-Amoah LK, Owusu-Sekyere JD. The effects of four sources of irrigation water on soil chemical and physical properties. Asian Journal of Plant Science. 2011;10(1):92 – 96. ISSN 1682 – 3974/ DOI: 10.3923/ajps.2011.92.96 ©

2011 Asian Network for Scientific Information.
36. Abyhammer T, Fablin A, Nelson A, Henfrindison V. Askater Foringssystem Deiproject I: Tekniker Ochmojiligheter. (Production of wood ash, techniques and possibilities). 1994;341. In Swedish with English Summary).
37. Markikainen PN. Nitrification in two coniferous forest soils after different fertilizer treatments. Soil Biol. Biochem. 2002;16:577–882.
38. Nwite JC, Obalum SE, Igwe CA, Wakatsuki T. Properties and potential of selected ash sources for improving soil condition and rice yields in a degraded Inland Valley in Southeastern Nigeria. World Journal of Agricultural Sciences. 2011;7(3):304– 310. ISSN 1817 – 3047.
39. Opara-Nadi OA, Ezua BS, Wogu A. Organic manures and inorganic fertilizers addeded to an acid ultisol in Southeastern Nigeria: II. Effects on soil chemical properties and nutrient loss, In: proceedings of the 15th Annual Conf. SSSN, Kaduna, Nigeria; 1987.
40. Follet RF. Soil carbon sequestration and greenhouse gas mitigation. Soil Sci Soc Am J. 2010;74:345–346.
41. Bhagat RM, Verma TS. Impact of rice straw management on soil physical properties and wheat yield. Soil Sci. 1991;152:108-115.
42. Buresh RJ, Castillo EG, De Datta SK. Nitrogen losses in puddled soils as affected by timing of water deficit and nitrogen fertilization. Plant and Soil. 1993;157:197-206.
43. Buresh RJ, Reddy KR, van Kessel C. Nitrogen transformations in submerged soils. In Nitrogen in agricultural systems. (Eds JS Schepers, WR Raun). 2008;401-436. Agronomy Monograph 49. (ASA, CSSA, and SSSA: Madison, WI, USA).
44. Becker M, Johnson DE. Improved water control and crop management effects on lowland rice productivity in West Africa. Nutrient Cycling Agroecosystems. 2001; 59:119–127.
45. Sakurai T. Intensification of rainfed wetland rice production in West Africa: Present status and potential green revolution. Developing Economies. 2006;44:232–251.
46. Touré A, Becker M, Johnson DE, Koné B, Kossou DK, Kiepe P. Response of lowland rice to agronomic management under different hydrological regimes in an inland valley of Ivory Coast. Field Crops Research. 2009;114:304–310.
47. Kyuma K, Wakatsuki T. Ecological economy sustainability of paddy rice systems in Asia. In: Juo ASR, Russell DF. (eds.); Agriculture and Environment. Bridging Food production in developing countries. ASA special publication No. 60, 1995;139-159. ASA, CSSA, SSA, Wisconsin.
48. Greenland DJ. Sustainability of rice farming, CABI, Wallingford, and IRRI, Los Banõs, The Philippines; 1997.
49. Eswaran H, Almaraz R, Van den Berg E, Reich P. An assessment of the soil resources of Africa in relation to productivity. Geoderma. 1997;77:1–18.
50. Abe SS, Buri MM, Issaka RN, Kiepe P, Wakatsuki T. Soil fertility potential for rice production in West African lowlands. Japan Agricultural Research Quarterly. 2010;44: 343–355.
51. Errikson H. Short term effects of granulated wood ash on forest soil chemistry in Southwest and Northwest Sweden. Scandinvian J. Forest. Res. Supplement. 1998;2:43–55.
52. Serafinelion A. Wood ash: An alternative liming material for agricultureal soils. Soil Bullettin. 2002;35:80–95.
53. Dobermann A, Fairhurst T. Rice: Nutrient disorders and nutrient management. (Potash and Phosphate Institute and Potash and Phosphate Institute of Canada: Singapore and International Rice Research Institute: Los Baños, Philippines); 2000.
54. Wakatsuki T, Shinmura Y, Otoo E, Olaniyan DO. System for integrated watershed management of small inland valleys in West Africa. In: Institutional and Technical Options in the Development and Management of Small Scale Irrigation, Water Report No 17, FAO, Rome, 1998; 45–60.
55. Wakatsuki T, Buri MM, Oladele OI. West African green revolution by eco-technology and the creation of African SATOYAMA systems. Kyoto Working Papers on Area Studies No. 63 (G-COE Series 61). Center for Southeast Asian Studies, Kyoto, Japan. 2009;30. ISBN 978 4 901668 63 7. Available:http://www.humanosphere.cseas.kyoto-u.ac.jp/article.php/workingpaper61
56. Enwezor WO, Ohiri AC, Opuwaribo EE, Udo EJ. A review of soil fertilizer use of

crops in Southeastern zone of Nigeria. Fertilizer Procurement and Distribution Department, Lagos, Nigeria; 1988.

57. Igwe CA, Akamigbo FOR, Mbagwu JSC. Physical properties of soils of Southeastern Nigeria and the role of some aggregating agents in their stability, Soil Sci. 1995;160:431–441.
58. Abe SS, Wakatsuki T. Ecotechnology – A tiger for a rice green revolution in Sub-Saharan Africa: Basic concept and policy implications. Outlook on Agriculture. 2011;40(3):221–227. DOI: 10.5367/oa.2011.0049
59. IRRI (International Rice Research Institute). Physical measurements in rice soils: the Japanese methodologies. Los Baños (Philippines): IRRI; 1987.
60. Barron J, Rockstrom J, Hatibu N, Gichuki F. Dry spell occurrence and maize yields for two locations in semi-arid East Africa. Agricultural Forest and Meteorology (in press). 2003;117(1–2):23–37.
61. Mupangwa W, Love D, Twomlow S. Soil–water conservation and rainwater harvesting strategies in the semi-arid Mzingwane Catchment, Limpopo Basin, Zimbabwe, Phys. Chem. Earth, 31 15-16. 2006;893-900.
62. Makurira H, Mul ML, Vyagusa NF, Uhlenbrook S, Savenije HHG. Evaluation of community-driven smallholder irrigation in dryland South Pare Mountains, Tanzania: A case study of Manoo micro dam. Phys. Chem. Earth, 32(15-18). 2007;1090-1097.
63. Buri MM, Issaka RN, Wakatsuki T, Kawano N. Improving the productivity of lowland soils for rice cultivation in Ghana: The role of the 'sawah' system. Journal of Soil Science & Environment Management. 2012;3(3):56–62.

23

Soil Physical Responses of a Compacted Sports Field Following Various Core Aerification Techniques

P. J. Brown[1], J. L. Atkinson[2], L. B. McCarty[1*] and V. L. Quisenberry[1]

[1]Clemson University, Clemson, SC, 29634, USA.
[2]11550 North Meridian Street, Suite 600, Carmel, IN 46032, USA.

Authors' contributions

This work was carried out in collaboration between all authors. Authors LBM, VLQ and PJB designed the study and wrote the protocol. Author JLA wrote the first draft of the manuscript and managed the literature searches. Author PJB conducted the in field research, data collection, and analysis. All authors read and approved the final manuscript.

Editor(s):
(1) Ying Ouyang, United States Department of Agricultural Forest Service, USA.
Reviewers:
(1) Edward Missanjo, Malawi College of Forestry and Wildlife, Malawi.
(2) Arturo Perez-Vazquez, Colegio de Posgraduados, Campus Veracruz, Mexico.

ABSTRACT

Aims: Soil compaction is detrimental to turfgrass health and potentially hazardous to users of sports field participants. Previous research has evaluated numerous core aeration programs and their effects on water infiltration and thatch levels in a myriad of soil series under various management programs. The goal of this study was to identify detailed soil responses of a highly compacted, multi-purpose sports field to different soil aerification techniques while minimizing the negative impact of core removal on turfgrass quality.

Study Design: The study was designed as a randomized complete block with 4 replications and a 2 × 2 factorial treatment arrangement.

Place and Duration of Study: The study was conducted over two years on Clemson University's band practice field, Clemson, SC, USA.

Methodology: The study evaluated deep (17.8 cm) and shallow (7.6 cm) tine core aerification and number of yearly aerification events on several soil and turf parameters. Extracted cores were either removed or incorporated back into the plots.

**Corresponding author: E-mail: philipb@clemson.edu*

Results: Little effect on turfgrass quality was observed due to mowing height (2.65 cm) masking any reduction in turfgrass density. Deep tine aerification lowered bulk density 5% in the first year compared to shallow tine aerification. In year one, infiltration was increased 29% after deep tine core aeration and 34% when cores were removed after aerification.
Conclusion: Treatment effects were not as drastic as studies conducted in sand based soils, reflecting the necessity of a perpetual soil aerification program in clay-based soils.

Keywords: Soil compaction; turfgrass; bulk density; clegg impact value; turfgrass quality; aerification.

ABBREVIATIONS

BD: Bulk density; CIV: Clegg impact value; TQ: Turfgrass quality.

1. INTRODUCTION

Soil compaction is a potentially serious problem for turfgrass managers and users of sports fields. The soil physical properties of a compacted soil can adversely influence plant growth, efficiency of an irrigation program [1] and decrease player safety as surface cushioning will not be adequate to prevent injury during use. Compaction is most problematic in high traffic areas [2].

Soil compaction is the reduction of macro-porosity by the pressing together of soil particles, resulting in a more dense soil mass with less aeration porosity, increased bulk density, increased soil strength, and altered pore size distribution [1]. Compaction also often results in destruction of soil structure [3]. In turfgrass systems, soil compaction discourages root and shoot growth, slows replenishment of carbohydrate reserves, and eventually causes a decline in overall plant health.

In commercial turfgrass, soil aerification is typically performed with vertically operating hollow tine core aeration units, which selectively remove soil cores from the soil profile [3]. More recently, soil aerification with solid tines has gained in popularity as they typically cause less surface disruption and have lower labor costs due to less post-aerification cleanup compared to hollow tine aerification [3]. However, solid tine core aeration may create an impermeable layer at the bottom of the aerification zone and directly adjacent to coring holes [4].

Previous research has evaluated numerous core aeration programs and their effects on water infiltration and thatch levels in a myriad of soil series under various management programs. Detailed soil responses to various cultivation techniques such as deep tine and shallow tine aerification have received limited attention. The goal of this study was to identify detailed soil responses to different soil aerification techniques in a highly compacted, multi-purpose sports field while minimizing the negative impact of core removal on turfgrass quality.

2. MATERIALS AND METHODS

2.1 Site Description

A two-year field study was conducted between 2002 & 2003 to evaluate various aerification techniques on soil physical properties on the Clemson University band practice field in Clemson, SC. The field presented a worst-case-scenario combination of frequent soil compression from a variety of sources and a Cecil soil series [Cecil sandy clay loam (fine, kaolinitic, thermic Typic Kanhapludults)] that is highly susceptible to compaction, particularly after catastrophic disruption to existing soil structure which likely occurred during shaping of the field surface and subsequent daily use. Prior to this study, no known core aerification was performed on this field. The field was exposed to daily band practice and was also used as an overflow parking lot on home football games. The site was planted with common bermudagrass (*Cynodon dactylon* L.) and maintained by the Clemson University Facilities and Landscape Service Department. In the absence of rain, the field was irrigated with ~1.25 cm every 3 d, mowed once weekly at 2.65 cm using a reel mower, fertilized with at 45 kg N ha-1 monthly during the growing season to total 270 kg N ha-1 annually, and limed as recommended by soil tests. The field was not overseeded to provide green turf during winter dormancy.

Treatments consisted of an nontreated control, deep hollow tine (17.8 cm depth, 1.6 cm diameter) aerification with a tractor mounted aerification unit (Southern Green, Zachary, LA 70791) with cores removed; deep hollow tine aerification without core removal; shallow hollow tine (7.6 cm depth, 1.3 cm diameter) aerification with a walk-behind aerification unit (Ryan Manufacturers, Lincoln, NE 68504) with cores removed; and, shallow hollow tine aerification without cores removed (Table 1). All treatments were applied on 5 x 5 cm centers. Treatments were made 2 d after an irrigation or rainfall event to ensure uniform soil moisture. Treatments were applied to the same plots for both years of the study.

For treatments with cores removed, cores were swept into the middle of the plot using a stiff shop broom then collected using a flat-head shovel and discarded. In treatments with cores re-incorporated, cores were allowed to dry on the soil surface and then crumbled to reincorporate back into the soil profile using a shop broom in several directions until cores disintegrated into aerification holes. The design was a randomized complete block with 4 replications and a 2 × 2 factorial treatment arrangement.

2.2 Variables Measureds

2.2.1 Turfgrass quality

Visual turfgrass quality (TQ) ratings were based on color, shoot density, and uniformity of stand using a scale of 1 to 9 with 1 representing dead turf and 9 representing dark green, dense turf. Visual TQ was evaluated monthly from May to October during each year.

2.2.2 Bulk density

Soil BD was measured 15 d after the final treatment by removing a 154.4 cm3 soil core between 2 and 8 cm below the turf surface from each plot using an A-145 Soil Core Sampler (ELE International, Lake Bluff, IL). Cores were dried in a forced air oven for 48 h at 105ºC, then weighed (g). Bulk density (g cm^{-3}) was determined by dividing dry soil core mass by total soil core volume.

2.2.3 Surface hardness

Surface hardness was assessed 14 d after each aerification event using a Clegg impact soil tester (Lafayette Instrument Co., Lafayette, IN) where a 2.25 kg weight was dropped from a height of 45 cm. Upon impact with the ground, the tester measured deceleration of the weight, or Clegg impact value (CIV), to indicate a relative surface hardness value. The CIVs were multiplied by 10 to convert to gmax (peak deceleration in gravities). Measurements were made 2 d following irrigation to ensure uniform soil moisture content. Two readings were taken per plot during each assessment and averaged before statistical analysis.

2.2.4 Infiltration rate

Infiltration rate was measured using a double ring infiltrometer (IN5-W, Turf-Tec International, Tallahassee, FL) with an outer ring diameter of 30.5 cm, and inner ring diameter of 15.2 cm. Measurements were taken 15d (+5 d) after the final treatment and 2 d following irrigation to ensure uniform soil moisture. Rings were randomly placed within each plot and inserted 2 cm below soil surface. Water was added in both rings to the top of the infiltrometer. Infiltration (cm h-1) was calculated as the drop in water level after 10 min from an initial height of 8 cm in the inner ring and converted to cm hr-1.

Table 1. Treatment list comparing deep (17.8 cm) and shallow (7.6 cm) hollow tine core aerification with cores removed or incorporated. Treatments were applied either 1, 2, or 4X per year in Clemson, South Carolina

Aerification depth	Cores removed	Aerification timing
cm		
17.8	yes	June
17.8	no	June
17.8	yes	June & Sept.
17.8	no	June & Sept.
17.8	yes	June, July, August, & Sept.
17.8	no	June, July, August, & Sept.
7.6	yes	June
7.6	no	June
7.6	yes	June & Sept.
7.6	no	June & Sept.
7.6	yes	June, July, August, & Sept.
7.6	no	June, July, August, & Sept.

2.3 Statistical Analysis

All data were analyzed using ANOVA to test for overall treatment effects and interactions, and Fisher's protected LSD to test for specific differences among pairs of treatment levels. Analyses focused on the effects of aerification depth and aerification frequency, and interactions of these on parameters measured. All tests were performed at a significance level of 0.10. This level was chosen to avoid type II errors that could occur due to the inherent variability of soil measurements [5]. The General Linear Model procedure of SAS [6] was used for all calculations.

3. RESULTS AND DISCUSSION

Treatment effects of core aeration depth, frequency, and core removal or incorporation varied between measured parameters and years (Table 2). Because of main effect-by-year interaction, main effects will be discussed for each parameter, separated by year, and further interaction effects discussed where appropriate.

3.1 Turf Quality

A certain degree of disruption to TQ should be expected to improve soil physical properties through core aeration. Frequent core aeration may reduce the amount of time a field is in acceptable playing condition, often during periods of heavy use. Minimizing surface disruption is an important consideration when developing a balanced, effective core aeration program.

Much of the previous research evaluating core aeration programs was conducted on turfgrass stands maintained at <0.66 cm (0.25 in) mowing height [3,7-10] Typically at lower mowing heights, reducing the number of core aerations per growing season limits thinning of turf cover and subsequent healing period necessary to return to pre-core removal turfgrass density [7]. The taller mowing height in this study (2.65 cm) helped mask any reduction in turfgrass density following core aeration. Variation between experiments in turfgrass stands of similar mowing height may be attributed in part to differences in turfgrass cultivar, soil fertility, water, or light availability, among other differences in environmental conditions effecting speed of recovery, turfgrass color, and density.

In this study, TQ values do not necessarily reflect TQ at the end of the experiment; rather, they reflect mean TQ across the experiment duration. In year one, significant treatment effects ($P < 0.05$) were limited to treatments core aerated 4 times per year (Table 3). In these treatments, deep tine core aeration improved TQ 4.7% compared to shallow tine core aeration. Additional variance in TQ values could not be attributed to other treatment effects (Table 2).

In year two, TQ was 2-3% higher in treatments aerified once yearly compared to treatments aerified 2 or 4 times yearly and the untreated

Table 2. Analysis of variance among aerification depth (17.8 vs. 7.6 cm), core removal, and aerification events (0, 1, 2 or 4X), on turfgrass quality and soil physical properties in Clemson, SC

Source	df	Year 1				Year 2			
		Turfgrass quality	Bulk density	Surface hardness	Infiltration	Turfgrass quality	Bulk density	Surface hardness	Infiltration
Depth (D)	1	ns†	*	*	*	ns	ns	*	ns
Cores Removed (CR)	1	ns	*	ns	*	ns	*	*	*
Aerification Events (AE)	3	ns	*	ns	*	*	*	*	*
D x CR	1	ns	*	*	ns	ns	*	*	*
D x AE	3	*	*	ns	*	*	ns	*	ns
CR x AE	3	ns	*	*	ns	ns	*	*	*
D x CR x AE	3	*	*	*	ns	*	*	*	ns

**Significance at α = 0.10 level, † ns, not significant*

(Table 4). Frequent removal of turf cover through core aerification can result in an overall decrease in season-long TQ if turfgrass density is a consideration in this measurement [7]. Treatment effects of aerification depth and core removal or incorporation were not significant in year two (Table 2).

In previous research, the overall effect of core aeration on turfgrass quality has been inconsistent. Turf quality improved from core aeration in some studies [3,11-13] and declined in others [14,15]. Numerous factors are responsible for the overall effect of core aeration on turfgrass quality. Site history and soil type determine the amount of aerification necessary while turf cultivar, fertilization, irrigation, and other cultural practices determine speed of recovery after core aeration. Short-term studies may not fully capture long-term effects of core aeration on turfgrass quality. Depending on existing conditions, multiple years of a consistent soil aerification program may be necessary before appreciable improvements to turfgrass quality will be attained. Additional evaluations of long-term core aeration programs are necessary to separate the short-term effects of core aeration on turfgrass density from the long-term impact of core aeration on overall plant health.

3.2 Bulk Density

In a compacted soil with high bulk density, gas exchange and water holding capacity are low, water binds tightly to micro pores and the resistance roots face to growth is increased which discourages rhizosphere development for more efficient utilization of soil available water and nutrients [1]. In a clay to clay-loam soil, a bulk density value of 1.0 to 1.6 g cm^{-3} is recommended to balance water retention with drainage and soil gas exchange [16]. Core aeration aims to reduce bulk density of compacted soil by reducing the amount of soil mass per given soil volume.

Table 3. Turfgrass quality, soil bulk density, surface hardness, and infiltration responses to deep tine (17.8 x 1.6 cm) and shallow tine (7.6 x 1.3 cm) aerification in all plots receiving four aerifications yearly in Clemson, SC

Aerification depth	Year1				Year2			
	Turfgrass quality†	Soil bulk density	Surface hardness	Infiltration	Turfgrass quality	Soil bulk density	Surface hardness	Infiltration
cm	1 - 9	g cm^{-3}	g_{max}‡	mm hr^{-1}	1 - 9	g cm^{-3}	g_{max}	mm hr^{-1}
17.8	7.02a§	1.16b	79.24a	21.75a	6.82a	1.24a	95.39c	14.25a
7.6	6.70b	1.27a	81.49a	21.75a	6.89a	1.25a	100.24a	17.25a
Nontreated	6.79ab	1.28a	79.29a	14.81b	6.88a	1.28a	98.82b	17.63a
p-value	0.06	0.05§	<0.01	1.00	0.61	0.55	<0.01	0.37
LSD (0.1)	0.28	0.09	2.35	5.79	0.23	0.05	0.21	5.56

†Visual turfgrass quality ratings were based on a scale of 1 to 9 where 1 = brown dead turfgrass and 9 = green healthy turfgrass, ‡Relative surface hardness value quantifies deceleration of 2.25 kg weight dropped from height of 45 cm, §Means followed by the same letter within a column are not significantly different at the 0.10 significant level

Table 4. Turfgrass quality, soil bulk density, surface hardness, and infiltration responses to one, two, or four core aeration events per year in Clemson, SC

Aerification event yr^{-1}	Year 1				Year 2			
	Turfgrass quality	Soil bulk density	Surface hardness	Infiltration	Turfgrass quality	Soil bulk density	Surface hardness	Infiltration
	1 - 9†	g cm^{-3}	g_{max}‡	mm hr^{-1}	1 - 9	g cm^{-3}	g_{max}	mm hr^{-1}
0	6.79a	1.28a§	79.29a	14.81a	6.88ab	1.28a	98.82ab	9.88a
1	6.91a	1.23ab	79.68a	17.25a	7.02a	1.29a	98.29b	9.83a
2	6.96a	1.24ab	79.60a	16.69a	6.80b	1.27ab	100.09a	10.01a
4	6.86a	1.21b	79.29a	14.81a	6.86b	1.25b	97.87b	9.78a
p-value	0.75	0.74	0.14	0.02	0.14	0.14	0.07	0.35a
LSD (0.1)	0.25	0.06	2.35	3.74	0.16	0.03	0.15	3.75

†Visual turfgrass quality ratings were based on a scale of 1 to 9 where 1 = brown dead turfgrass and 9 = green healthy turfgrass, ‡Relative surface hardness value quantifies deceleration of 2.25 kg weight dropped from height of 45 cm, § Means followed by the same letter within a column are not significantly different at the 0.10 significant level

Bulk density values are often inconsistent from year-to-year as a reflection of changing levels of traffic, cultivation, and soil moisture content throughout the year. Also for these reasons, treatment effects in this study were inconsistent between years (Table 2). In year one, deep tine core aeration reduced BD ~5% compared to the untreated and shallow tine core aeration (1.28 and 1.26 to 1.20 g cm^{-3}, respectively). Deep tine core aeration did not further reduce BD in year two; however, BD remained below pre-study levels. Shallow tine core aeration did not affect BD compared to the untreated in either year (Table 5).

Within treatments, removal or incorporation of cores back into the soil profile after core aeration did not affect BD in year one. In year two, core removal reduced bulk density 5% from 1.33 to 1.26 g cm^{-3} compared to incorporating cores back into the soil profile (Table 6). Although incorporation of soil back into the profile following core aeration does not drastically reduce the amount of soil mass per given soil volume per se, re-incorporated soil is likely loosely packed and a portion of removed soil will invariably remain on the soil surface. The additional reduction in soil mass by core removal following core aeration may not be sufficient to consistently reduce bulk density compared to reincorporating loosely packed soil material into core aeration holes.

Treatment effects were more consistent when analyzed by number of core aerations performed per year. Treatments with four core aerifications per year lowered bulk density 5% compared to the untreated in year one from 1.28 to 1.21 g cm^{-3} and 2% in year two from 1.28 to 1.25 g cm^{-3} (Table 4). Multiple core aerations have been necessary to reduce bulk density relative to the untreated in similar studies [7,10]. Only performing one core aeration within a year may not disrupt a large enough percentage of total soil volume to impact BD. Calculated at a soil depth of 17.8 cm, deep tine core aeration (17.8 cm) impacted 8% of total soil volume and shallow tine core aeration (7.6 cm) impacted only 2% of total soil volume. Impacting a higher percentage of total soil volume through multiple core aerations may be necessary to significantly affect soil bulk density, especially in dense clay-based soils.

Table 5. Turfgrass quality, soil bulk density, surface hardness, and infiltration responses to deep tine (17.8 x 1.6 cm) and shallow tine (7.6 x 1.3 cm) aerification in Clemson, SC

Aerification depth	Year 1				Year 2			
	Turfgrass quality	Soil bulk density	Surface hardness	Infiltration	Turfgrass quality	Soil bulk density	Surface hardness	Infiltration
cm	1 - 9†	g cm^{-3}	g_{max}‡	mm hr^{-1}	1 - 9	g cm^{-3}	g_{max}	mm hr^{-1}
17.8	6.92a	1.20b‡	77.50b	19.13a	6.83a	1.27a	97.06c	15.25a
7.6	6.89a	1.26a	80.34a	18.00ab	6.96a	1.27a	100.04a	15.88a
Nontreated	6.79a	1.28a	79.29ab	14.81b	6.88a	1.28a	98.82b	17.63a
p-value	0.76	0.08§	<0.01	0.12	0.25	0.7	<0.01	0.51
LSD (0.1)	0.25	0.05	2.35	3.36	0.14	0.03	0.13	3.32

†Visual turfgrass quality ratings were based on a scale of 1 to 9 where 1 = brown dead turfgrass and 9 = green healthy turfgrass, ‡Relative surface hardness value quantifies deceleration of 2.25 kg weight dropped from height of 45 cm, § Means followed by the same letter within a column are not significantly different at the 0.10 significant level

Table 6. Turfgrass quality, soil bulk density, surface hardness and infiltration response to removing or reincorporating cores removed during core aeration in Clemson, SC

Cores removed	Year 1				Year 2			
	Turfgrass quality	Soil bulk density	Surface hardness	Infiltration	Turfgrass quality	Soil bulk density	Surface hardness	Infiltration
	1 - 9†	g cm^{-3}	g_{max}‡	mm hr^{-1}	1 - 9	g cm^{-3}	g_{max}	mm hr^{-1}
Nontreated	6.79a	1.28a‡	79.29a	14.81b	6.89a	1.29ab	98.82b	17.25a
No	6.92a	1.23b	79.19a	17.25ab	7.00a	1.33a	100.08a	11.25b
Yes	6.89a	1.23b	78.65a	19.88a	7.04a	1.26c	96.50c	16.50a
p-value	0.76	0.79§	0.82	0.05	0.77	0.04	<0.01	0.07
LSD (0.1)	0.25	0.05	2.38	3.34	0.2	0.06	0.2	4.68

†Visual turfgrass quality ratings were based on a scale of 1 to 9 where 1 = brown dead turfgrass and 9 = green healthy turfgrass, ‡Relative surface hardness value quantifies deceleration of 2.25 kg weight dropped from height of 45 cm, § Means followed by the same letter within a column are not significantly different at the 0.10 significant level

3.3 Surface Hardness

Clegg impact values (CIV) are a surface barometer of soil strength. A firm surface will slow water infiltration and effect playability and user safety by reducing the amount of surface cushioning. The degree of surface hardness is a cumulative effect of soil texture, density, and moisture level, among other factors.

In this study, deep tine core aeration reduced surface hardness 3% compared to shallow tine core aeration; however, no treatment effects were significant compared to the untreated in year one (Table 5). By year two, deep tine core aeration reduced surface hardness 2% compared to the untreated and shallow tine core aeration increased surface hardness 2% compared to the untreated (Table 5). In heavier soils, core aeration has the potential to increase soil strength in the region directly adjacent to the coring hole [4]. Shallow tine core aeration may have produced too little mechanical fracture within the soil profile to overcome any localized increase in soil strength.

Core removal or incorporation did not affect surface hardness in year one; however, in year two core removal reduced surface hardness 4% compared to plots with cores reincorporated (Table 6). In plots core aerated once per year, core removal reduced surface hardness 5% and 4% in year one and two, respectively (Table 7).

The number of core aerations performed per year did not drastically affect surface hardness (Table 4), although previous research has shown surface hardness to be reduced by increasing number of core aeration events [7]. The limited effect of core aeration frequency and other main effects in this study may be related to soil type. In general, surface hardness has decreased in response to core aeration, although the majority of these studies were conducted on sand-based golf course putting greens with inherently less soil strength than clay-based soils [3,7,8,10,17]. The structural integrity of clay-based soil is likely too great to induce a rapid change in surface hardness in response to a single season of core aeration. A consistent, long-term core aeration program may be necessary to realize effects of core aeration on surface hardness in clay-based soils.

3.4 Infiltration Rate

In clay-based soils, water availability is dependent on adequate surface water infiltration after an irrigation or rainfall event to prevent excessive water loss through surface runoff. Clay soil particles can form a tight matrix at the soil surface that prevents water infiltration, especially in compacted soils. Disruption of this matrix through core aeration aims to lower the tortuosity water must overcome at the soil surface by creating channels for water to quickly infiltrate into the soil profile.

Infiltration speed increased an average of 29% after deep tine core aeration and 34% if cores were removed compared to the untreated in year one (Tables 5 and 6). In year two, treatment effects were less pronounced. Incorporation of cores back into the soil profile after core aeration slowed infiltration speed 35% compared to the untreated. Further treatment effects were not significant (Table 2).

The limited effect of core aeration on surface infiltration in this study may be related to soil texture. In sand based systems, surface water infiltration has typically increased following core aeration [7,8,17]. Although core aeration

Table 7. Turfgrass quality, soil bulk density, surface hardness and infiltration response to removing or reincorporating cores removed during core aeration in treatments receiving on core aerification yearly in Clemson, SC

Cores removed	Year1				Year2			
	Turfgrass quality	Soil bulk density	Surface hardness	Infiltration	Turfgrass quality	Soil bulk density	Surface hardness	Infiltration
	1 - 9†	g cm^{-3}	g_{max}‡	mm hr^{-1}	1 - 9	g cm^{-3}	g_{max}	mm hr^{-1}
No	6.91a	1.25a‡	81.54a	16.88a	7.00a	1.33a	100.08a	11.25b
Yes	6.91a	1.22a	77.82c	17.63a	7.04a	1.26b	96.50b	16.50a
p-value	1	0.71§	0.02	0.84	0.77	0.04	<0.01	0.07
LSD (0.1)	0.25	0.12	2.70	6.07	0.20	0.06	0.20	4.68

†Visual turfgrass quality ratings were based on a scale of 1 to 9 where 1 = brown dead turfgrass and 9 = green healthy turfgrass, ‡Relative surface hardness value quantifies deceleration of 2.25 kg weight dropped from height of 45 cm, § Means followed by the same letter within a column are not significantly different at the 0.10 significant level

opens large channels to facilitate movement of water through the soil surface, clay-based soils are likely more sensitive to disruptions in soil structure than sand based systems. Soil fracture and motion from core aeration tines smearing soil particles along the inside of core aeration holes may prevent water from reaching existing channels within the soil structure that facilitate water movement through the soil profile, effectively slowing water infiltration.

The data produced by this study suggests deep tine core aeration and core removal following aeration may be necessary to improve surface water infiltration in heavy soils. Shallow tine core aeration may not penetrate the soil to an adequate depth to encourage movement of water past the uppermost compacted horizon to less compacted horizons. Further, core removal decreases the tortuosity water encounters in newly formed channels while core reincorporation introduces pockets of unstructured soil which may resist water movement.

Continued core aeration of heavier soils at a consistent depth may further slow surface water infiltration as compacted pockets accumulate at the lower end of the aerification zone, influencing the speed at which water can infiltrate [3]. Formation of a compaction zone was not observed in this study; however, long term evaluation of core aeration programs could further the understanding of this effect.

4. CONCLUSION

Although the data presented here does not reflect a drastic nor immediate response in plant health or soil physical properties after a short-term core aeration program, it is overall supportive of an aerification program that utilizes frequent core aeration to improve soil physical properties in heavily compacted, clay-based soils.

Improvements to soil physical properties after core aeration are often short-lived and require regular aerification to maintain high quality turf. A perpetual core aeration program is recommended to provide sustainable improvements to both TQ and soil physical properties over several growing seasons. In addition, knowledge of the particular soil system under aerification will aid in determining the correct course of action for improvement of soil physical properties.

Most soil aerification studies, including this study, are limited in scope by study duration. Continuing research is needed to fully appreciate the effect of long-term soil aerification programs on soil physical properties and overall plant health, especially as new equipment and technology emerge.

COMPETING INTERESTS

Authors have declared that no competing interests exist.

REFERENCES

1. O'Neil KJ, Carrow RN. Perennial ryegrass growth, water use, and soil aeration status under soil compaction. Agron J. 1983;75: 177-180.
2. Swartz WE, Kardos LT. Effects of compaction on physical properties of sand-soil-peat mixtures at various moisture contents. Agron J. 1963;55:7-10.
3. Murphy JA, Rieke PE, Erickson AE. Core cultivation of a putting green with hollow and solid tines. Agron J. 1993;85:1-9.
4. Petrovic AM. The effects of vertical operating hollow tine (VOHT) cultivation on turfgrass soil structure. Ph.D. diss., Michigan State Univ., East Lansing; 1979.
5. Weicko G, Carrow RN, Karnock KJ. Turfgrass cultivation methods: Influence on soil physical, root/shoot and water relationships. Int Turf Soc Res J. 1993;7: 451-457.
6. SAS Institute. Users guide: Statistics. SAS Inst., Cary, NC; 2004.
7. Atkinson JL, McCarty LB, Bridges WC. Effect of core aerification frequency, area impacted, and topdressing rate on turf quality and soil physical properties. Agron J. 2012;104:1710-1715.
8. Bunnell BT, McCarty LB, Hill HS. Summer cultivation effects on a sand based creeping bentgrass golf green. Int Turf Res J. 2001;9:3-9.
9. Murphy JA, Rieke PE. Comparing hollow and solid-tine cultivation. USGA Green Section Record. 1990;28:7-10.
10. Murphy JA, Rieke PE. High-pressure water injection and core cultivation of a compacted putting green. Agron J. 1994; 86:719-724.
11. Carrow RN, Johnson BJ, Burns RE. Thatch and quality of tifway bermudagrass turf in relation to fertility and cultivation. Agron J. 1987;79:524-530.

12. McWhirter EL, Ward CY. Effect of vertical mowing and cultivation on golf green quality. Rep. 2. Mississippi Agric. For. Exp. Stn. Starkville, Mississippi; 1976.
13. Murray JJ, Juska FV. Effect of management-practices on thatch accumulation, turf quality, and leaf spot damage in common Kentucky bluegrass. Agron J. 1977;69:365-369.
14. Cooper RJ, Skogley CR. An evaluation of several topdressing programs for Agrostis palustris huds. and *Agrostis canina* L. putting green turf. In R.W. Sheard, editor, Proceedings of the 4th International Turfgrass Research Conference, Guelph, Ontario. 198;129-136.
15. Weston JW, Dunn JH. Thatch and quality of meyer zoysia in response to mechanical cultivation and nitrogen fertilization. In F. Lemaire, editor, Proceedings of the 5th International Turfgrass Conference, Avignon, France. 1985;449-457.
16. McCarty LB. 2011. Best golf course management practices. 3rd ed. Upper Saddle River, New Jersey: Pearson Education, Inc; 2011.
17. McCarty LB, Gregg MF, Toler JE. Thatch and mat management in an established creeping bentgrass golf green. Agron J. 2007;99:1530-1537.

24

Evaluation of Zeolite Based Nitrogen Nano-fertilizers on Maize Growth, Yield and Quality on Inceptisols and Alfisols

A. Manikandan[1*] and K. S. Subramanian[2]

[1]ICAR - Central Institute for Cotton Research, Nagpur, India.
[2]Tamil Nadu Agricultural University, Coimbatore, India.

Authors' contributions

This work was carried out in collaboration between both authors. Author AM designed the study, performed and wrote the first draft of the manuscript with the guidance of author KSS. Both authors read and approved the final manuscript.

Editor(s):
(1) Dionisios Gasparatos, Soil Science Laboratory, Faculty of Agriculture, Aristotle University of Thessaloniki, Greece.
Reviewers:
(1) Linda Mondini, Tuscia University, Italy.
(2) Muhammad Rafi Qamar, University of Sargodha, Pakistan.

ABSTRACT

Nano-fertilizer technology is designed to deliver nutrients in a regulated pattern in correspondence with the crop demand thereby nutrient use efficiency can be improved without associated ill-effects. This present study hypothesize that nano-zeolite possesses extensive surface area and its coating or blending with the conventional nitrogenous fertilizers can regulate the release of nitrogen. This serves as an excellent slow release fertilizer that assists in sustained release of nitrogen that commensurate with crop growth without associated environmental harm. The newly developed intercalated N nano-fertilizer formulations had been tested using maize as a model system. The fate of N in the soil system was examined in order to gain insights of nano-fertilizer in promoting productivity of crops with higher N use efficiency. Response of maize plants to the fabricated fertilizers were tested in two greenhouse experiments of two distinct soil textures (Inceptisol – Periyanayakkan palayam soil series – clay loam and Alfisols - Irugur soil series- sandy loam). The grain N content of nanozeourea on inceptisol (Control: 0.26%; Treated: 0.32%) and alfisols (Control 0.48; Treated 0.76%) were higher consistently. The response was more pronounced in alfisol than

**Corresponding author: E-mail: poonamani223@gmail.com*

inceptisol. The growth, yield, quality and nutrient uptake were consistently higher for nanozeourea treatment than conventional urea.

Keywords: Alfisols; inceptisols; nano-fertilizer; nitrogen; maize yield; slow release; zeolite.

1. INTRODUCTION

Fertilizers are indispensable in agricultural production system. Application of fertilizers started in the country in 1960's which closely coincided with the introduction green revolution when fertilizer responsive varieties have been inducted in Indian agriculture. Despite fertilizer application remarkably increased the grain growth, the yields of several crops got plateaued due the low fertilizer response ratio, imbalanced fertilization, low organic matter, increased intensities of multi-micronutrient deficiencies across the country. Within the set of challenges faced by the soil scientists, imbalanced fertilization is one of the most critical factors to be considered for N management. Since N fertilization exhibits universal response in crops besides low price of urea due to decontrol (subsidized rate), farmers started using nitrogenous fertilizers particularly urea heavily which has led to the current NPK ratio of 8.2: 3.2: 1 while optimal ratio is stipulated as 4:2:1. This is very serious issue causing nitrate pollution in ground water and eutrophication in aquatic system. This necessitates to develop slow release fertilizers to regulate the nitrification processes thereby N availability be sustained during the crop period. Nanotechnology deals with manipulation of materials at atom level, molecular and macromolecular scales, which operate at 100 nm, their properties significantly, differed from at a large scale [1,2]. It increases surface area to volume ratio altering mechanical, thermal and catalytic properties of materials. Nanotechnology has potential to develop slow release efficient fertilizers [3]. As we are all aware that clays are the key factor in deciding the soil fertility due to the fact that the surface area is huge and it adsorptive sites assists in retention and release of nutrients.

One gram of montmorillonite can be spread to a dimension of 30-40 m^2 [4], if the same clay reduced by ball milling process and its dimension got reduced to nano-metre that increase the surface area by several fold of 750 m^2 g^{-1} [5]. Extensive studies have been done to exploit "top-down approach" wherein the particles are subjected to intense grinding to reach the desirable size of 1-100 nm. In order to stabilize the particles, surfactants have been used. The surfactant had an added advantage of changing the surface charge of the particle which will allow the nano-particles to hold anionic nutrients (eg. NO_3^-). This approach had been widely exploited to synthesize nano-fertilizers carrying nitrogen [6-10]. These nano-fertilizer formulations have a potential to increase nutrient use efficiencies under the greenhouse conditions [11]. Nano-fertilizers and nano composites can be used to control the release of nutrients [12] from the fertilizer granules so as to improve the nutrient use efficiency while preventing the fixation or loss of nutrients to the environment [13] and supply with range of nutrients in desirable proportions [14]. Zeolite and nanoporous zeolite used as a slow release fertilizer in farming [15,16]. Zeolite incorporated urea, potassium sulphate and calcium hydroxyapatite as a slow release nano-fertilizer increased availability for 60 days [17]. Synthesised clay polymer nutrient nanocomposite using crystalline and non-crystalline components of soil clays increased biomass yield [18]. Nanocomposite fabricated using nanoclays and zeolite for maize as a slow release fertilizer which regulated N availability up to 45-49 days [5]. Therefore, responses of maize plants to zeolite based nano N fertilizer formulations were undertaken on Inceptisols and Alfisols.

2. MATERIALS AND METHODS

Fine size natural zeolite (clintoptilolite) was procured from GM Chemicals, Allahabad. On the basis of zeolite forms (micro, nano) loading efficiency, the intercalation and impregnation of urea was done to produce novel nano-nitrogen fertilizer using both adsorbents [3]. Experiments were conducted in two types of soils (light and heavy textured i.e., Alfisols - Irugur soil series-sandy loam and Inceptisol – Periyanayakkan palayam soil series – clay loam) to study the response of maize (NK-6240) to different fertilizer formulations during 2012 and 2013 at Tamil Nadu Agricultural University. Initial soil samples were analyzed for their physical and chemical properties (Table 1).

Table 1. Initial characteristics of the experimental soils

Properties	Inceptisols	Alfisols
	Coimbatore black soil	Bhavanisagar red soil
Clay (%)	42	15.6
Silt (%)	14.7	5.5
Fine sand (%)	16.2	46
Coarse sand (%)	21.6	31.8
Textural class	Clay loam	Sandy loam
pH (1:2.5)	8.52	8.4
EC (dSm^{-1})	0.2	0.06
CEC (cmol (p+) kg^{-1})	23.2	10.9
Bulk Density (gcm^{-3})	1.44	1.62
Particle Density (gcm^{-3})	2.65	2.61
Porespace (%)	52.3	32.1
Water holding capacity (%)	47.3	27.9
Total nitrogen (%)	0.09	0.03
Organic carbon (mg kg^{-1})	4200	3420
Available nitrogen (Kg ha^{-1})	154	148
Available phosphorus (Kg ha^{-1})	20	35
Available potassium (Kg ha^{-1})	638	907

The following 5 treatments were adopted in both soil conditions. The recommended dose of N for hybrid maize @ 250 kg N ha^{-1} was followed for all the five treatments. The P & K were applied in the form of single super phosphate and murate of potash at the recommended dose of 75 and 75 kg P_2O_5 and K_2O, respectively.

- T1 Urea alone
- T2 Zeolite + Urea (1:1 ratio – Physical mixing at equal proportion on w/w basis)
- T3 Nano Zeolite + Urea (1:1 ratio - Physical mixing at equal proportion on w/w basis)
- T4 Zeourea (1:1) – Intercalated
- T5 Nanozeourea (1:1) – Intercalated

Plant height was measured at 30, 60, 90 and harvest stage. Total chlorophyll content in leaves was estimated by SPAD meter (Minolta SPAD52 meter) at 60 DAS [19] prior to emergence of tassel. Root length was determined by measuring the length of root from the base of the stem to the tip of the lengthiest root and expressed in cm. Dry matter production was estimated at post- harvest stage. The uptake of N was obtained by multiplying the dry matter with the nutrient content concentrations. The units expressed as g per pot.

2.1 Plant Nutrient Analysis

One g of digested samples were made up to 50 mL using distilled water and stored for further nutrients analysis. The nitrogen content of the plant samples were determined by the standard procedure. About 10 mL of the di acid digest was taken and transferred to micro kjeldahl to distill the ammonia. 2% boric acid with 2-3 drops of double indicator was used to collect evolved ammonia and titrated against 0.02 N sulphuric acid. Blanks were maintained without adding sample. Crude protein was estimated by using the formula N (%) content × (6.25) conversion factor.

2.2 Yield Components

The number of days for the emergence of tassel and silking were recorded. The mean length of cob from bottom to the top was measured and expressed in cm. The mean girth of the cob at maximum girth was measured and expressed in cm. The mean number of grain rows cob^{-1} was counted from the cob obtained from the sample plants and expressed. The total number of grains in each row of the cobs of sample plant was counted and mean was worked and expressed as number grains row^{-1}. The weight of individual cob was recorded after drying and the mean weight was arrived at and expressed in g cob^{-1}.The data collected from various experiments were subjected to analysis by variance (P = .05) with mean separation by least significant difference (LSD) as well as by DMRT [20].

3. RESULTS AND DISCUSSION

3.1 Plant Height (cm)

Maize plants fertilized with new fertilizer formulation derived from physical blending of urea with zeolite or nano-zeolite or fusing through intercalation processes significantly increased the plant height especially at the active growth stage of the crop (60 & 90 DAS). Despite these new fertilizer formulations increased the growth but the level of significance was not observed during early stage and harvest. The treatment difference was quite obvious in plants fertilized with nanozeourea. The best treatment registered a plant height of 144.8 cm while conventional urea fertilization produced 131.8 cm at 90 DAS. Other treatments exhibited comparable plant heights. Similar trend of results were recorded in Alfisol with the exception that the values are higher by 15-20% in all the treatments (Table 2).

Plant height was considered as indicator of growth, differed with application of fertilizer source. Irrespective of soil, fertilizer application, the plant height increased over the vegetative growth period. This trend of growth and results are in corroboration with the findings of [21] sugarcane growth and yield significantly increased by the adding zeolite incorporated with chemicall fertilizer and [22] zeolite application (@ 20 g and 40 g) had taken minimum days for emergence of unifoliate first, second, sixth trifoliate leaves and plant height of soyabean on allophanic soil. The mixing of zeolite with wolverine sand improved the corn growth [23]. Consumption of 2 g zeolite in 1 kg soil had important effects on the leaf area and the diameter of maize, in comparison with control treatment it could effect on the yield [24]. The positive results obtained on maize plant height and grain yield on natural zeolite [25].

3.2 SPAD Value

The SPAD readings of urea-fertilized maize plants were compared with urea blended with zeolite or nano-zeolite. All plants had comparable in SPAD Meter readings and were statistically non-significant (Table 3).

SPAD meter reading on chlorophyll content at a tassel growth stage varied with amount of fabricated nitrogenous fertilizer fed and light intensity of maize micro climate of growth period. Cholorophyll content was determined with Minolta SPAD 52 meter of non-destructive method. Analogous findings observed strong correlation between dry matter yields with the amount of total N in tropical maize [26]. Similar readings recorded in [27] non-destructive chlorophyll meters have been used to predict N fertilizer requirement of maize.

Table 2. Effect of zeolite based N fertilizers on DAS of maize plant height (cm)

Treatments	Inceptisols				Alfisols			
	30	60	90	115	30	60	90	115
T_1- Urea	34.57	91.43	131.8	144.3	34.0	102.1	153.7	162.4
T_2- Zeolite +Urea	35.14	89.28	127.4	142.4	41.6	107.0	153.1	160.4
T_3- Nanozeolite + Urea	35.28	89.28	130.4	142.4	32.1	101.1	150.2	158.0
T_4- Zeourea	37.57	94.57	130.8	146.3	37.6	109.4	158.8	165.8
T_5-Nanozeourea	40.86	94.86	144.8	152.3	37.3	110.8	167.4	178.1
S.Ed	3.53	3.98	3.86	3.84	3.24	7.70	5.19	4.75
CD (0.05)	NS	NS	7.89	NS	NS	NS	NS	9.70

DAS- Days after sowing

Table 3. Effect zeolite based N fertilizers on maize SPAD Value, root length, Dry Matter Production (DMP), tasselling and silking

Treatments	SPAD value		Root length(cm)		DMP (g)		Tasseling (Days)		Silking (Days)	
	I	A	I	A	I	A	I	A	I	A
T_1- Urea	39.32	36.73	56.8	43.5	120.7	105.2	71.4	60.9	83.4	66.1
T_2- Zeolite +Urea	37.97	38.76	53.3	60.3	142.5	115.7	68.6	56.1	78.4	60.9
T_3-Nanozeolite+Urea	39.01	38.18	56.8	59.3	144.7	118.6	71.3	61.0	82.9	65.4
T_4- Zeourea	38.52	38.21	59.2	61.9	146.7	123.6	72.4	58.6	83.4	63.4
T_5-Nanozeourea	38.45	39.17	65.9	67.4	151.4	130.1	71.1	56.0	85.0	60.1
S.Ed	1.33	1.70	5.00	5.12	10.46	9.03	2.55	2.32	2.48	2.12
CD (.05)	NS	NS	NS	10.46	NS	NS	NS	NS	NS	4.34

I-Inceptisols, A-Alfisols

3.3 Root Length (cm)

Maize plants fertilized with either urea or new fertilizer formulations did not produce any statistical significance on root length in inceptisols (Table 3). However, nano-urea-fertilized plants had the highest values for root length and the value was 9.1% higher than control. On the other hand, in alfisol, nanozeourea fertilization increased the root length to the tune on 23.8% higher in comparison to urea-fertilized plants. The root length increase may be the effect of nanozeolite and as a consequence to physical contact of nanozeourea with root surface. Similar findings reported [28] that the highest increase of weight/length ratio of maize primary root was obtained for treatment with micronized clinoptilolite at the lowest dose of 0.1%, while a decrease was found for treatment with granular clinoptilolite, attesting the role of different clinoptilolite particles size on root development.

3.4 Total Dry Matter (g)

The dry matter production of maize plants did not differ significantly among the treatments especially at the harvest stage in both inceptisol and alfisol (Table 3). The highest dry matter yield obtained from nanozeourea treated soil may be attributed to the increased N availability due to reduced ammonia loss. Similar findings were recorded by [29] found that the activity of water after adding nano-materials was increased and N, P and K were absorbed into the plants with the absorbing of water, thus the dry matter production was also increased. Similar findings[30] revealed that the highest dry matter yield (29.25 g pot^{-1}) was obtained from gelatin + Cu coated urea and (26.50 g pot^{-1}) micronutrient coated urea treated pots, closely followed by (25.5 g pot^{-1}) Palm stearin + Cu coated urea, (26 g pot^{-1}), Agar + Cu coated urea and (20.75 g pot^{-1}) Cu coated urea. The uncoated urea produced the least dry matter yield (19 g pot^{-1}) at the same level of N application. Results indicated that N, P and K-enriched zeolite was an adequate slow-release nutrient source for plants. The enrichment of zeolite with chemical fertilizer (KNO_3, K_2HPO_4, H_3PO_4) improved the dry matter production and nutrient accumulation after successive crops of Lettuce, Tomato, Rice and Andropogon grass [31].

3.5 Days Taken for Tasseling and Silking

The number of days taken to tassel or silking did not differ significantly in an inceptisol. The number of days to tassel was in the range of 68 to 72 days and silking in the range of 78-85 days. One of the striking response was in nanozeo-urea fertilized plants which took 85 days to silk which is 2 days longer than urea fertilized plants. In alfisol, the days to attain tasseling did not change significantly. But, there was a remarkable difference between red and black soils. Maize plants grown in alfisol had a early tasseling by 10-14 days in comparison to inceptisol. Such trend was also seen in silking. The conventional urea fertilized plants took 66 days while nanozeourea took 60.1 days and zeolite blended urea fertilized plants registered 60.9 days (Table 3).

3.6 Nitrogen Content (%)

In inceptisol, nitrogen contents of roots, shoots and grains were significantly influenced by fertilization with various forms of zeolite based N fertilizers. The highest N content was registered in roots of maize plants fertilized with nanozeourea (0.32%) while urea fertilized plants had only 0.26%. On the hand, zeourea fertilized plants had the highest N content of 0.78% which is significantly different from the rest of the treatments. Highest N content was registered in grains harvested from nanozeourea fertilized maize plants. Even in alfisol, N contents in maize plants were significantly different for various fertilizer formulations. Roots had the highest N uptake in zeourea or nanozeourea but they were significantly different from urea fertilized or physically blended urea with zeolite. Nitrogen contents in stover and grains were significantly higher in nanozeourea than other treatments. In grains, nanozeourea fertilized maize plants had 28% higher N content (0.76%) in comparison to urea fertilized plants (0.48%) (Table 4).

Nanozeourea fertilizer application caused a highly significant increase in total nitrogen content among the treatments on maize plant parts. Total nitrogen content was recorded as higher under both pot conditions regardless of soil. Both soils showed low total nitrogen content towards the root and leaves under all the treatments and higher N uptake was recorded on grain and stover of nanozeourea. The slow release pattern might be the responsible factor for enhanced nitrogen uptake. Similar results were reported in low land rice [7], sugarcane [21] and the total N uptake in maize[30] plants was obtained highest to lowest as micronutrient coated urea >gelatin + Cu coated urea > Palm stearin + Cu coated urea > Cu coated urea >

Agar + Cu coated urea > uncoated urea, which which were 762, 676, 523, 491, 324 mg pot^{-1}, respectively. Zeolites mixed with chemical fertilizers helped to retain nutrients in root zone and, hence, improving the long term soil quality by enhancing nutrient absorption [32] and [33] application of Nano-chelate levels Increased accumulation of iron, potassium and decreased accumulation of sodium, nitrate and nitro in spinach leaves.

3.7 Yield Parameters

In inceptisol, grain yield was significantly influenced by various N formulations while 100 grain weight was not altered significantly. Zeourea and nanozeourea fertilized maize plants registered significantly higher grain yield per plant. Despite test weight was not significant, higher grain yield was registered in nanozeourea treatment which is comparable to zeourea. Lowest grain yield was registered in zeolite + urea or nano-zeolite + urea treatments. In alfisol, the highest grain yield of 254 g per plant was registered in treatment that received recommended dose of nanozeourea which was 38% higher than conventional urea fertilization. Even zeo-urea fertilized treatment had increased the grain yield to the tune of 31.5%. Same set of treatments had also registered higher values for 100 grain weight (Table 5).

3.8 Quality Parameters - Crude Protein

Crude protein content of maize grains was the highest in nanozeo urea fertilized plants regardless of light textured or heavy textured soils. Crude protein in the best treatment was 4.9% and 4.7% in black and red soils, respectively (Table 5).

The crude proteins recorded in nanozeo urea treated plants in black and red soils were 26.1% and 36.1%, higher than urea fertilized treatments. The lowest crude proteins recorded in the treatment that received zeolite blended with urea. On yield components the significant differences were observed (Days taken for Tasseling and Silking, hundred grain weight and grain and stover yield) of maize. It may be the effect of slow release and controlled release of nitrogen from the by nanozeourea application and availability of nitrogen throughout crop growth period. The combined application of zeolite and chemical fertilizer increased growth and yield of apple per tree and per unit area. This is in agreement with the findings of [21] application of zeolite and chemical fertilizer

Table 4. Effect of zeolite based N fertilizers on maize N content (%)

Treatments	Inceptisol			Alfisols		
	Root	Stover	Grain	Root	Stover	Grain
T_1- Urea	0.26	0.34	0.58	0.25	0.25	0.48
T_2- Zeolite +Urea	0.22	0.30	0.53	0.25	0.23	0.52
T_3- Nanozeolite + Urea	0.25	0.26	0.62	0.19	0.23	0.52
T_4- Zeourea	0.27	0.78	0.62	0.28	0.23	0.59
T_5-Nanozeourea	0.32	0.51	0.78	0.28	0.32	0.76
S.Ed	0.02	0.09	0.04	0.02	0.02	0.06
CD (0.05)	0.05	0.18	0.09	0.06	0.04	0.13

Table 5. Effect of zeolite based N fertilizers on maize yield and quality parameters

Treatments	Inceptisol			Alfisols		
	Grain yield (g)	100 grain wt(g)	Crude protein (%)	Grain yield (g)	100 grain wt(g)	Crude protein (%)
T_1- Urea	268	27.8	3.62	156	25.8	3.00
T_2- Zeolite +Urea	232	28.2	3.32	203	25.4	3.25
T_3- Nanozeolite + Urea	238	28.0	3.85	133	25.7	3.22
T_4- Zeourea	295	29.3	3.90	173	27.1	3.70
T_5-Nanozeourea	291	29.8	4.90	254	29.4	4.70
S.Ed	23.01	1.11	0.28	27.59	1.27	0.41
CD (0.05)	47.00	NS	0.57	56.36	2.60	0.83

improved the sugarcane yield and sweetness (Brix). The nano-subnano composites of kaoline were used as the cementing and coating materials of slow/controlled release fertilizer [34]. The nutrient loaded zeolite (zeopro) improved growth, yield and quality of cucumber and tomato [35]. The mixture of compost and zeolite was enhanced the growth characters, essential oil yield and chemical composition of Achillea plants [36]. Nitrogen fertilizer management of Eggplant and nano-iron chelate foliar spray improved yield and yield components [37].

4. CONCLUSION

Response of maize plants to the fabricated fertilizers were tested two distinct soil textures (Inceptisol – Periyanayakkan palayam soil series – clay loam and Alfisols - Irugur soil series-sandy loam). The grain N content of nanozeourea on inceptisol (Control: 0.26%; Treated 0.32%) and alfisols (Control 0.48; Treated 0.76%) were higher consistently. The response was more pronounced in alfisol than inceptisol. The growth, yield, quality and nutrient uptake were consistently higher for nanozeourea treatment than conventional urea.

ACKNOWLEDGEMENTS

First author gratefully acknowledge UGC, New Delhi and GOI for the financial support of this Ph.D. Research fellowship award "Smart delivery of N in plant system through nano-fertilizers"

COMPETING INTERESTS

Authors have declared that no competing interests exist.

REFERENCES

1. Lal R. Soils and India's food security. J. Ind. Soc. Soil Sci. 2008;56(2):129-138.
2. Sharma PD. Nutrient management – Challenges and options. J. Ind. Soc. Soil Sci. 2008;55(4):395-403.
3. Komarneni S. Potential of nanotechnology in environmental soil science. Proc of 9th International Conf East and Southeast Asia Federation of Soil Science Societies, (Korean Society of Soil Science and Fertilizers, Seoul) Oct 27-30. 2009;16-20.
4. Macht F, Eusterhues K, Pronk GJ, Totsche KU. Specific surface area of clay minerals: Comparison between atomic force microscopy measurements and bulk-gas (N_2) and -liquid (EGME) adsorption methods. App. Clay Sci. 2011;53(1):20-26.
5. Sharmila Rahale C. Nutrient release pattern of nano – Fertilizer formulations, Ph.D. Thesis, Tamil Nadu Agricultural University, Coimbatore; 2010.
6. Li Z. Use of surfactant-modified zeolite as fertilizer carriers to control nitrate release. Microporous and Mesoporous Materials. 2003;61:181–188.
7. Aljabri M. Nano technology of zeolite mineral for slow release nitrogen of urea fertilizer on vertisols-paddy soil. Proceeding of the International Conference on Materials Science and Technology (ICMST - 2010), Jakarta, Indonesia, 19-23 October. 2010;205-210.
8. Bhardwaj D, Tomar R. Use of surface modified inorganic nano materials as slow release nitrogen fertilizer. Behnassi M, et al. (eds.), Sustainable Agricultural Development, Springer Science + Business Media B.V. 2011;171-184.
9. Pereira EI, Minussi FB, Cruz CCT, Bernardi ACC, Ribeiro C. Urea–montmorillonite-extruded nanocomposites: A novel slow-release material. J. Agric. Food Chem. 2012;60:5267–5272.
10. Mohanraj J. Effect of nano-zeolite on nitrogen dynamics and greenhouse gas emission in rice soil eco system. M. Tech. Thesis; Tamil Nadu Agricultural University, Coimbatore; 2013.
11. Subramanian KS, SharmilaRahale C. Synthesis of nano-fertiliser for regulated release of nutrients. In: Biomedical applications of nanostructured materials (Rajendran V, Hillebrands B, Prabu P, Geckeler KE) Macmillan Publishers India Ltd. New Delhi. 2010;57-62.
12. DeRosa MC, Monreal C, Schnitzer M, Walsh R, Sultan Y. Nanotechnology in fertilizers. Nat. Nanotechnology. 2010;5: 91.
13. Subramanian KS, Tarafdar JC. Nanotechnology in soil science. Proc. Platinum Jubilee Symposium Soil Science, (IARI, New Delhi) December 22-25. 2009; 326-334.
14. Datta SC. Nanoclay research in agriculture environment and industry. National Symposium on 'Applications of Clay Science: Agriculture, Environment and

Industry', NBSS & LUP, Nagpur. 2011;71-73.

15. Gholizadeh A. Zeolitic slow release fertilizer: A brief review. Iran International Zeolite Conference (IIZC'08) April 29 - May1, Tehran –Iran. 2008;1-2.
16. Ramesh K, Biswas AK, Somasundaram J, Subbarao A. Nanoporous zeolite in farming: Current status and issues ahead. Curr. Sci. 2010;99(6):760-764.
17. Kottegoda N, Munaweera I, Madusanka N, Karunaratne V. A green slow-release fertilizer composition based on urea-modified hydroxyapatite nanoparticles encapsulated wood. Curr. Sci. 2011; 101(1):73-78.
18. Sarkar S. Synthesis of clay polymer-nutrient nanocomposites and their characterisation with respect to water holding capacity and nutrient release behaviour. Ph.D. Thesis 2011, Indian Agricultural Research Institute, New Delhi.
19. Markwell J, Osterman JC, Mitchell JL. Calibration of the Minolta SPAD-502 leaf chlorophyll meter. Photosyn. Res. 1995; 46:467–72.
20. Panse VG, Sukhatme PV. Statistical Methods for Agricultural Workers. ICAR Publications, New Delhi, 1985;1-21.
21. Junrungreang S, Limtong P, Wattanaprapat K, Patsarayeangyong T. Effect of zeolite and chemical fertilizer on th change of physical and chemical properties Latya soil series for sugarcane. 17th WCSS, 14-21 August 2002, Thailand. 1897;1-7.
22. Khan M, Kim KW, Mingzhi W, Lim BK, Lee WH, Lee JY. Nutrient-impregnated charcoal: An environmentally friendly slow-release fertilizer. Environmentalist. 2008; 28:231–235.
23. Ippolito JA, Tarkalson DD, Lehrsch GA. Zeolite soil application method affects inorganic nitrogen, moisture and corn growth. Soil Sci. 2011;176(3):136-142.
24. Siuki AK, Kouchakzadeh M, Riahi H, Sirdari ZZ. Investigating the effects of naural zeolite clinoptilolite of natural trend of maize growth. Iran International Conference; 2012.

    Available:http://www.civilica.com/EnPaper-zeolite01-zeolite01_109.html
25. Mahrokh A, Zendehdel M. Penological stage, plant height and grain yield of maize affected using clinoptilolite zeolite natural. Conference 3rd Iran International Zeolite Conference (IIZC 2012) 6-7 Jun, Arak, Iran. 2012;182-183.
26. Lafitte HR, Edmeades GO. Improvement for tolerance to low soil N in tropical maize. I. Selection criteria. Field Crops Res. 1994; 39:1–14.
27. Ciganda V Gitelson A, Schepers J. Non-destructive determination of maize leaf and canopy chlorophyll content. J. Plant Physiol. 2009;166:157–167.
28. Trinchera A, Rivera CM, Rinaldi S, Salerno A, Rea E, Sequi P. Granular size effect of clinoptilolite on maize seedlings growth. Open Agr. J. 2010;4:23-30.
29. Liu AX, Liao ZW. Effects of nano-materials on water clusters. J. Anhui Agric. Sci. 2008;36:15780-15781. (In Chinese).
30. Junejo N, Khanif MY, Dharejo KA, Hamid HA, Abdu A. Evaluation of coated urea for ammonia volatilization loss, nitrogen mineralization and microsite pH in selected soil series. Afr. J. of Biotechnol. 2012; 11(2):366-378.
31. Bernardi ACC, Monte MBM, Paiva PRP, Werneck CG, Haim PG Barros FS. Dry matter production and nutrient accumulation after successive crops of Lettuce, Tomato, Rice, and Andropogon grass in a substrate with Zeolite. R. Bras. Ci. Solo. 2010;34:435-442.
32. Mumpton FA. Uses of natural zeolites in agriculture and industry. Proc. Natl. Acad. Sci. USA. La Roca Magica. 1999;96(7): 3463-3470.
33. Vattani H, Keshavarz N, Baghaei N. Effect of sprayed soluble different levels of iron chelate nano-fertilizer on nutrient uptake efficiency in two varieties of spinach (Varamin88 and Virofly). Intl. Res. J. App. Basic Sci. 2012;3:2651-2656.
34. Liu XM, Feng ZB, Zhang FD, Zhang SQ, He XS. Preparation and testing of cementing and coating nano-subnanocomposites of slow/controlled – release fertilizer. Agri. Sci. China. 2006; 5(9):700-706.
35. Lisa R. Effects of zeolites on the growth of cucumber and tomato seedlings. Bachelor (Horticulture) project, Swedish University of Agriculture, Alnarp. 2007;1-41.

36. Harb EMZ, Mahmoud MA. Enhancing of growth, essential oil yield and components of Yarrow plant (*Achillea millefolium*) grown under safe agriculture conditions using zeolite and compost. 4th Conference on Recent Technologies in Agriculture, 3-5, Nov. Cairo, Egypt. 2009;586-592.

37. Bozorgi HR. Study effects of N fertilizer management under nano iron chelate foliar spraying on yield and yield components of Eggplant (*Solanum melongena*). ARPN J. Ag. Biolog. Sci. 2012;7(4):233-237.

25

Study on Genetic Variability and Character Association of Aromatic Rice (*Oryza sativa* L.) Cultivars

Most. Fatema Kaosar Mishu[1], Md. Waliur Rahman[1*], Mohammad Abul Kalam Azad[1], Bhabendra Kumar Biswas[1], Md. Aminul Islam Talukder[2], Md. Omar Kayess[1], Md. Rafiqul Islam[1] and Md. Rakibul Alam[1]

[1]Department of Genetics and Plant Breeding, Faculty of Agriculture, Hajee Mohammad Danesh Science and Technology University, Dinajpur, Bangladesh.
[2]Department of Agricultural Chemistry, Faculty of Agriculture, Hajee Mohammad Danesh Science and Technology University, Dinajpur, Bangladesh.

Authors' contributions

This work was carried out in collaboration between all authors. All authors read and approved the final manuscript.

Editor(s):
(1) Sławomir Borek, Faculty of Biology, Department of Plant Physiology, Adam Mickiewicz University, Poland.
Reviewers:
(1) Willem Jansen van Rensburg, Vegetable and Ornamental Plant Institute, South Africa.
(2) Guillermo Castañón_Nájera, Universidad Juárez Autónoma de Tabasco, Tabsco, Mexico.

ABSTRACT

An experiment was conducted to evaluate correlation between some traits and yield components of six aromatic rice varieties and also to determine the most effective factors on its yield. The experiment was conducted in a randomized complete block design (RCBD) with 3 replications. Among six variteits Badshabhog was the topmost yielder followed by Kataribhog, Chinigura, and Radhunipagal, Begunbichi and Kalozira. Analysis of variability parameters revealed that the phenotypic coefficients of variation were higher than genotypic coefficient of variation for all the characters studied. The spikelets panicle^{-1} and 1000-grain weight showed high heritability and effective tillers hill^{-1} showed low heritability. Yield has positive significant correlation with days to maturity, spikelet length and 1000-grain weight. Negative significant correlation found with plant height and sterility percentage and non-significant correlation with other characters.

**Corresponding author: E-mail: waliur.hstu@gmail.com*

Keywords: Rice; yield performance; variability; correlation.

1. INTRODUCTION

Rice is the principal food crop and most extensively grown cereal in the tropical and subtropical regions of the world and a major source of nutrition for about two-thirds of world's population [1] and it is the staple food consumed by more than half of the people, three billion of which are Asian [2].

It is the staple food for the people of Bangladesh and playing an important role in the national economy of many developing countries [3,4]. Bangladesh is the fourth largest producer and consumer of rice in the world where rice provides 75% of the calories and 55% of the protein in the average daily diet [5]. In Bangladesh most of the aromatic rice cultivars are of traditional type, photoperiod-sensitive and are grown during Aman season in the rainfed low land ecosystem [6]. Due to high prices and export potentiality, the production of aromatic rice is becoming popular in Bangladesh [7]. Generally, the price of the aromatic rice is three to four times greater than that of the non-aromatic or ordinary rice varieties in national and international markets. Aroma which is being appreciated by many people and represents a high value added trait is the major feature of these aromatic rice varieties [8]. The aroma in rice is merely due to the charisma of some volatile chemicals [9] and the most vital one is the 2-acetyl -1-pyrroline [10]. Therefore, rice needs attention toward improvement in its cooking qualities as well as several biochemical and morphological characteristics [11]. But unfortunately, the aromatic rice often has undesirable agronomic characters, such as low yield, susceptibility to pests and diseases, and strong shedding [12]. Study on yield contributing characters involves greater importance of fixing up characters that influence yield [13,14].

In breeding programs designed to increase the yield potential of a crop plant, an understanding of the mode of inheritance of the yield components, the correlations among them and the relationship between the components and yield is necessary for a better selection of breeding procedures for developing high-yielding varieties [15,16]. Significant improvement in the efficiency of crop breeding programs through the use of proper selection indices where Interrelationships between yield and its contributing components are existed [17,18]. It is useful in the selection of several traits simultaneously influencing yield through the correlation coefficient analysis [19]. For better insight into cause and effect relationship between different pairs of traits the association in amalgamation with path analysis can effectively use [20]. By genotypic and/or phenotypic coefficients of correlation the association among traits may be measured depending on the types of studied materials and the kind of experimental design used [21]. In aromatic rice, genotypic evaluation can identify the relationship of yield components with grain yield and the information of these relationships can be helpful to find superior aromatic rice genotypes [22]. The present investigation was carried out to understand the association of grain yield with yield contributing traits to determine the inter-correlation between the traits at genotypic level to obtain path values, showing the extent of direct and indirect effects of various biometrical traits upon grain yield.

2. MATERIALS AND METHODS

2.1 Plant Material

Six aromatic rice genotypes namely Radhunipagol, Chinigura, Kataribhog, Begunbichi, Kalozira, Badshabhog were used in this investigation. The seeds were collected from the Genetics and Plant Breeding laboratory of Hajee Mohammad Danesh Science and Technology University, Dinajpur, Bangladesh.

2.2 Experimental Design and Growing Condition

All of six genotypes are raised in small plots (3×2 m^2) with three replications laid out on Randomized Complete Block Design (RCBD). Recommended rice production practice of Bangladesh Agricultural Research Council [23] was followed. The experimental site is situated in the tropical climatic zone, characterized by heavy rainfall during the months from May to September and minimum rainfall during winter season. The soil of the experimental field was sandy loam in texture having a pH of around 6.2 Each seedbed was prepared with 2 m length and 1 m width and recommended doses of manure and fertilizers were applied. A 50 cm drainage channel was allowed between two adjacent seedbeds to drain out excess water whenever needed. The grains were soaked in water for 24 hours and then incubated in a moist jute sack for

48 hours with a view of quick germination. The beds were irrigated to a depth of 2-3 cm after establishing the roots. Excess water was occasionally drained off to attain water depth of 5 cm that partially controlled weeds, and the remaining weeds were removed to raise healthy and vigorous seedlings.

2.3 Data Collection of Agronomic Traits

Data were collected on plant height (cm), tillers hill^{-1} (maximum vegetative stage), effective tillers hill^{-1}, days to 50% flowering, days to maturity, panicle length (cm), racillae panicle^{-1}, spikelets panicle^{-1}, sterility percentage, flag leaf angle (0°), flag leaf length (cm), spikelet length (mm), length-breadth ratio, 1000-grain weight (g) and yield (t ha^{-1}). After the appearance of 70% of the panicles, days to flowering have been recorded as soon as possible. Number of tillers was recorded after emerging the fertile panicles and grain development. The plant height was measured from ground level to the tip of the longest panicle. At maturity, panicles were harvested separately and placed individually in an envelope. The panicles were taken out of the envelopes and air-dried for one week at room temperature. Fertility of panicle, 1000 grain-weight and grain yield plot^{-1} was estimated after that.

2.4 Data Analysis

The collected data were analyzed by using the statistical program SPSS (Version 16.0, 2007) for analysis of variance (ANOVA) and mean separation.

3. RESULTS AND DISCUSSION

Performance of different characters in six aromatic cultivars is shown in Table 1. The highest value for plant height was observed in Begunbichi which was however statistically different from other varieties. The lowest value was exhibited in Kataribhog. Kalozira found to produce highest value for tillers hill^{-1} (maximum vegetative stage) while the lowest value was exhibited Radhunipagol.

Effective tillers hill^{-1} was observed maximum in Badshabhog which was however statistically different from other varieties and the lowest value was exhibited Radhunipagol that is also different from other varieties

The variety Kalozira required highest value for days to 50% flowering and days to maturity while Begunbichi required the lowest period for 50% flowering and days to maturity.

Panicle length was observed highest in Kataribhog and varied significantly from other varieties while the lowest value was exhibited Chinigura.

Spikelets panicle^{-1}, length of flag leaf and 1000-grain weight were observed maximum in Radhunipagol and varied significantly from other varieties. The lowest value of spikelets panicle^{-1} was exhibited in Kalozira while lowest length of flag leaf and 1000-grain weight was found in Begunbichi.

Highest yield was found in Badshabhog which was however statistically different from other varieties and the lowest was in Kalozira.

Variability of different characters of the rice varieties are shown in Table 2. Analysis of variability parameters revealed that the phenotypic coefficient of variation were higher than genotypic coefficient of variation for all the characters studied. In case of plant height, the difference between phenotypic and genotypic variance was low indicating minimum environmental influence on this trait. The character showed moderate genotypic and phenotypic coefficient of variation, high heritability in broad sense, moderate genetic advance and moderate genetic advance in percentage of mean.

The difference between phenotypic and genotypic variance was also low in case of number of tillers hill^{-1}, effective tillers hill^{-1}, days to 50% flowering, days to maturity, panicle length, rachillae panicle^{-1}, spikelets panicle^{-1}, sterility percentage, flag leaf length, spikelet length, 1000-grain weight, and yield. The characters showed moderate genotypic and phenotypic coefficient of variation, high heritability in broad sense, moderate genetic advance and moderate genetic advance in percentage of mean. These indicate that the influence of environmental condition low on expression of genes controlling the characters. In other words, the expression of these traits is mainly due to the genetic constituents rather than environmental influence. Therefore selection for improvement of these traits might be beneficial. Similar observations were reported for one or more characters [24-26].

Table 1. Mean performance of different characters in six aromatic cultivars

Cultivars	Plant height (cm)	Tillers/hill	Effective tillers/hill	Days to 50% flowering	Days to maturity	Panicle length (cm)	Racillae/ panicle	Spikelets/ panicle	Sterility percentage	Flag leaf angle(^{0})	Flag leaf length (cm)	Spikelet length (mm)	Length-breadth ratio	1000 grain weight (g)	Yield (ton ha^{-1})
Radhunipagol	148.1b	12.53 d	9.76 b	94.33 a	124.0 a	26.32ab	13.47 a	327.0 a	12.58 bc	26.19 b	37.26a	6.43 c	4.62 a	11.68 a	3.34a
Chinigura	136.2c	16.17 b	11.00 ab	95.33 a	124.7 a	24.30 d	10.02 c	224.3 c	12.00 bc	17.73 d	29.00c	6.37 c	3.87 b	10.34 d	3.43ab
Kataribhog	133.7c	14.33 c	10.76 ab	94.67 a	124.7 a	28.45 a	14.00 a	316.7 ab	14.11 ab	24.31 c	34.43b	6.59 b	4.51 a	9.48 e	3.72 a
Begunbichi	153.3a	15.20 bc	11.21 ab	90.33 b	121.0 b	26.75 b	11.99 b	203.9 d	12.19 bc	29.43 a	26.62d	6.42 c	4.55 a	9.23 f	3.16 b
Kalozira	146.4b	17.77 a	11.99 a	95.67 a	125.7 a	25.72 c	13.25 a	163.0 e	10.27 c	25.10 bc	26.58d	6.10 d	3.73 c	11.21 b	3.13 b
Badshabhog	136.4c	16.20 b	12.05 a	95.33 a	124.0 a	24.50 d	10.60 c	305.3 b	16.00 a	29.40 a	33.40b	6.73 a	4.57 a	10.61 c	3.74 a
LSD	2.61	1.40	1.71	1.39	1.99	o.82	0.99	13.99	2.81	1.18	1.60	0.08	0.12	0.21	0.41
Range	133.7-153.3	12.53-17.77	9.76-12.05	90.33-95.67	121-125.7	24.30-28.45	10.02-14.00	163-327	10.27-16.00	17.73-29.43	26.58-37.26	6.10-6.73	3.73-4.62	9.23-11.68	3.13-3.74

Mean values having same letter did not differ significantly when tested against DMRT at 0.05 level of probability

Table 2.Estimation of genetic variability parameters on fifteen characters in aromatic rice

Characters	Genotypic variance (σ^2g)	Phenotypic variance (σ^2p)	GCV (%)	PCV (%)	Heritability (%)	GA	GA in 1% mean
Plant height (cm)	63.00	65.07	5.57	5.67	96.81	16.08	11.30
Tillers/hill	3.04	3.64	11.35	12.42	83.60	3.28	21.34
Effective tillers/hill	0.42	1.31	5.84	10.28	32.26	0.76	6.83
Days to 50% flowering	3.77	4.36	2.06	2.22	86.51	3.72	3.95
Days to maturity	2.13	3.33	1.18	1.47	63.99	2.41	1.94
Panicle length(cm)	2.30	2.51	5.84	6.19	91.83	3.0	11.53
Rachillae /panicle	2.56	2.86	13.11	13.85	89.54	3.12	25.55
Spikelets /panicle	4681.97	4741.08	26.66	26.82	98.75	140.06	54.57
Sterility (%)	3.07	5.47	13.64	18.19	56.26	2.71	21.07
Flag leaf angle (◦)	18.44	18.866	16.93	17.13	97.74	8.75	34.50
Flag leaf length(cm)	19.55	20.33	14.16	14.44	96.18	8.93	28.62
Spikelet length(mm)	0.045	0.047	3.29	3.36	95.74	0.42	6.64
Spikelet L/B	0.15	0.16	9.15	9.30	96.89	0.80	18.55
1000-grain wt(g)	0.90	0.91	9.12	9.18	98.47	1.94	18.63
Yield(t/ha)	0.08	0.13	8.61	10.88	62.58	0.48	14.04

GCV= Genotypic coefficient of variation, PCV=Phenotypic coefficient of variation, GA=Genetic Advance

Table 3. Phenotypic correlation (r_{ph}) of different characteristics of aromatic rice

	PH	T/H	ET/H	DF	DM	PL	R/P	Sp/P	St. P	FLA	FLL	Sp. L	L/B	TGW	YLD
PH	1														
T/H	-0.127	1													
ET/H	-0.145	0.909*	1												
DF	-0.671	-0.287	0.176	1											
DM	-0.565	0.303	0.127	0.953**	1										
PL	0.114	-0.492	0.407	-0.373	0.180	1									
R/P	0.255	-0.407	0.389	-0.056	0.165	0.832*	1								
Sp/P	-0.429	-0.749	0.527	0.179	0.037	0.261	0.184	1							
St. P	-0.558	-0.264	0.078	-0.097	-0.131	0.704	0.239	-0.760	1						
FLA	-0.466	-0.136	0.224	-0.478	-0.537	0.232	0.233	0.168	0.369	1					
FLL	-0.350	-0.767	-0.606	-0.298	0.190	0.237	0.300	0.964**	0.597	0.103	1				
Sp.L	0.508	0.397	0.072	0.049	-0.266	0.082	0.233	0.789	0.976**	0.325	0.607	1			
L/B	0.093	0.742	0.402	0.465	-0.582	0.420	0.199	0.951**	0.687	0.652	0.639	0.763	1		
TGW	0.130	-0.050	0.151	0.589	0.553	-0.429	0.111	0.502	-0.223	-0.078	0.331	0.336	0.234	1	
YLD	-0.862*	-0.192	0.819	0.820	0.948**	0.641	0.163	0.755	-0.886*	-0.033	0.630	0.854*	0.409	0.980**	1

PH=Plant Height, T/H=Tillers/hill, ET/H=Effective tillers/hill, DF=Days to 50% flowering. DM=Days to maturity, PL=Panicle Length, R/P=Racillae/panicle, Sp/P=Spikelets /panicle, St.P=Sterility percentage, FLA=Flag leaf angle, FLL=Flag leaf length, Sp.L=Spikelets length, L/B=length-breadth ratio, TGW=Thousand grain weight and YLD=Yield.

**Significant at 0.05 of probability, **Significant at 0.01 of probability*

Yield is an important trait being affected by various interrelated traits. Therefore planning a protocol for increased yield might not be effective until other yield attributes influencing directly or indirectly are taken under consideration. The character association among the considered genotypes is shown in Table 3 above. Tillers hill^{-1} at the maximum vegetative stage showed the negative association with the plant height (r= -0.127). The effective tillers hill^{-1} showed the negative association with the plant height (r= -0.145). It had a significant association with the tillers hill^{-1} (r=0.909*).

The days to 50% flowering showed the negative association with the plant height (r= -0.671) and tillers hill^{-1} (r= -0.287). It had positive association with the effective tillers hill^{-1} (r=0.176).

The days to maturity showed the negative insignificant association with the plant height (r= -0.565). It had positive association with the tillers hill^{-1} (r=0.303) and effective tillers hill^{-1} (r=0.127). It showed highly positive association with the days to 50% flowering (r=0.953**).

The panicle length showed the positive association with the plant height (r=0.114), effective tillers hill^{-1} (r= 0.407) and days to maturity (r= 0.180). It had negative association with the tillers hill^{-1} (r= -0.492) and days to 50% flowering (r= -0. 373). The rachillae panicle^{-1} showed the positive association with the plant height (r= 0.255), effective tillers hill^{-1} (r= 0.389) and days to maturity (r= 0.165). It had negative association with the tillers hill^{-1} (r= - 0.407), and days to 50% flowering (r= -0.056). It showed positive association with the panicle length (r=0.832*). The spikelets panicle^{-1} showed the positive association with the effective tillers hill^{-1} (r= 0.527), days to 50% flowering (r= 0.179), days to maturity (r= 0.037), panicle length (r=0.261) and rachillae panicle^{-1} (r=0.184). It had negative association with the plant height (r= -0.429) and tillers hill^{-1} (r= - 0.749). The sterility percentage showed the positive association with the effective tillers hill^{-1} (r= 0.078), panicle length (r=0.704) and rachillae panicle^{-1} (r=0.239). It had negative association with the plant height (r= -0.558), tillers hill^{-1} (r= - 0.264), days to 50% flowering (r= -0.097), days to maturity (r= -0.131) and spikelets panicle^{-1} (r= -0.760).

Flag leaf length showed the positive insignificant association with the days to maturity (r= 0.190), panicle length (r=0.237), rachillae panicle^{-1} (r=0.300), sterility percentage (r=0.597) and flag leaf angle (r=0.103). It had negative association with the plant height (r= -0.350), tillers hill^{-1} (r= - 0.767), effective tillers hill^{-1} (r= -0.606) and days to 50% flowering (r= -0.298). It showed highly positive association with the spikelets panicle^{-1} (r= 0.964**).

1000-grain weight showed the positive association with the plant height (r=0.130), effective tillers hill^{-1} (r=0.151), days to 50% flowering (r=0.589), days to maturity (r=0.553), rachillae panicle^{-1} (r=0.111), spikelets panicle^{-1} (r= 0.502), flag leaf length (r=0.331), spikelet length (r=0.336) and length-breadth ratio (r=0.234). It had negative association with the tillers hill^{-1} (r= -0.050), panicle length (r= -0.429), sterility percentage (r= -0.223) and flag leaf angle (r= -0.078).

Yield showed the positive association with the effective tillers hill^{-1} (r=0.819), days to 50% flowering (r=0.820), panicle length (r= 0.641), rachillae panicle^{-1} (r=0.163), spikelets panicle^{-1} (r= 0.755), flag leaf length (r=0.630), and length-breadth ratio (r=0.409). It had negative association with the tillers hill^{-1} (r= -0.192) and flag leaf angle (r= -0.033). It showed highly positive association with the days to maturity (r=0.948**), spikelet length (r=0.854*) and thousand grain weight (r=0.980**). It had negative association with the plant height (r= -0.862*) and sterility percentage (r= -0.886*).

Positively correlated results indicated that increase of one trait will result in increase of the correlated trait and negative association of traits indicates that increase one character will decrease the negatively correlated character. Therefore, selection would be effective for increasing grain yield with simultaneous consideration of panicle length, spikelets panicle^{-1}, effective tillers/hill and 1000-grain weight. Previous report also indicated that harvest index, days to maturity and effective tillers hill^{-1}, 1000-grain weight; flag leaf area and panicle length should be given emphasis on the selection criteria for genetic improvement of rice yield [27].

4. CONCLUSION

The results of various characters studied in this experiment suggested that some good characters do exist in the local cultivars. Genetic variability parameters revealed that plant height, number of tillers hill^{-1}, effective tillers, days to maturity, panicle length, rachillae panicle^{-1}, spikelets panicle^{-1}, sterility percentage, flag leaf length, spikelet length, 1000-grain weight, and yield can be improved by proper selection. The desirable characters may be transfer in the modern varieties by using appropriate breeding techniques for developing superior varieties of rice and play an important role in meeting the global demand.

COMPETING INTERESTS

It is declared that the authors have no competing interests as defined by the publisher, or other interests that might be perceived to influence the results and/or discussion reported in this article.

REFERENCES

1. Vaughan DA, Morishima H, Kadowak K. Diversity in the *Oryza* genus: Current opinion. Journal of Plant Biology. 2003;6: 139-146.
2. Waghmare CR, Guhey A, Saxena R, Kulkarni AA, Agrawal K. Genetic divergence of morpho- physiological traits in rainfed early rice genotypes. Agricultural Science Digest. 2008;28(3):198–200.
3. Faruq G, Zulaani KN, Jennifer AH, Subha B, Zulqarnain M, Nazia AM, Mohammad O. Evaluation of kernel elongation ratio and aroma association in global popular aromatic rice cultivars in tropical environment. African Journal of Agricultural Research. 2010;5(12):1515–1522.
4. Trans DV. Closing the rice yield gap for food security. p.27-41. In: Peng S, Hardy B, (eds.). Rice Research for food security and poverty alleviation. Progressing of the Intl. Rice Res. Conf. 31 March-3 April, 2000. Los Banos, IRRI, Philippines. 2001; 692-693.
5. Bhuiyan NI, Paul DNR, Jabbar MA. Feeding the extra millions by 2025 challenges for rice research and extension in Bangladesh. 2002;9.
6. Das SK, Singh J, Tripathy M, Mishra D. Association of quantitative traits and path analysis in medium land rice. Environmental Ecology. 1996;14(1):99-102.
7. Dipti SS, Hossain ST, Bari MN, Kabir KA. Physico-chemical and cooking properties of some fine rice varieties. Pakistan Journal of Nutrition. 2002;1(4):188-190.
8. Dela CN, Khush GS. Rice grain quality evaluation procedures. In: Aromatic rices. Singh RK, Singh US, Khush GS, (Eds). Oxford & IBH Publishing Co. Pvt. Ltd., New Delhi, India. 2000;16-28.
9. Yajima I, Yanajm T, Nakamuram M, Sakakibara H, Hayaski K. Volatile components of cooked Kaorimai scented rice. Agricultural and Biotechnological Chemistry. 1979;43:2425-2430.
10. Buttery RG, Ling LC, Juliano BO, Turnbaugh JC. Cooked rice aroma and 2-acetyl-1- pyroline. Journal of Agricultural and Food Chemistry. 1983;31(4):823-826.
11. Faruq G, Mohamad O, Hadjimand K, Craig AM. Inheritance of kernel elongation in rice. International Journal of Agricultural Biology. 2004;5:813-815.
12. Berner DK, Hoff BJ. Inheritance of scent in American long grain rice. Crop Science. 1986;26:876–878.
13. Prasad B, Patwary AK, Biswas PS. Genetic variability and selection criteria in fine rice (*Oryza sativa* L.). Pakistan Journal of Biological Science. 2001;4(10):1188-1190.
14. Kole PC, Hasib KM, Correlation and regression analysis in scented rice. Madras Agricultural Journal. 2008;95(1-6): 178-182.
15. Ahmad A, Saleem M. Path coefficient Analysis in *Zea mays* L. Int. J. Agri. Bio. 2003;5(3):245-248.
16. Khodambashi M, Bitaraf N, Houshmand S. Generation mean analysis for grain yield and its related traits in lentil. J. Agric. Sci. Technol. 2012;14:609-616.
17. Mohammadi SA, Prasanna BM, Singh NN. Sequential path model for determining interrelationships among grain yield and related characters in Maize. Crop Sci. 2003;43:1690-1697.
18. Kashiani P, Saleh G, Abdullah NAP, Abdullah SN. Variation and genetic studies on selected sweet corn inbred lines. Asian J. Crop Sci. 2010;2:78-84.
19. Menkir A. Genetic variation for grain mineral content in tropical-adapted maize inbred lines. Food Chem. 2008;110:454-464.
20. Jayasudha S, Sharma D. Genetic parameters of variability, correlation and path coefficient for grain yield and physiological traits in rice (*Oryza sativa* L.) under shallow lowland situation. Electronic J. Plant Breed. 2010;1:1332-1338.
21. Sadek SE, Ahmed MA, Abd El-Ghaney HM. Correlation and path coefficient analysis in five parents inbred lines and their six white maize (*Zea mays* L.) single

crosses developed and grown in Egypt. J. Applied Sci. Res. 2006;2:159-167.

22. Tahir M, Khan DR, Zada A, Khan A, Wadan D, Said A, Marwat KB. Genetic variability of plant and yield characters in rice. Sarhad Journal of Agriculcure. 2002; 18(2):207-210.
23. BARC (Bangladesh Agricultural Research Council). Fertilizer Recommendation Guide. Bangladesh Agricultural Research Council, Dhaka, Bangladesh. 2005;60-70.
24. Hossain MA, Haque ME. Genetic variability and path analysis in rice genotypes. Bangladesh J. Pl. Breed. Genet. 2003; 16(1):33-37.
25. Sadhukhan RN, Chattopadhyay P. Variability and character association between yield attributes and grain quality in aromatic rice. Journal of Interacademicia. 2000;4(4):494-497.
26. Biswas PS, Prasad B, Dewan SBA. Variability, character association and path analysis of rice (*Oryza sativa* L). Bangladesh Journal of Plant Breeding and Genetics. 2000;13:19-25.
27. Pandey VR, Singh PK, Verma OP, Pandey P. Inter-relationship and path coefficient estimation in rice under salt stress environment. International Journal of Agricultural Research. 2012;7(4):169-184.

26

Evaluation of Soil Aggregate Stability under Long Term Land Management System

Olaoye J. Olawale[1*], S. T. Abu[2] and Ogunwole O. Dorcas[2]

[1]Department of Soil Science, Ahmadu Bello University, P.M.B 1044, Zaria, Nigeria.
[2]Department of Crop Science, Ahmadu Bello University, P.M.B 1044, Zaria, Nigeria.

Authors' contributions

This work was carried out in collaboration between all authors. All authors read and approved the final manuscript.

Editor(s):
(1) L.S. Ayeni, Adeyemi College of Education, Ondo State, Nigeria.
(2) Junhong Bai, School of Environment, Beijing Normal University, Beijing, China.
Reviewers:
(1) Bassey Udom, University of Port Harcourt, Nigeria.
(2) Anonymous, South Dakota State University, USA.
(3) Yi Zhang, Tsinghua University, China.
(4) Nyatuame Mexoese, Ho Polytechnic, Ho, Ghana.

ABSTRACT

The influence of land management on soil structural stability of National Animal Production Research Institute (NAPRI) farm Shika Zaria in Northern ecological zone of Nigeria was studied. Soil samples were collected from depth of 0-15 cm in fields under three perennial pasture fields not subjected to tillage for over 20 years since establishment using an adjacent continuously cultivated field as reference which has also been under cultivation for over 20 years. The long-term field experiment had four treatments: Three perennial pasture Digit grass *"Digitaria smutsii"*, Signal grass *"Bracheria decumbens"*, Gamba grass *"Andropogon gayanus"* fields and a continuously cultivated field with 12 samples collected from each plot. The highest dry aggregate stability means 47.40, 42.20, and 38.60 were observed under the perennial grasses land management in the 2-5mm aggregate class. The highest water stable aggregate stability means was also observed under the three perennial pastures following the same trend B.D>A.G>D.S 36.02, 35.18, and 33.68 as observed dry aggregate stability mean values in the 2-5 mm aggregate class range. The land management system under the *Bracheria decumbens* has 2.07, 0.99 as the highest values for dry mean weight diameter, wet geometric mean diameter, while 0.57 and 1.00 were respectively

**Corresponding author: E-mail: olawalejohn2013@yahoo.com*

observed as the highest value for wet mean weight diameter and dry geometric mean diameter under the *Digitaria smutsii* land management system Fields under the pasture grasses are higher in organic matter and total nitrogen. The highest C/N ratio value of 1:7.9 was also observed under the pasture grasses fields. The data obtained from the experiment showed that land management under pasture grasses increases the organic matter content, total nitrogen content and consequently increases the structural stability of the soil on long-term basis.

Keywords: Soil; land; mean weight diameter.

1. INTRODUCTION

Land use and management practices often have significant effects on soil properties, especially on soil organic carbon (SOC) dynamics, aggregation and aggregate stability, which are important indicators of soil physical quality [1]. Erodibility as a dynamic property of a soil depends on the stability of soil aggregates and the percentage of coarse primary particles that are resistant to erosion. The formation and maintenance of stable aggregates is the essential feature of the soil tilth, a qualitative term used to describe highly desirable physical condition in which soil is optimally loose, friable and contains a porous assemblage of stable aggregates. This in turn permits free percolation of water and air, enhance the ease of cultivation as well as germination, and emergence of seedlings and growth of roots through the soil pores.

Evaluating the impact of agricultural practices on agroecosystems functions is essential to determine the sustainability of management systems [2], which cover the productivity, economic, social, and environmental components of land use systems [3]. However, decision making for sustainable land management could be improved by tools that provide integration and synthesis of soil research results, management priorities, and environmental concerns [4].

Soil quality is considered a key element used to evaluate sustainable land management in agroecosystems [5] through identification of soil quality indicators [6]. Soil quality, encompasses both soil's productive and environmental capabilities [7], has two parts: An intrinsic part covering a soil's inherent-capacity for crop growth and a dynamic part influenced by the soil management. Soil structure is a crucial soil physical property used to infer soil quality [8]. The degradation of soil structure should be balanced and or exceeded by regeneration in order to have a sustainable soil management [9].

Conservation agriculture (CA), defined as minimal soil disturbance and permanent soil cover (live mulch) combined with crop rotations, was found to be more sustainable cultivation system for the future than those conventionally practiced cultivation system, because conservation agriculture can recover soil functioning through improving water infiltration, reducing erosion, increasing soil organic matter content, and improving soil surface aggregates [10].

Rapid population growth in Nigeria requires additional farmlands for food production. One way to expand the cropland is by cutting down the forests and converting pasturelands to the croplands. This results in destruction of natural ecosystems and reduction of the current or future capacity of soil productivity. In Nigeria, land use change is one of the major factors causing soil degradation [11]. The problem is further aggravated by the intensive cultivation on fragile soil types, which are characterized by low activity clay, low organic matter, low water holding capacity, weak aggregation, high bulk density and poor buffering capacity [12]. Adoption of any management practice that would improve the quality of such soils is imperative in guaranteeing good environmental conditions and high productivity.

2. MATERIALS AND METHODS

2.1 Soil Sampling

The study area is located in National Animal Production Research Institute (NAPRI) farm at Shika Zaria (Latitude 11°00'S and 11°12'N and Longitude 007°33'E) in the Northern Guinea Savanna ecological zone of Nigeria. The mean annual rainfall of the area is 1150 mm usually from May to September ending and a mean daily temperature of 24°C.

Four land use system closely located to each other were selected for this study. The first three were cropped with three different species of perennial grasses namely: Digit grass "*Digitaria smutsii* (DS)", Signal grass "*Bracheria decumbens* (BD)", and Gamba grass "*Andropogon gayanus* (AG)" without tillage (No-tillage) for over fifty years, while the fourth land use system was continuously cultivated (CC) land cropped with maize and cowpea as sole or combined for 20 years. Twelve auger soil samples were collected from each of the four plots at the depth 0-15 cm. The samples were air-dried and prepared for analyses. Water stable aggregate fractions, dry aggregate size fractions, total organic carbon, and total nitrogen were determined using standard methods. Number Cruncher Statistical System "NCSS" computer package was used to analyze the data and the means were compared using Tukey-Kramer Multiple-Comparison Test. Significance differences were reported for 5% probability level.

2.2 Laboratory Analysis

2.2.1 Dry sieving

Dry aggregate size distribution was determined after air drying by mechanical sieving into different size fractions as described by [13]. The same number and size of sieves (2 mm, 1 mm, 0.5 mm, 0.25 mm, 0.05 and <0.05 mm) were used throughout. The weight of aggregate sizes collected from each sieve size were taken and recorded.

2.2.2 Wet sieving

The wet sieving method of Kemper and Rosenau [14] with a set of sieves of 2, 1, 0.5, 0.25, 0.05 and <0.05 mm diameter was used. The mean weight diameterboth of wet and dry sieved samples was calculated by the relationship:

$$MWD = \sum_{i=1}^{n} \bar{X}_i W_i$$

MWD = Mean weight diameter (mm)

W_i = Proportion of each size class with respect to the total sample

$\bar{X}_i$ = Mean diameter of the class

2.3 Total Organic Carbon

The soil organic carbon was determined by the Walkley-Black wet oxidation method.

2.4 Total Nitrogen

Total nitrogen was analyzed by micro Kjeldahl digestion method [15].

3. RESULTS AND DISCUSSION

3.1 Aggregate Size Distribution

The distribution of dry aggregate size class under various land management systems The dry aggregate size obtained in the 2-5 mm, 1-2 mm and 0.5-1 mm aggregate class range are significantly higher under the three pasture grasses compared with the values obtained under continuous cultivation. This may be attributed to the high root biomass, high microbial activities, and high organic carbon content in the soil under the perennial grasses. On the other hand the values obtained for the 0.25-0.5, 0.05-0.25 and < 0.05 mm aggregate size classes are significantly higher under continuous cultivation compared to the values obtained under the three perennial grasses except that CC and BD maintained statistical par on <0.05 mm-sized aggregate. This may be attributed to the collapse of larger aggregates into smaller sized aggregate caused by continuous conventional tillage [16-18] which results from low root biomass, reduced microbial activities, and low organic carbon content in the soil. Franzluebber and Arshad [19] found that macroaggregates (> 0.25 mm) and mean weight diameter were greater under zero tillage (harrowing following harvest) than under conventional tillage (one fall tillage with a cultivator followed by two cultivation in the spring prior to seeding) in coarse-textures soils.

3.2 Water Stable Aggregates

The distribution of water stable aggregate size class under various land management system. The proportion of 2 mm-5 mm, 1 mm-2 mm and 0.5 mm-1 mm water stable aggregate (WSA) size classes were significantly higher under the three pasture grasses compared with the values obtained under continuous cultivation. This also may be attributed to the high root biomass, high microbial activities, and high organic carbon content in the soil under the perennial grasses. On the other hand the values obtained for the 0.25 mm-0.5 mm, 0.05 mm-0.25 mm and < 0.05 mm aggregate classes are significantly higher under continuous cultivation compared to the values obtained under the three perennial grasses land management. This is attributed to the reduction in the size of soil aggregate formed under continuous cultivation which results from low root biomass, reduced microbial activities, and low organic carbon content in the soil.

More pronounced negative effects of continuous cultivation are found in macro-aggregate size classes (>0.25 mm). Tisdall and Oades [20] and Gupta and Germida [21] stated that fungal biomass is a key factor in the formation of macroaggregates and emphasized the sensitivity of macro-aggregate size classes and their nutrient and microbial content (1.00 to 0.25 mm) to long term cultivation, while micro-aggregate size classes are stabilized by organic matter. However, our findings also show a significant (*p*=0.05) impact of long term cultivation on the micro-aggregate size classes (<0.25 mm).

3.3 Aggregate Stability Indices

Significantly higher values of wet and dry mean weight diameter (MWD) and geometric mean diameter (GMD) were observed under the three pasture grasses than in continuous cultivation land management (Table 3.0). This is attributed to high root biomass, microbial activities and organic carbon content of the soil and this implies that the soil under the perennial pasture grasses land use presents greater macro-aggregate stability, higher water permeability and is more effective in mitigating wind and water erosion [22] than the other land uses. Detachability index (Di) show a significantly higher value under continuous cultivation compared with the three perennial grasses. This is as a result of decrease in size of the aggregates, low organic carbon content and reduced microbial activities in the soil. The low values of MWDw under CC land use relative to the land use under perennial pasture grasses showed that continuous mechanical soil disturbance weakens macro-aggregate stability [23] thus resulting to detachment of smaller aggregates and particles from macroaggregates. This also points to the fact that greater proportion of large WSA shift to small macro-and microaggregates with cultivation which induces significant reduction in MWD. Since soil stability decreases as the MWD decreases, the result obtained in this study indicate that macro-aggregate stability against wind erosion decreased in the order: BD>AG>DS>CC, while macro-aggregate stability against water erosion decreased in order: DS>BD>AG>CC.

3.4 Total Organic Carbon, Total Nitrogen and C:N Ratio

Total organic carbon (TOC), total nitrogen (TN) and C:N ratio (see below Table 4) did not show significant difference between the perennial pasture grasses fields, but the values were significantly higher than that of CC. Lack of differences between the perennial pasture grasses in TOC, TN and C:N ratio may reflect long-term stability of the fields.

Table 1. Percentage weight of dry sieving aggregate fractions under various land management systems

Treatment	2-5	1-2	0.5-1	0.25-0.5	0.05-0.25	<0.05
				mm		
AG	42.2b	19.8a	13.4ab	10.8b	12.0b	1.7b
BD	47.4a	17.4a	12.5ab	10.1b	9.6b	3.1ab
CC	11.5c	8.5b	11.9b	19.8a	45.4a	2.9b
DS	38.6b	20.3a	15.0a	12.5b	9.1b	4.4a

AD= Andropogon gayanus, BD= Bracaria decumbens, CC= continuous cultivation, DS= digitaria smuthsii. means followed by the same alphabets are not significantly different at p= 0.5 probability

Table 2. Percentage weight of wet sieving aggregate fractions under various land management systems

Treatment	2-5 mm	1-2 mm	0.5-1 mm	0.25-.5 mm	0.05-.25 mm	<0.05 mm
AG	35.1a	26.9a	14.0a	10.2000b	8.7667b	4.9
BD	36.0a	26.8a	16.7a	9.1792b	7.4417b	3.9c
CC	3.6b	6.9c	8.7b	16.4458a	26.8917a	37.5a
DS	33.7a	21.9b	16.4a	11.4667b	9.0708b	7.5b

AD= Andropogon gayanus, BD= Bracaria decumbens, CC= continuous cultivation, DS= digitaria smuthsii. means followed by the same alphabets are not significantly different at p= 0.5 probability

Table 3. Dry and wet mean weight diameter, dry and wet geometric mean diameter and detachability index under various land management systems

Treatment	dMWD	wMWD	wGMD	dGMD	Di
			mm		
	1.94b	0.55c	0.9429d	1.0008b	0.2862b
BD	2.07a	0.60b	0.9934b	1.0013a	0.2892b
CC	0.76d	0.50d	0.9922c	0.9950d	0.6791a
DS	1.83c	0.63a	0.9937a	1.0005c	0.3464b

AD= Andropogon gayanus, BD= Bracaria decumbens, CC= continuous cultivation, DS= digitaria smuthsii, dMWD= dry mean weight diameter, wMWD= wet mean weight diameter, wGMD= wet geometric mean diameter, dGMD= dry geometric mean diameter. Means followed by the same alphabets are not significantly different at p= 0.5 probability

Table 4. Total organic carbon and total nitrogen under various land management systems

Treatment	TOC(g/kg)	TN(g/kg)	C/N ratio
AG	9.5767a	1.2175a	1:7.9a
BD	9.5091a	1.2300a	1:7.7a
CC	6.8158b	1.1608b	1.5.9b
DS	9.2933a	1.2208a	1:7.6a

AD= Andropogon gayanus, BD= Bracaria decumbens, CC= continuous cultivation, DS= digitaria smuthsiitc= total carbon content, TN= total nitrogen content. means followed by the same alphabets are not significantly different at p= 0.5 probability

The higher total organic carbon under the perennial pasture grasses than under CC may be as a result of the differences in aggregation, high root biomass, litter return [24,25] and microbial activities. Šimansky et al. [26] also stated that the highest average TOC content is found under minimum tillage (9.5767 g/kg) compared with conventional tillage (6.8158 g/kg), which suggests that less or non-disturbance of soil favours accumulation of great amount of organic matter in the soil. The lower organic C and total N for the long-term conventionally tilled field was probably a result of high organic matter decomposition enhanced by disruption of aggregates [27]. Soil C:N ratio was significantly lower in the CC field, compared to the grasses probably due to the higher degree of humification and favour N mineralization.

Although nitrogen mineralization is expected to contribute an important part of the available N to plants under CC, the low N content reflects high leaching process due to continuous conventional tillage. However, the higher C:N ratios at the surface under no-tilled grasses fields compared to that of CT also indicated that even though substantial humification had taken place, the organic matter in the NT fields is still less humified and more carbon-enriched compared to the plots where CT has been applied.

4. CONCLUSION

Measurements of soil aggregate stability and distribution are strongly influenced by an interaction between soil sample pre-treatment, aggregate size class selected and sieving methodology.

Compared with continuous cultivation and recognizing the different forces of disintegration studied, the land management system (zero tillage) as practiced under the three perennial grasses fields improved soil structure by increasing soil organic matter, nitrogen content and increasing the proportion of larger aggregates and the stability of soil when subjected to changes in wetness as well as wind erosion.

ACKNOWLEDGEMENT

The authors thank Lawal Bashir, Professor J.D Olarewaju and Mr A.B Shobayo for stimulating us in reviewing this important topic.

COMPETING INTERESTS

Authors have declared that no competing interests exist.

REFERENCES

1. Shrestha BM, Singh BR, Sitaula BK, Lal R, Bajracharya RM. Soil aggregate- and particle-associated organic carbon under

different land uses in Nepal. Soil Science Society of America Journal. 2007;71(4): 1194-1203.

2. Liebig MA, Varvel G, Doran JW. A simple performance- based index for assessing multiple agroeco-system functions. Agron. J. 2001;93:313-318.
3. Smith AJ, Dumanski J. A framework for evaluating sustainable land management. Can. J. Soil Sci. 1995;75:401–406.
4. Andrew SS, Carrol CR. Designing a decision tool for sustainable agro-ecosystem management Soil quality assessment of a poultry litre management case study. Ecol. Applic. 2001;11(6):1573-1585.
5. Carter MR. Soil quality for sustainable land management: Organic matter and aggregation interactions that maintain soil functions. Agron. J. 2002;94:38–47.
6. Skukla MK, Lal R, Ebinger M. Soil quality indicators for the North Appalachian experimental watersheds in Coshocton Ohio. Soil Sci. 2004;169:195-205.
7. Wander MM, Bollero GA. Soil quality assessment of tillage impacts in Illinois. Soil Sci. Soc. Am. J. 1999;63:961-971.
8. Sánchez-Marañón M, Soriano M, Delgado G, Delgado R. Soil quality in Mediterranean Mountain environments effects of land use change. Soil Sci. Soc. Am. J. 2002;66:948-958.
9. Munkholm LJ, Schjonning P. Structural vulnerability of a sandy loam exposed to intensive tillage and traffic in wet conditions. Soil Till. Res. 2004;79:79-85.
10. Hobbs PR. Conservation agriculture: What is it and why is it important for future sustainable food production? J. Agric. Sci. 2007;1995;145:127-137.
11. Lal R. Global overview of soil erosion. In, Soil and water science: Key to understanding our global environment. Soil Sci. Soc. Am. Special Publ. 41, Madison, WI. 1994;39–51.
12. Oikeh SO, Kling JG, Okoruwa AE. Nitrogen fertilizer management effects on maize grain quality in West African moist savanna. Crop. Sci. 1998;38:1056-1061
13. Bremmer JM, Mulvancy CS. Nitrogen-Total. Methods of soil analysis. Part 2. 2nd ed. Agron. Monogr. 9 ASA and SSSA, madison, WI. *In* A.L. page et al. (ed.). 1982;595-624.
14. Chepil WS. A compact rotary sieve and importance of dry sieving in physical soil analysis. Soil sci. Sco. Ams. Proc. 1962; 26:4-6.
15. Kemper WD, Rosenau RC. Aggregate stability and size distribution. in A. Klute, ed. Methods of soil analysis. Part 1 2nd ed. American Society of Agronomy, Madison, WI. 1986;425–442.
16. Berntsen R, Berre B. Fracturing of soil clods and the soil crumbling effectiveness of draught tillage implements. Soil Till. Res. 1993;28:79–94.
17. Munkholm LJ, Kay BD. Effect of water regime on aggregate-tensile strength, rupture energy and friability. Soil Sci. Soc. Am. J. 2002;66:702-709.
18. Dexter AR, Birkas M. Prediction of the soil structures produced by tillage. Soil Till. Res. 2004;79:233–238.
19. Franzluebbers AJ, Arshad MA. Soil microbial biomass and mineralizable C of water-stable aggregates. Soil Sci. Soc. Am. J. 1997;61:1090-1097.
20. Tisdall JM, Oades JM. Organic matter and water stable aggregates in soils. J. Soil Sci. 1982;33:141–163.
21. Gupta VV, Germida JJ Distribution of microbial biomass and its activity in different soil aggregate size classes as affected by cultivation. Soil Biol. Biochem. 1988;20:777-786.
22. Colazo JC, Buschiazzo DE. Soil dry aggregate stability and wind erodible fraction in a semiarid environment of Argentina. Geoderma. 2010;159:228-236.
23. Lal R, Kimble J, Stewart BA. World Soils as a sink for radio-active gases, in: Soil management and greenhouse effect, edited by: Jorgensen BB, Lewis, London. 1995;1–7.
24. Goma-Tchimbakala J, Makosso S. Soil organic matter and biological changes in a natural-to planted forest succession: Terminalia superba plantatioins grown in deforested plot of Congo. J. applied sci. 2008;8:4346-4353.
25. Qiang-Sheng Wu, X. Ren-Xue, and Z. Ying-Ning. Improved soil structure and citrus growth inoculation with three arbuscular micorrhiza fungi under drought stress. J. Soil Bio. 2008;44(1):122-128.
26. Šimansky V, Tobiašova E, Chlpik J. Soil tillage and fertilization of Orthic Luvisol and

their influence on chemical properties, soil structure stability and carbon distribition in waterstable macro-aggregates. Soil Tillage Res. 2008;100:125–132.
ISSN 0167-1987.

27. Hassink J. Density fractions of soil macroorganic matter and microbial biomass as predictors of C and N mineralization. Soil Biol. Biochem. 1995; 27:1099- 1108.

27

Impact of Farm Yard Manure and Goat Manure as Organic Amendment on Physico-chemical and Physical Properties of Theri Soil

E. Indira[1*] and B. Annadurai[2]

[1]Manonmaniam Sundaranar University, Tirunelveli-627 012, Tamilnadu, India.
[2]Department of Physics, Aditanar College of Arts and Science, Tiruchendur-628216, Tamilnadu, India.

Authors' contributions

This work was carried out in collaboration between both authors. Author EI designed the study, wrote the protocol and wrote the first draft of the manuscript. Author EI managed the literature searches; analyses of the study performed the spectroscopy analysis and managed the experimental process. Author BA identified the species of plant. Both authors read and approved the final manuscript.

Editor(s):
(1) Scd Victoria Anatolyivna Tsygankova, Department for Chemistry of Bioactive Nitrogen-Containing Heterocyclic Compounds, Institute of Bio-organic Chemistry and Petrochemistry, Plant Physiology, NAS, Ukraine.
Reviewers:
(1) Prakash M. Munnnoli, SDM College of Engineering and Technology, India.
(2) Petigrosso, Lucas Ricardo, National University of Mar del Plata, Argentina.

ABSTRACT

The study on soil quality assessment in red sandy soils (chemical Theri soil) of Punnai Nagar (Kachanavillai) in Tuticorin district, Tamilnadu, south India was carried out during 2014(Oct) to 2015(Jan). Investigations were carried out on the surface (15 cm depth) of soil samples with the objective to assess the available nutrients and their relationship with various physico-Chemical properties. The soil has low available N=78 kg/ha, low organic carbon OC=0.19%, medium P=5 kg/ha, K=104 kg/ha and Fe, Mn, Zn, Cu are takes a value of 0.369 ppm, 0.985 ppm, 4.861 ppm and 1.996 ppms. Theri soil subject to higher rate of erosion resulting in low nutrient and minerals. The permeability of water in theri soil is high and is not suitable for agriculture which is presently considered as a wasteland. Hence, the fertility and productivity of theri soil may be improved on sustainable basis using Organic farming. The treatments of this study were combination of Farmyard manure (FYM) and Goat manure. The experiment was laid out in Pot

Corresponding author: E-mail: indiraesaki76@gmail.com

culture with 3 replications. All the amendments were applied and wetting with water regularly for 90 days without allowing them to get dried. After 90 days measurements were made. The result suggest that the use of organic amendment will increases NPK, Water holding capacity, Pore space, EC, Organic Carbon and decreased pH, bulk density, Particle density, Hydraulic conductivity. These changes in soil environment improved the soil fertility without affecting soil health.

Keywords: Theri soil; FYM; goat manure; water holding capacity; reclamation; physico-chemical properties.

ABBREVIATIONS

1. *W.H.C- Water holding capacity;*
2. *Vol.Exp- Volume Expansion;*
3. *Bulk.Den- Bulk density;*
4. *Th- Theri soil;*
5. *Hyd.Cond- Hydraulic Conductivity.*
6. *OC- Organic Carbon.*
7. *EC- Electrical Conductivity*
8. *FYM- Farmyard Manure*
9. *pH- Potential of Hydrogen*
10. *FG- Farmyard+ Goat manure*

1. INTRODUCTION

The coastal red sandy dunal soils of Tuticorin and Tirunelveli districts of Tamilnadu are locally called theri soils. They occur in the Tuticorin and Tirunelveli districts of Tamilnadu with an extend of 20,171 hectares [1]. Tuticorin district has the highest theri area (16,978 ha). The southern and eastern parts of two districts adjoining the coastal area of gulf of manner are degraded due to wind erosion and moving sand dunes. The high velocity wind blows from the Western Ghats during the months of June-July scoop the sand and moves it towards Bay of Bengal situated in the east, resulting in the formation of sand dunes. This type of transported soil is locally called as theri soil (Th). Wind speed up to 66 Km/hr has recorded. The colour span of theri soils is red due to the presence of iron compounds and the aerobic conditions prevailing confirm the presence of haematite mineral in the soil. Haematite is in the dust form to adhere on the soil particle uniformly [2]. Yadav et al. [3] inferred that both topography and drainage were responsible for the different colours in the associated red soils. Theri soils are poor in nutrient status, water holding capacity, organic carbon content and fertility status. Soil fertility is determined by the presence or absence of micro and macro nutrients. The availability of nutrients is particularly sensitive to changes in soil environment. The texture of Theri land profile soils were mostly sandy and loamy sand. Due to this coarser texture theri soils had excessive drainage, poor moisture holding capacity, poor nutrient status etc. These observations are in line with the inferences of Mayalagu [4]. Addition of organic matter brings improvement of soil structure, favourable change in total porosity, and increase in water holding capacity, soil aeration and permeability [5]. The detail study of physical, physico-chemical character, available macro and micro nutrients of soil determines the soil potential which is essential for better scientific utilization of land use planning for crop growth. Keeping in view of above points the detailed micro level study was conducted to assess the physico-chemical characteristics of theri soil.

2. MATERIALS AND METHODS

Soil samples (Th) were collected at Punnai nagar (Kachanavilai) that is located at 19 km west from Tiruchendur in Tuticorin district of Tamil Nadu. A soil sample was collected in the experimental site by making 'V' shaped cut at a depth of 15 cm air dried ground, mixed pass through 2 mm sieve and analyzed for their physico-chemical and physical properties. A pot experiment was conducted during Rabi season during 2014- 2015. The soil was sandy clay loam in texture, neutral in reaction having pH range of 7.63 and low in organic carbon 0.19%, available NPK values are 84 kg/ha, 5 kg/ha and 104 kg/ha. EC was found to be below $1.0 dSm^{-1}$. The experiment was laid out in a pot culture with ten treatments replicated thrice. The pots were arranged under the screen house according to steel and Torrie [6]. In the screen house study, 1 litre (1000 c.c) of theri soil thoroughly mixed with different percentage of farm yard and goat manure mixture in a clay pot. (2 mm sieve powder). Ten different combinations, namely theri soil (Th)+10% of FG (FYM + Goat manure); Th+20% of FG; Th+30% of FG; Th+40% of FG; Th+50% of FG; Th+60% of FG; Th+70% of FG; Th+80% of FG; Th+90% of FG; Th+100% of FG. These different combinations were thoroughly ameliorated mechanically before use. For example Th+10% of FG mean 100 cc of Fym+Goat manure was mixed with 1000 cc of theri soil. The volume of the soil is fixed. It was found that 100 cc of FG is equal to 35 g. It was also found that 100 cc of theri soil is equal to 165 g. These mixtures were allowed to settle for a period of 90 days by wetting with water regularly and without allowing them to get dried. After 90 days measurements were made on the physical and physico-chemical properties, such as pH, EC, NPK, particle density, bulk density, porosity, water holding capacity, volume expansion, organic carbon content and hydraulic conductivity.

2.1 Methods Used for Analysis

The soil samples were characterized for important physico-chemical properties using standard procedures. Bulk density, Particle density, Water holding capacity, Porosity, Volume expansion by KeenRaczkowski(KR) Box model [7], Soil pH and EC was determined in 1:2.5 soil water suspensions with help of glass electrode pH meter with a digital display and with a null balance conductivity meter correspondingly [8], Organic Carbon determined Chromic acid wet oxidation method [9]. Available "N" alkaline permanganate method [10]. The available "P" was estimated by Flame photometer [11] and available "K" determined by Ammonium acetate method [12]. Finally soil samples were collected from pots and air dried in shade processed and screened through a 2mm sieve. After sieving all the samples were packed in polythene bags for analysis.

3. RESULT S AND DISCUSSION

Application of FYM and Goat manure with theri soil at various proportions show a regular increase of all oxidizable organic carbon fractions except in control shown in (Table 2). Application of FYM enhances the OC content of the soil. Increased OC values range from 0.55% to 0.71% (Table 2) with the mean value of 0.62% was observed. Addition of FG with theri soil decreases the pH, Bulk density, Particle density, Hydraulic conductivity but increases the EC, Water holding capacity, Volume expansion and Porosity (Table1). The bulk density of soil ranges from 1.23-1.77 g/cm^3, generally bulk density of 1.5-1.6 g/cm^3 is critical for root growth of the most of the plants [13]. But theri soil has bulk density of 1.73 g/cm^3 which shows less convenience for root growth. The decrease in bulk density might be due to higher organic matter content of the soil (figure) with the application of FYM and Goat manure, this result was already inferred by Gopal Reddy and Surya narayan Reddy [14]. The nitrogen content increases as the dosage of the amendment increases. It ranges from 84 kg/ha-112 kg/ha (Table2). Srikanth et al. [15] showed that the decreased in soil pH with the application of FYM. The reduction in soil pH was due to increased CO_2 concentration in soil arising from the microbial decomposition of FYM & Goat manure. The CO_2 in contact with water forms Carbonic acid, which reacts with native $CaCO_3$ of soil to bring Ca in soil solution, resulted in to lower value of soil pH. Reduction in soil pH and ESP by the application of FYM due to microbial decomposition was also reported by [16]. FG treated theri soil shows a pH values ranges from 7.34-7.11 (Table 2). The pH value above 9 is undesirable and pH value below 4.5 is good. Here in all treatments the porosity ranges from 40%-51% (Table 1).

This range of porosity slightly hinders the growth of plants. The neutral pH value is obtained for all FG treatments. So this is most favourable for the cultivation purpose. The EC values vary from 0.9dsm^{-1} to 1.1dsm^{-1} (Table 2) higher than the control soil, but in all treatments EC do not exceed the threshold limit. The desired increase of water holding capacity ranged from 26.13% to 33.8% (Table1) will improve the ability to supply the nutrients to the soil. The hydraulic conductivity controlled from very rapid stage to moderate rapid stage ranges from190 mm/hr to 102 mm/hr shown in (figure) this enhances soil nutrients. In all the treatments water holding capacity increases and attains the maximum value of 33.82% (Table1) and the hydraulic conductivity reduces to the minimum value of 102.3 mm/hr (Table 1). There is an inverse relationship between porosity and bulk density. Total pore space is typically increased by addition of organic matter [17,18].

The FYM & Goat manure in addition to its ameliorative effect being a store house of almost all the nutrients required for plant growth and improved soil environment via improvement in physio-chemical properties of soil. Thus on addition of FYM & Goat manure, increases the nutrient status due to mineralization of native as well as applied nutrients through organic fertilisers in addition to its own nutrient content. The significant increase in available 'N' content of soil due to application of FYM may be attributed to mineralization of 'N' by FYM in soil.

Application of gypsum, FYM and sulphuric acid according to Hussain et al. [19] improved the bulk density and porosity and has contributed to enhanced water permeability and hydraulic conductivity. So the use of organic amendment helps to build up soil humus and beneficial microbes, besides improvement of soil physical properties.

Table 1. Effect of various mixtures of farmyard and goat manure on physical properties of theri soil

Treatment	Bulk density. g/cm^3	Particle density g/cm^3	Porosity, %	W.H.C %	Vol. Exp, %	Hyd. Cond.mm/hr
Th+10%FG	1.77	2.967	40.02	26.13	4.36	190.22
Th+20%FG	1.60	2.854	43.83	26.63	4.56	183.23
Th+30%FG	1.57	2.76	43.15	26.91	4.65	177.12
Th+40%FG	1.46	2.583	43.65	27.29	5.82	165.03
Th+50%FG	1.40	2.550	45.63	27.59	6.22	154.74
Th+60%FG	1.42	2.500	46.89	28.71	6.33	155.15
Th+70%FG	1.43	2.350	51.21	28.31	7.09	142.51
Th+80%FG	1.33	2.300	50.97	31.72	8.40	127.00
Th+90%FG	1.27	2.222	51.47	33.24	8.54	120.54
Th+100%FG	1.23	2.050	51.07	33.82	8.56	102.29
Control	1.73	2.857	39.13	25.12	3.9	210.00

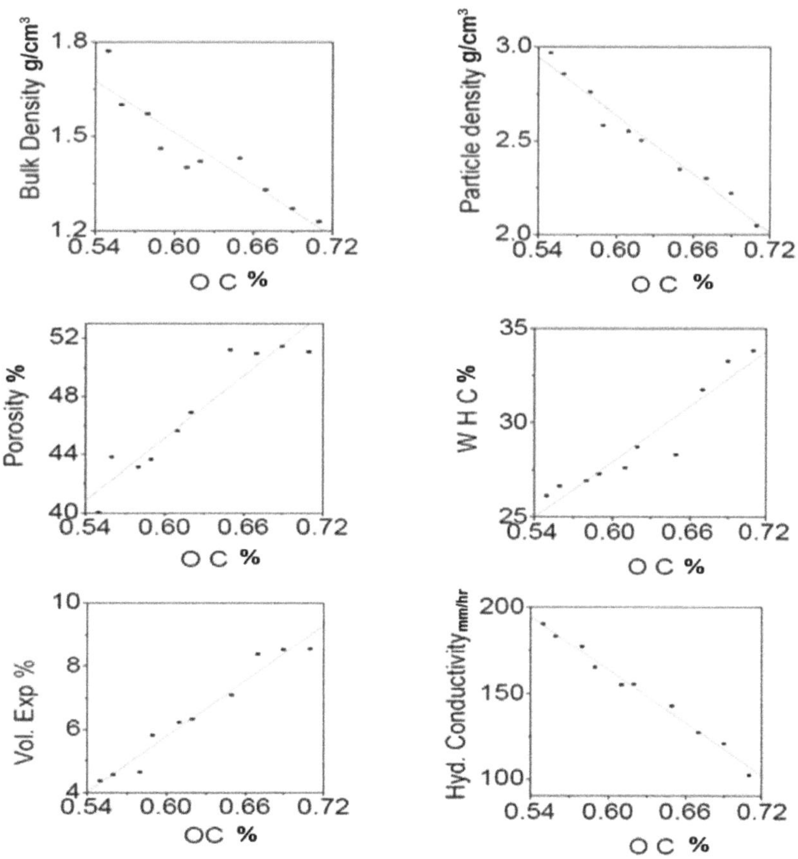

Fig. 1. Relation between organic carbon and physical properties of theri soil

Table 2. Effect of various mixtures of farmyard and goat manure on chemical properties of theri soil

Treatment	pH	EC dsm^{-1}	OC %	N Kg/ ha	P Kg/ ha	K Kg/ ha
Th+10%FG	7.34	0.946	0.55	84	26.72	384
Th+20%FG	7.31	1.045	0.56	89	28.63	389
Th+30%FG	7.28	1.052	0.58	91	29.46	394
Th+40%FG	7.25	1.058	0.59	94	31.35	395
Th+50%FG	7.23	1.073	0.61	96	34.56	411
Th+60%FG	7.21	1.078	0.62	98	39.42	419
Th+70%FG	7.18	1.085	0.65	101	41.25	422
Th+80%FG	7.15	1.089	0.67	104	43.84	426
Th+90%FG	7.13	1.092	0.69	109	46.73	429
Th+100%FG	7.11	1.106	0.71	112	50.22	432
Control	7.63	0.84	0.19	78	10.8	315

4. CONCLUSION

In the era of ever increasing population, it is vital to utilise each and every unit of soil scientifically and get the best out of it. In this context the uncultivated sand dune areas have to be identified and examined through scientific methods. This examination shows the poor pedogenesis due to constant movement of theri soil. There was poor retention of nutrients, loose and friable in nature, high Bulk density, high particle density, high pore space, low water holding capacity, low volume expansion due to severe wind erosion. So to improve this poor nutrient soil Organic farming (FYM+Goat manure) has to be introduced here, to increase the level of soil organic matter. Organic farming enhances water-retention capacity of the soil and hence the amount of water needed for irrigation is reduced. Further, organic farming can mitigate the climate change as it reduces emission of greenhouse gases, especially nitrous oxide, because no chemical nitrogen fertilizers are used. In the present study may be associated with the supply of essential nutrients by continuous mineralization of organic manures enhanced inherent nutrient supplying capacity of the soil and its favourable effect on soil physical, Physico-chemical and biological properties and to meet the nutrient requirement of crops for sustainable agriculture.

COMPETING INTERESTS

Authors have declared that no competing interests exist.

REFERENCES

1. Jawahar D, Studies on the sand dunes (theri) in the coastal belt of V. O. Chidambaranar and Nellai Kattabomman districts, Tamilnadu Agricultural University Ph.D. Thesis, Coimbatore, Tamilnadu; 1996.
2. Subramanian V. Land use planning for management of agricultural resources, final report, TNAU, Killikulam, T.N. South India. 2004;45-46.
3. Yadav BR, Gupta RN, Singh BD. Catenary relationship existing among the soils of the lower Vindhyan Plateau in Uttar Pradesh. J. Indian Soc. Soil Sci. 1977;25:253-259.
4. Mayalagu K, Sree Ramulu US. N, P and K (Total and available) status of major series of red soils of Coimbatore district, Tamil Nadu. Madras Agric. J. 1983;70:51-54.
5. Gorbunov NI. Effects of method of soil properties. Agrochimiya. 1980;50:93-100.
6. Steel RGD, Torrie JH. Principles and procedures of statistics. Second Edition. McGraw-Hill Book Co. New York; 1980.
7. Keen BA, Raczkowski H. Relation between the clay content and certain physical properties of soil. Journal of Agricultural Science. 1921;11:441-449.
8. Jackson MC. Soil chemical analysis. Prentice Hall Pvt, Ltd. New Delhi; 1973.
9. Walkley A, Black JA. An examination of the Degt Jareft method for determination of soil organic matter and proposed modification of chromic acid titration method. Sci. Soil. 1934;37:29-38.
10. Subbaiah BV, Asija GC. A rapid procedure for the determination of available nitrogen in soils. Current Science. 1956;25:259-262.
11. Olsen SR, Cole CU, Watanabe FS, Deen LA. Estimation of available phosphorus in soil by extracting with sodium bicarbonate, USDA circular 939, Washington; 1954.

12. Hanway JJ, Heidel. Soil analysis methods as used in lowa state college of soil testing laboratory. Iowa Agric. 1952;57:1-31.
13. Ayres KW et al. Soil morphology and soil physical properties. II. Mechanical impedance moisture retention and movement. Can. J. Soil. Sci. 1973;53:9-19.
14. Gopal Reddy B, Suryanarayan Reddy MJ. Indian Soc. Soil sci. 1998;46(3):474-476.
15. Srikanth KC, Srinivasa murthy A, Siddaramappa R, Rama Krishna Parama VR. Direct and residual effects of enriched composts FYM, Vermi compost and fertilizers on properties of an alfisol. J. Indian Soc. Soil Sci. 2000;48:494-499.
16. Deshveer CL, Singh A, Mali GC. Effect of partial substitution of inorganic fertilizers by FYM on the productivity of wheat. Annals of Plant and Soil Research. 2002; 4:341-342.
17. Barzegar AR, Yousefi A, Daryashenas A, Plant and soil. 2002;247:295-301.
18. Dorado J, Zancada MC, Almendras G, Lopez-Fando C. J. Plant Nutrition and Soil Sci. 2003;166:31-38.
19. Hussain N, Hassan G, Arshadullah M, Mujeeb F. Evaluation of amendments for the improvement of physical properties of sodic soil. Int. J. Agric. and Biology. 2001;3(3):319-322.

28

Effect of Brassinolide on the Growth of Mustard Crops Grown in Semi-arid Tropics of Nizamabad

P. Latha[1] and B. Vidya Vardhini[2*]

[1]Department of Botany, Govt. Degree College, Armoor, Nizamabad, India.
[2]Department of Botany, Telangana University, Dichpally, Nizamabad -503322, India.

Authors' contributions

This work was carried out in collaboration between both authors. Author BVV is the research supervisor and author PL is pursuing her Ph. D under author BVV. Author BVV designed the study, wrote the protocol and wrote the first draft of the manuscript. Author PL managed the literature searches; analyses of the study performed the spectroscopy analysis and managed the experimental process. Both the authors read and approved the final manuscript.

ABSTRACT

The effect of Brassinolide (BL) sprayed in three concentrations viz., 0.5 μM, 1.0 μM and 2.0 μM on the shoot growth, root growth and foliar growth (plant fresh weight, leaves per plant and leaf area) of mustard plants grown in the semi-arid tropics of Nizamabad was studied. The soil in Nizamabad district is saline land black soil wherein the plants usually experience drought and saline stresses. All the three concentrations of BL stimulated the shoot growth, root growth as well as the foliar growth of mustard plants. BL at 0.5μM was found most effective in increasing the shoot and root growth. The promotion of shoot, root and foliar growth is an indicator that BL mitigated the negative effect of the semi-arid conditions of the soil.

Keywords: Brassinolide; foliar growth; mustard; root growth; shoot growth.

**Corresponding author: E-mail: drvidyavardhini@rediffmail.com*

ABBREVIATIONS

Brassinolide=BL; Brassinosteroids =BRs; 24-Epibrassinolide = 24-EpiBL; 28-Homobrassinolide =28-Homo BL.

1. INTRODUCTION

Brassinosteroids (BRs) are a new type of polyhydroxy steroidal phytohormones with significant growth-promoting influence [1]. Mitchell et al. [2] discovered BRs which were later extracted from the pollen of *Brassica napus* L. by Grove et al. [3]. BRs can be classified as C_{27}, C_{28} or C_{29} BRs according to the number of carbons in their structure [4]. However, Vardhini et al. [5] reported that brassinolide (BL), 28-homobrassinolide (28-HomoBL) and 24-epibrassinolide (24-EpiBL) are the three bioactive BRs being widely used in most physiological and experimental studies. The work with BR biosynthetic mutants in *Arabidopsis thaliana* [6] and *Pisum sativum* [7] have provided strong evidences that BRs are essential for plant growth and development and BR- signaling plays a positive in plant growth and development viz., spatiotemporal control of BR pathways in plant development employing microscope lens turret to study the pleiotropic phenotypes of the BR mutants at a higher magnification [8].

Rao et al. [9] stated that BRs are a new group of plant growth hormones that perform a variety of physiological roles like growth, seed germination, rhizogenesis, senescence, etc. and also confer resistance to plants against various abiotic stresses. Though, BRs were initially identified based on their growth promoting activities, subsequent physiological and genetic studies revealed additional functions of BRs in regulating a wide range of processes, including source/sink relationships, seed germination, photosynthesis, senescence, photomorphogenesis, flowering and responses to different abiotic and biotic stresses [4].

Nizamabad district experiences a tropical dry and wet season with most of the rainfall in June to October. It usually experiences erratic rain fall. The soil is saline land black soil which is deep loamy to clay loam, moderately drained, neutral to alkaline in nature. The reduction of growth of many plants by salinity and drought usually effects on dry matter production, ionic relations, metabolic variations, physiological processes, water contents. The semi-arid condition directly poses a threat to the overall yield of the plants as they usually experience drought and saline stresses.

Mustard (*Brassica juncea*) is an oil yielding plant that belongs to the family Brassicaceae which is grown throughout the world. The plant has tiny seeds which are usually used as a condiment. Mustard oil is extracted from the seed which are used for cooking, massaging etc. Some kinds of mustard plants have edible leaves that are known as *mustard greens*. It is a well-established fact from time immemorial that plants are the vital components of dietary food chains in which they provide almost all the essential mineral and organic nutrients to humans [10]. The present study is undertaken to understand the effect of application of BL on the growth of mustard plants in terms of shoot, root and foliage grown in the semi - arid soils of Nizamabad.

2. MATERIALS AND METHODS

2.1 Chemicals and Plant Material

Brassinolide (BL: double) is a commercially available brassinosteroid was procured from Bahar Agrochem & Feeds Pvt. Ltd, Ratnagiri, Maharastra, India, Ltd. It is marketed by Godrej Agrovet Ltd., Hyderabad, Andhra Pradesh, India. Brassinolide (Double) consists of 0.1% of brassinolide, 2.0% of emulsifier and 97.9% of solvent IPA.

Mustard (*Brassica juncea* var tulasi) was purchased from National Seeds Corporation, Hyderabad, Andhra Pradesh, India.

2.2 Shoot and Root Growth

Mustard seeds were sown in clay pots containing fresh sieved black soil mixed with farmyard manure. Plants were grown under natural day length. BL was supplied to the plants as foliar spray at three different concentration levels viz., 0.5 µM, 1.0 µM and 2.0 µM on 35th, 40th, 45th and 50th day (from the day of sowing). Growth parameters in terms of shoot weight diameter and length were recorded on the 55th day. Growth parameters in terms of root weight, diameter and length were recorded on the 55th day.

2.3 Foliar Growth

Foliar growth was recorded in terms of plant fresh weight, leaves per plant and leaf area per plant were recorded on 55th day. Average leaf area per plant was determined with the help of the equation developed by Kemp [11].

$$A = (L \times B \times 0.9) \times n,$$

Where A= leaf area, L= leaf length, B=leaf breadth, 0.9= constant factor and n= number of leaves.

The data is represented in terms of Mean±S.E (n=9).

3. RESULTS

Exogenous application of BL resulted in substantial increase in shoot growth of mustard plants as reflected in increases in length, fresh weight and diameter of the shoots (Table 1.). All the three concentrations applied viz., 0.5µM, 1 µM and 2 µM increased the shoot growth of mustard plants grown in semi-arid soils of Nizamabad over control plants. BL at 0.5µM was found most effective in substantial increase in shoot growth compared to the other two concentrations as well as control plants.

The mustard plants grown in semi-arid soils of Nizamabad and treated with foliar application of BL showed increased root growth in terms of length, fresh weight and diameter (Table 2.). All the three concentrations of BL applied viz., 0.5 µM, 1 µM and 2 µM increased the shoot growth of mustard plants grown in semi-arid soils of Nizamabad over control plants. 0.5 µM Conc. of BL exhibited maximum enhanced root growth compared to the other treatments as well as untreated controls.

Foliar application of BL exhibited improved foliage in terms of plant fresh weight, leaves/plant and leaf area of mustard plants grown in semi-arid tropics of Nizamabad (Table 3.). All the three concentrations applied viz., 0.5 µM, 1 µM and 2 µM increased the shoot growth of mustard plants grown in semi-arid soils of Nizamabad over control plants. Application of BL at 0.5 µM conc. was more pronounced in increasing the foliar growth of mustard plants compared to the other two treatments as well as control plants.

Table 1. Effect of brassinolide on the shoot growth in *Brassica junceae*

Treatment (µM)	Shoot weight (gm/fr. wt.)*	Shoot diameter (cm)*	Shoot length (cm)*
0.5	8.093±1.925	6.400±0.213	61.66±3.179
1.0	6.610±0.100	6.427±0.019	57.10±1.000
2.0	4.510±0.238	5.098±0.036	47.66±7.881
Control	2.001±1.041	4.330±0.169	32.00±4.582

**Values are Mean±S.E. (N=9), (gm/fr. wt.) = gram/fresh weight; (cm) = centimeters*

Table 2. Effect of brassinolide on the root growth in *Brassica junceae*

Treatment (µM)	Root weight (gm/fr. wt.)*	Root diameter (cm)*	Root length (cm)*
0.5	1.130±0.430	4.850±0.087	11.02±0.288
1.0	0.830±0.100	4.370±0.134	10.50±0.288
2.0	0.710±0.058	4.435±0.131	10.33±0.033
Control	0.440±0.145	3.320±0.150	8.83±4.582

**Values are Mean±S.E. (N=9), (gm/fr. wt.) = gram/fresh weight; (cm) = centimeters*

Table 3. Effect of brassinolide on the foliar growth of *Brassica junceae*

Treatment (µM)	Plant fresh weight (gm/fr. wt.)*	Number of leaves/plant	Leaf area (sq. cm)*
0.5	9.550±2.224	12.33±1.850	483.1±11.26
1.0	8.403±1.315	11.26±1.650	411.3±9.90
2.0	5.510±1.043	10.66±2.060	398.2±10.10
Control	2.991±1.041	7.33±1.520	133.1±9.91

**Values are Mean±S.E. (N=9), (gm/fr. wt.) = gram/fresh weight; (sq cm) = square centimeters*

4. DISCUSSION

It is a well-established fact that BRs are a new group of plant growth regulators which play a positive role in the growth and development of plants. BRs increased the height of two field-grown inbred lines of maize (*Zea mays* L.) during the vegetative and early reproductive phases of plant development during the early weeks after their application [12]. Fariduddin et al. [13] reported that cucumber seedlings sprayed with 10^{-8}, or 10^{-6} M of 28-homoBL exhibited improved growth at the 30-day stage after treatment plants were exposed for 18 h to chilling temperature (10/8°C, 5/3°C). BL applied to the sunflower seedlings significantly increased the hypocotyl growth under different light qualities [14]. Even in the present study, application of BL significantly increased the shoot growth in mustard plants grown in the semi –arid soils of Nizamabad.

Vardhini et al. [15] reported that application of BRs increased the qualitative growth of radish (*Raphanus sativus* L. var Pusa chetki long) in terms of root fresh weight, root dry weight, oxalic acid, ascorbic acid, folic acid and niacin. Hayat et al. [16] reported that 28-homoBL and 24-epiBL increased the growth of *Lycopersicon esculentum*. Imbibition of 24-epiBL to pea (*P. sativum* L.) cv. Climax seeds subjected to sodium chloride stress exhibited increased germination, embryo axis length and most of the aspects of shoot and root growth at seedling stage [17]. Earlier studies clearly emphasized that external supplementation/application of BRs improved plant growth in the case of wheat [18] and groundnut [19] which is in tune with the present study where foliar application of BL substantially increased the root growth of mustard plants.

The enhanced shoot and root growth of BL treated mustard plants was associated with increased foliar growth. Arteca and Arteca [20] also reported that BRs not only induce exaggerated growth in hydroponically grown *A. thaliana*, but also control the proliferation of its leaf cells [21]. Kudryashova et al. [22] reported that supplementation of 24-epiBL played a pivotal role in *in vitro* regeneration of highbush blueberry (*Vaccinium corymbosum* L., cv. Brigitta blue).

5. CONCLUSION

The present study reveals that application of BL on mustard plants as foliar spray promoted the shoot, root and foliar growth of mustard plants grown in semi-arid soils of Nizamabad. The soils of Nizamabad are saline and dry in nature inhibiting the growth of plants. BRs have the ability to promote growth of plant under stressful conditions. The present study reveals a new insight that application of BL overcame the negative effect of the semi-arid conditions of the soil (reflected in the control plants) and promoted the shoot, root and foliar growth (reflected in the BL-treated plants) of mustard.

COMPETING INTERESTS

Authors have declared that no competing interests exist.

REFERENCES

1. Vardhini BV, Anjum NA. Brassinosteroids make plant life easier under abiotic stresses mainly by modulating major components of antioxidant defense system. Frontiers in Environmental Sciences. 2015;2:67. DOI: 10.3389/fenvs.2014.0006.
2. Mitchell JW, Mandava NB, Worley JE, Plimmer JR, Smith MV. Brassins: A family of plant hormones from rape pollen. Nature. 1970;225:1065-1066.
3. Grove MD, Spencer GF, Rohwededer WK, Mandava NB, Worlet JF, Warthen Jr JC, Steffens GL, Flippen-Andersen JL, Cook Jr JC. Brassinolide, a plant promoting steroid isolated from *Brassica napus* pollen. Nature. 1979;281:121-124.
4. Vardhini BV. Brassinosteroids, Role for amino acids, peptides and amines modulation in stressed plants - A review. In: Anjum NA, Gill SS, Gill R, editors. Plant Adaptation to Environmental Change: Significance of Amino Acids and their Derivatives, United Kingdom: CAB International of Nosworthy Way, Wallingford OX10 8DE. 300-316; 2013.
5. Vardhini BV, Anuradha S, Rao SSR. Brassinosteroids - New class of plant hormone with potential to improve crop productivity. Indian J. Plant Physiol. 2006;11:1-12.
6. Tao YZ, Zheng J, Xu ZM, Zhang XH, Zhang K, Wang GY. Functional analysis of ZmDWF1, a maize homolog of *Arabidopsis* brassinosteroids biosynthetic DWF1/DIM gene. Plant Sci. 2004;167:743-751.
7. Nomura T, Nakayama N, Reid JB, Takeuchi Y, Yokota T. Blockage of

brassinosteroid biosynthesis and sensitivity cause dwarfism in *Pisum sativum*. Plant Physiol. 1997;113:31-37.
8. Fábregas N, Caño-Delgado AI. Turning on the microscope turret: A new view for the study of brassinosteroid signaling in plant development. Physiol Plant. 2014;151:172-183.
9. Rao SSR, Vardhini BV, Sujatha E, Anuradha S. Brassinosteroids – new class of phytohormones. Curr Sci. 2002;82: 1239-1245.
10. Grusak MA, Dellapenna D. Improving the nutrient composition of plants to enhance human nutrition and health. Annu Rev Plant Physiol Plant Mol Biol. 1999;50:133-161.
11. Kemp CD. Methods of estimating leaf area of grasses from linear measurement. Ann Bot. 1960;24:491-512.
12. Hola D, Rothova O, Kocova M, Kohout L, Kvasnica M. The effect of brassinosteroids on the morphology, development and yield of field-grown maize. Plant Growth Regul. 2010;61:29-43.
13. Fariduddin Q, Yusuf M, Chalkoo S, Hayat S, Ahmad A. 28- Homobrassinolide improves growth and photosynthesis in *Cucumis sativus* L. through an enhanced antioxidant system in the presence of chilling stress. Photosynthetica. 2011;49:55-64.
14. Kurepin LV, Joo SH, Kim SK, Pharis RP, Back TG. Interaction of brassinosteroids with light quality and plant hormones in regulating shoot growth of young sunflower and *Arabidopsis* seedlings. J Plant Growth Regul. 2012;31:156-164.
15. Vardhini BV, Sujatha E, Rao SSR. Studies on the effect of brassinosteroids on the qualitative changes in the storage roots of radish. Bulg J Agric Sci. 2012;18:63-69.
16. Hayat S, Yadav S, Wani AS, Irfan M, Ahmad A. Comparative effect of 28-homobrassinolide and 24-epibrassinolide on the growth, carbonic anhydrase activity and photosynthetic efficiency of *Lycopersicon esculentum*. Photosynthetica. 2011;49:397-404.
17. Shahid MA, Pervez MA, Balal RM, Mattson NS, Rashid A, Ahmad R, Ayyub CM, Abbas T. Brassinosteroid (24-epibrassinolide) enhances growth and alleviates the deleterious effects induced by salt stress in pea (*Pisum sativum* L.). Aust J Crop Sci. 2011;5:500-510.
18. Sairam RK. Effect of homobrassinolide application on plant metabolism and grain yield under irrigated and moisture – stress conditions of two wheat varieties. Plant Growth Regul. 1994;14:173-181.
19. Vardhini BV, Rao SSR. Effect of brassinosteroids on growth, metabolite content and yield of *Arachis hypogaea*. Phytochemistry. 1998;48:927-930.
20. Arteca JM, Arteca RN. Brassinosteroid-induced exaggerated growth in hydroponically grown *Arabidopsis* plants. Physiol Plant. 2001;112:104-112.
21. Nakaya M, Tsukaya H, Murakami N, Kato M. Brassinosteroids control the proliferation of leaf cells of *Arabidopsis thaliana*. Plant Cell Physiol. 2002;43:239-244.
22. Kudryashova OA, Volotovich AA, Vasilevskaya TI, Varavina NP, Rupasova ZA, Khripach VA. Effects of 24-epibrassinolide on in vitro micropropagation of highbush blueberry. Russ J Plant Physiol. 2012;59:586-593.

29

Micronutrient Composition of Soils and Supported *Solanum scabrum* in Bafut Sub-division (North-West Cameroon)

Margaret Awah Tita[1*], Primus Azinwi Tamfuh[2], Pierre Wotchoko[1], Alice Magha[1] and Rebecca Bienkaa Azinwi[2]

[1]Higher Teachers' Training College, University of Bamenda, P.O.Box 20, Bamenda, Cameroon.
[2]Faculty of Science, University of Yaoundé 1, P.O.Box 812, Yaoundé, Cameroon.

Authors' contributions

This work was carried out in collaboration between all authors. All authors read and approved the final manuscript.

Editor(s):
(1) Marco Trevisan, Institute of Agricultural Chemistry and Environmental Research Centre Biomass, Faculty of Agriculture, Catholic University of the Sacred Heart, Italy.
Reviewers:
(1) Zafar Iqbal Khan, University of Sargodha, Pakistan.
(2) Rajaram Pandurang Dhok, Savitribai Phule Pune University, Pune, India.
(3) Anelia Marais, South Africa.

ABSTRACT

Soil-plant metal interaction is vital in plant metabolism and on effects related to deficiency and toxicity, yet has received very little attention in terms of research in developing countries. The aim of the present study was to investigate the composition of Zn, Cu, Mn and Fe in soils and grown huckleberry (*Solanum scabrum)* in Bafut Sub-division (North-West Cameroon). Soils and huckleberry samples were collected in North and South Bafut in the rainy and dry seasons. The micronutrient levels in *Solanum scabrum* and soils were measured by Atomic Absorption Spectrometry. The results revealed that the metal levels of the soils were within the permissible range for agricultural soils. In the vegetables, only Fe and Mn, the most represented metals, were within the adequate and normal levels for optimum plant growth and human consumption, while Cu and Zn contents suggested a dietary intake deficiency for the inhabitants of the study area. The soil-metal transfer factors were globally less than 1 (0.03 to 0.95), lowest for Cu and highest for Fe. Between sites, Fe contents were fairly higher in South Bafut soils, while Zn, Mn and Cu were higher in North Bafut soils. These differences could be explained by the variation in the nature of parent

**Corresponding author: E-mail: mtitawa2000@yahoo.com*

rock in both localities and the slightly acidic soil pH which influences the solubility, mobility and bioavailability of soil nutrients. Between seasons, there was no significant difference (P<.05) between metal concentrations of soils and *Solanum scabrum,* except for Cu in the dry season vegetables. The present work suggests that *S. scabrum* could serve as a cheaper source of micronutrients especially Fe and Mn for the local inhabitants of Bafut and beyond.

Keywords: Solanum scabrum; soil; micronutrients; bioavailability; transfer factor; Bafut.

1. INTRODUCTION

Huckleberry (*Solanum scabrum*) is one of the most popular cash crops and most grown traditional vegetables in the western part of Cameroon [1]. It is an annual plant that is easy to cultivate, easy to cook and very palatable. *S. scabrum* serves as a traditional dish for most villages in the North West Region of Cameroon and it is mostly used in social gatherings and ceremonies. It is an indigenous vegetable that forms the basis of human nutrition: It is a source of vitamins and micronutrients required in minute quantities. These micronutrients have been given special attention worldwide due to the principal role that they play in numerous metabolic activities as well as harmful effects associated with deficiency and toxicity levels when their concentrations exceed certain limits [2-5]. The concentrations of micronutrients in vegetables depend on their intrinsic characteristics (varieties, maturity, genetics, age, etc) and environmental conditions (soils, geographical locations, season of the year, water quality, fertilizer use, etc) as well as on handling and processing techniques [6]. Vegetable contamination through metals is a serious threat because of the high toxicity, bioaccumulation and biomagnification of these metals once in the food chain [7]. The metals often accumulate in soils due to weathering, environmental pollution and irrigation with metal-contaminated water where they can then be absorbed by plants [8-10].

Much work has already been published on *S. Scabrum* mainly on its global distribution and consumption [1], its nutritional value [11-14] and its medicinal value [10,15,16]. However, some aspects like the micronutrient concentrations and soil-plant transfer factors are very rare in Cameroon. This can be due to some constraints like the long and fastidious methodology, high number of samples or the the high cost of their analyses that require very advanced technological approaches often out of reach of most developing countries. Some local studies, nevertheless, revealed a tendency for vegetables to accumulate variable levels of metals in their tissues: [17] in some edible upland plants in the Upper Noun valley (West Cameroon), [18] on vegetables grown along the Nkoup River (Foumbot), [19] in some vegetables in Ngaoundéré, [9] on metal levels in huckleberry grown along the Mezan River valley (Bamenda) and [20] on nutritional evaluation of indigenous vegetables (Yaoundé). The aim of the present work was, firstly, to quantify the metal levels of Fe, Mn, Cu and Zn in soils and vegetables grown in Bafut sub-division and, secondly, to investigate the effect of some environmental factors on the accumulation of micronutrients in soils and *S. scabrum.* The results obtained will supplement the available data on the bioavailability of metals in vegetables commonly consumed in Cameroon and beyond.

2. MATERIALS AND METHODS

2.1 Study Area

Bafut Sub-division is located in the north of Mezam division (North West region of Cameroon) between latitude 06°05′-06°11′ N and longitude 09°58′-10°11′ E (Fig. 1) [21]. It is situated at about 20 km northwest of Bamenda and covers an area of roughly 340 km^2. The mean annual precipitation is 2657.2 mm, with a long rainy season from March to November and a dry one from December to February. The mean annual temperature is 24°C, but maximum temperature attains 26.7°C in February while the minimum reaches 22.3°C between August and November [22]. The natural vegetation of Bafut is the grassland savannah, marked by grasses mixed with deciduous shrubs and stunted trees here and there, meanwhile the swampy valleys are dominated by raffia bushes and palm trees [23]. This natural vegetation is strongly modified by human activities mainly farming and demographic pressure imposed by a rapidly increasing population [24]. On a hydrographic point of view, Bafut Sub-division is drained by River Mezam (main collector) and its tributaries. The main collector together with its tributaries form a sparse and more or less angular drainage pattern [25]. Geologically, Bafut sub-division is located on the Cameroon Volcanic Line and comprises three main geological formations: Volcanic rocks, metamorphic rocks and alluvial deposits. The metamorphic rocks, mainly gneiss and schist, outcrop mainly in the North especially on slopes of high plateaus and constitute the basement. The volcanic rocks, mostly basalts, outcrop in South Bafut [26]. The alluvial deposits cover the Mezam River valley. The distribution of soils in Bafut is conditioned chiefly by topography and climate. Thus, ferruginous and ferrallitic soils occur in the southern high plateaus. In the north, most of the hill slopes are covered by brunified soils; alluvial soils are abundant in the Mezam river valley. Hydromorphic soils are common in the swampy valleys [27]. The estimated population of 80.305 inhabitants (2005 census) is settled in three main zones [28]: The Mumala'a (heart of the country) at the Centre clustered around the Fon's palace, the Ntare (ridge area) to the South and the Mbunti (lower) to the North which descends abruptly to the Menchum valley. In Bafut sub-division, vegetable cultivation is all year round to ensure a continuous supply, but cultivation is intensified in the dry season when it is more lucrative. In the rainy season, vegetables are grown both in uplands and lowlands. In the dry seasons, they are grown only in lowlands close to the banks of the Mezam River, main water course which runs through the Bafut municipality and whose water serves for irrigation. Hand watering using watering cans is common but flood irrigation is also practiced. Most of the farmers usually have a home garden for their own local consumption and a garden for commercial consumption usually some distance away from the home. The study sites were selected in North Bafut at Mbekong and in South Bafut at Mambu (Fig. 1). Bafut and Mbekong are two outstanding vegetable production zones in Bafut. The Mambu plot was located in a swampy valley at Atimuluh quarter flooded mainly in the rainy season while the Mbekong plot was located

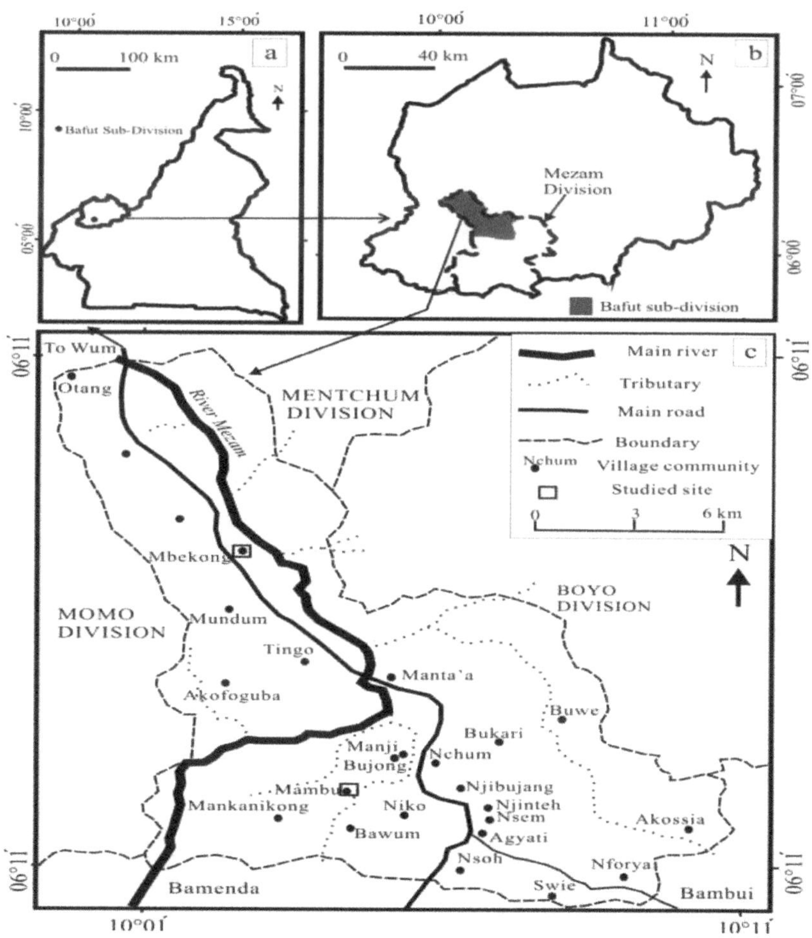

Fig. 1. Study area location (a) Cameroon map showing position of the Bafut sub-division; (b) Map of North West Region of Cameroon locating the Bafut Sub-division; (c) Map of Bafut Sub-division [21] showing the studied sites

along the Mezam main River Bank at the eastern side of the Bamenda-Wum road.

2.2 Methodology

2.2.1 Sample collection and pre-treatment

In the field, two plots (10 m x 10 m grid) were selected, one in Mambu and the other in Mbekong. Each was further divided into four equal sub-plots of 2.5 m by 2.5 m and planted with *S. scabrum* in March 2010 (rainy season) and in November 2010 (dry season). The *Solanum scabrum* shoots were harvested at six weeks after planting. The shoots of all vegetable plants from each plot were harvested by cutting stem at about 8 cm to the soil surface. The vegetables samples were randomly cut within each demarcated 2.5 m by 2.5 m sub-plot using a ceramic scissors to form a final composite sample of about 1 kg. All together four composite vegetable samples were obtained per 10 m x 10 m plot making a total number of 16 samples. In each sub-plot, after each vegetable sampling, five soil samples were also randomly collected at 0-20 cm depth (rooting zone) using a plastic shovel to avoid any metal contamination. These soil samples were mixed together to obtain a composite sample per sub-plot. All together, four composite soil samples were obtained per plot per season making a total numbers of 16 soil samples. Immediately after each collection, samples were stored in clean plastic bags and conveyed to the laboratory for further processing and analysis. In the laboratory, composite soil samples were air-dried at room temperature for one week and passed through a 2-mm polyethylene sieve to remove plant debris and pebbles. Afterwards, they were lightly crushed in an agate mortar into fine powder and passed through 0.149-mm nylon sieve. Vegetable samples were washed three times with distilled water to remove soil particles and were then oven-dried at 75°C to a constant weight for 48 hours. The dried vegetable samples were ground into fine powder in an agate mortar and passed through a 0.149-mm nylon sieve. The soil and vegetable sample powders were then stored in glass containers and preserved under ambient conditions pending analysis.

2.2.2 Laboratory analyses

2.2.2.1 Soil analyses

The soil analyses were comprised of physico-chemical analyses and determination of soil micronutrient levels. The physico-chemical analyses were done in the Laboratory of Soil Science and the Laboratory of Physico-chemistry of Mineral Materials (University of Yaoundé I) as well as at the International Institute for Tropical Agriculture (IITA) Nkolbissong, Yaoundé. The determination of soil micronutrient contents was done at the International Institute for Tropical Agriculture (IITA) of Nkolbisson (Yaoundé, Cameroon).

A battery of physicochemical analyses was performed. Thus, the soil moisture content was determined by noting the weight-loss of an air-dried sample, after subjecting it to an oven temperature of 105°C for 24 hours [29]. The bulk density (Db) was determined in reference to Archimedes' principle and particle density (Dp) was measured by pycnometer method [29]. Soil porosity was deduced from bulk density and particle density [29]. The particle size distribution was measured by Robinson´s pipette method [29]. The pH-H_2O was determined in a soil/water ratio of 1:2.5 using a glass pH-meter [30]. The organic carbon (OC) was measured by Walkley-Black method [31]. Organic matter (OM) was obtained from organic carbon (OC) using the Sprengel factor (OM= OC x 1.724) [32]. Total nitrogen (TN) was measured by the Kjeldahl method [33].

Determination of soil micronutrient levels was done by total digestion method [34]. Thus, aliquots of 0.5 g of dried soil samples were digested with HNO_3 acid, H_2O_2 and HCl acid mixture in the ratio 5:1:1 at 80°C until a clear solution was obtained. The solution was filtered through Whatman no. 42 filter paper and diluted to 50 ml with deionized water. The filtrates were then directly analyzed for Fe, Mn, Cu and Zn using atomic absorption spectrophotometry PG-900 Model, equipped with an air-acetylene flame and a hollow cathode lamp, under standard conditions using wavelengths and slit-widths specified for each element [35].

2.2.2.2 Vegetable analyses

The dried vegetables were digested with HNO_3 acid, H_2O_2 and HCl acid mixture (5:1:1 ratio), filtered, diluted to 50 ml and then analyzed for Fe, Mn, Cu and Zn by atomic absorption spectrophotometry PG-900 Model [35].

All soil and vegetable samples were analyzed along with a blank solution. Calibration was performed with standard solutions while precision and accuracy were checked by repeated analyses of sub-samples of the standards. The micronutrient concentrations in soils and vegetables were expressed in mg kg^{-1}. The metal transfer factors were calculated as the concentration of the metal in the edible part of vegetable relative to the concentration in the soil [36].

2.2.3 Statistical analysis

Statistical analysis was performed using the SPSS software program (SPSS Inc., Version 12.0). The data were analyzed by one-way analysis of variance (ANOVA). To detect the statistical significance of differences ($P<.05$) between means, the Tukey's test was performed.

3. RESULTS

3.1 Soil Physico-chemical Characteristics

The studied soils (0-20 cm depth) were dark grey (10 YR4/1) in South Bafut to dark brown (10 YR3/3) in North Bafut. They showed a heavy clayey texture in north Bafut and a clayey texture in South Bafut (Table 1). The hygroscopic water content ranged between 7.65 and 14.23, with slightly higher values in the south. Bulk density was 1.5 g/cm^3 in both zones but particle density was slightly higher in the south (2.6 gcm^{-3}) compared to the north. Total porosity was mediocre. The pH.H_2O was slightly acidic (5.6 and 6.3). Also, the pH.H_2O values of the dry season soils were slightly higher than those of the corresponding rainy season soils. The organic matter content was medium (2.57-3.58 %). The total nitrogen content was low to medium (0.05-0.11 %). The C/N ratio was very high (17-32). The Bafut soils fall within the medium fertility range [37,38].

3.2 Micronutrient Levels in Bafut Soils

Fe was the most abundant micronutrient in the soils and ranged between 213.0 and 342.1 mg kg^{-1} (Fig. 2). Between sites, Fe contents were significantly ($P<.05$) higher for soils from South Bafut; meanwhile for the lower values observed in the north, those of the rainy season were slightly higher (241.0 mg kg^{-1}) than those of the dry season (213.0 mg kg^{-1}).

Mn contents of the studied soils ranged between 113.0 and 264.0 mg kg^{-1} (Fig. 2; Table 2). However, no significant difference ($P<.05$) was observed between the different sites and seasons. The highest value was noted in the North (264.1 mg kg^{-1} in the dry season and 218.0 mg kg^{-1} in the rainy season). The lower contents in the South also showed a slightly lower value in the rainy season (113.0 mg kg^{-1}) relative to the dry season (141.0 mg kg^{-1}).

Zn contents in the two sites globally ranged from 82.0 to 118.0 mg kg^{-1} and no significant difference ($P<.05$) was noted between sites and seasons (Fig. 2; Table 2). However, the concentration of Zn in the soils was higher in the North than in South Bafut. In the north, zinc contents were slightly higher in the dry season (118.0 mg kg^{-1}) than in the rainy season (108.0 mg kg^{-1}). Zn contents of soils in the south were quite similar, though slightly higher in the dry season (94.0 mg kg^{-1}) than in the rainy season (82.0 mg kg^{-1}).

Cu contents in the Bafut soils ranged between 23.0 and 39.0 mg kg^{-1} (Fig. 2; Table 2). Generally, the Cu contents of the rainy season soils were slightly higher than those of the dry season for both sites. However, the Cu contents of soils sampled in North Bafut (28.0 mg kg^{-1} in the dry season and 39.0 mg kg^{-1} in the rainy season) were slightly higher than those of South Bafut (23.0 mg kg^{-1} in the dry season and 37.0 mg kg^{-1} in the rainy season). Nevertheless, there were no significant differences *($P<.05$)* in Cu contents between sites and seasons.

3.3 Levels of Micronutrients in *Solanum scabrum* grown in Bafut

Fe was the most concentrated metal in all the *S. scabrum* from the studied sites (Fig. 3; Table 3). In North Bafut, the highest Fe contents (202.3 mg kg^{-1}) appeared in the dry season *S. scabrum* as opposed to the rainy season ones (149.7 mg kg^{-1}). Globally, Fe contents in all other vegetable samples were significantly ($P<.05$) higher than VDS. In South Bafut, higher values were noted in the rainy season (192.3 mg kg^{-1}). Specifically in the South Bafut, Fe contents of VRS were significantly higher than those of VDS. In the north, there was no significant difference ($P<.05$) between Fe contents of rainy season and dry season vegetables.

The Mn levels in Bafut *S. scabrum* were slightly lower than those of Fe, varying between 58.3 and 81.3 mg kg^{-1} (Fig. 3; Table 3). Globally, only Mn contents of VDN were significantly ($P<.05$) higher relative to other sites and seasons. Thus, no significant difference was noted between VDS and VRS, meanwhile in the north VDN was significantly ($P<.05$) higher in Mn compared to VRN.

Table 1. Characteristics of the Bafut soils sampled at 0-20 cm depth (n =4)

Soil properties / Soil code	Moisture content (%)	Bulk density (gcm^{-3})	Particle density (gcm^{-3})	Porosity (%)	Particle size distribution (%)			Textural class	pH (H_2O)	OC (%)	OM (%)	TN (%)	C/N
					Sand	Silt	Clay						
SDS	14.23	1.5	2.5	40	25	14	61.00	Heavy clay	6.3	1.9	3.28	0.11	17
SRS	13.80	1.5	2.5	40	22	13	65.00	Heavy clay	6.1	1.6	2.76	0.06	27
SDN	8.23	1.5	2.6	42	40	14	46.00	Clayey	5.8	2.01	3.47	0.10	20
SRN	7.65	1.5	2.6	42	38	15	47.00	clayey	5.6	1.49	2.57	0.05	32

SDS: Dry season soil from South Bafut; SRS: Rainy season soil from South Bafut; SDN: Dry season soil from North Bafut; SRN: Rainy season soil from North Bafut

Table 2. Mean metal concentrations, standard deviation (in brackets) of soils in the Bafut Sub-division and permissible/critical levels for agricultural soils (n =4)

Soil micronutrient ($mg\ kg^{-1}$ dry weight) / Soil code	Fe	Cu	Mn	Zn
SDS	342.10^{b}	23.00^{a}	141.00^{a}	82.00^{a}
	(85.53)	(5.75)	(35.25)	(20.50)
SRS	296.80^{b}	37.00^{a}	113.00^{a}	94.00^{a}
	(74.20)	(9.25)	(28.25)	(23.50)
SDN	213.00^{a}	28.00^{a}	264.10^{a}	118.00^{a}
	(53.25)	(7.00)	(66.02)	(29.50)
SRN	241.00^{a}	39.00^{a}	218.00^{a}	108.00^{a}
	(60.25)	(9.75)	(54.50)	(27.00)
Permissible limits for agricultural soils [39]	50.00-250.00	50.00-140.00	15.00-500.00	150.00-300.00

SDS: Dry season soil from South Bafut; SRS: Rainy season soil from South Bafut; SDN: Dry season soil from North Bafut; SRN: Rainy season soil from North Bafut. Values in the same column followed by different letters are significantly different at P =.05

Zn contents of *S. scabrum* ranged from 28.4 to 43.6 mg kg^{-1} (Fig. 3; Table 3). There was no significant difference (*P*<.05) between Zn contents of *S. scabrum* from both sites and seasons, although *S. scabrum* VRS contents were fairly higher than VDS and VDN was slightly higher than VRN.

Cu contents of *S. scabrum* from Bafut were significantly low (0.78 and 1.99 mg kg^{-1}) compared to the other metals (Fig. 3; Table 3). Cu contents were significantly higher in the South (*P* <.05) than in the North. In the South, dry season *S. scabrum* showed no significant difference (*P*<.05) in Cu contents between seasons, although slightly higher (1.99 mg kg^{-1}) in the dry season than the rainy season ones (1.46 mg kg^{-1}). In the North, no significant difference (*P*<.05) was also observed between seasons, although higher Cu contents (0.78 mg kg^{-1}) were noted in the dry season compared to the rainy season (1.01 mg kg^{-1}).

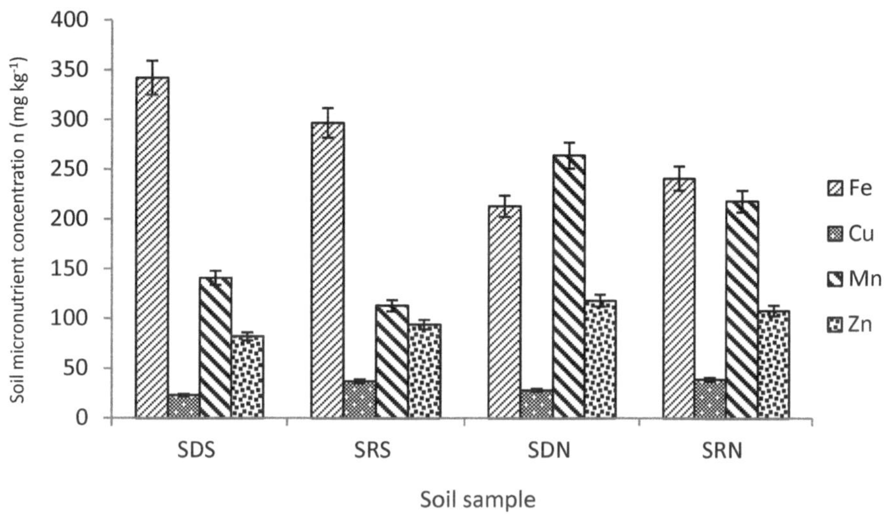

Fig. 2. Micronutrient concentrations (±standard error) in soils from Bafut Sub-Division
(SDS: Dry season soil sample from South Bafut; SRS: Rainy season soil sample from South Bafut; SDN: Dry season soil sample from North Bafut; SRN: Rainy season soil sample from North Bafut)

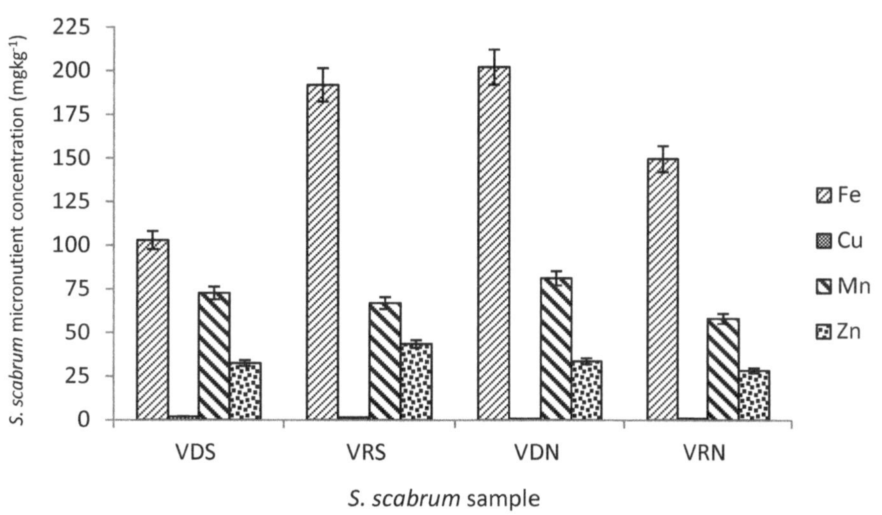

Fig. 3. Micronutrient concentrations (± standard error) in *S. scabrum* from Bafut Sub-Division
(VDS: Dry season S. scabrum samples from South Bafut; VRS: Rainy season S. scabrum sample from South Bafut; VDN: Dry season S. scabrum sample from North Bafut; VRN: Rainy season S. scabrum sample from North Bafut)

3.4 Linear Correlations and Metal Transfer Factors

The Pearson linear correlation test revealed positive (*P*<0.05) correlation coefficients between soil and vegetables micronutrient concentrations (Table 4). VRS versus SRS showed the highest correlation (r=0.99; p=.001) while the least was shown by VRN versus SRN (r=0.69; *P*=.002).

The soil-plant metal transfer factors in both sites and seasons were all less than 1 (0.3 to 0.95 for Fe, 0.28 to 0.59 for Mn, 0.26 to 0.46 for Zn and 0.03 to 0.09 for Cu) (Fig. 4; Table 5). Globally, except for Fe, transfer factors were higher in the south that in the north. Thus, Fe, specifically in VDN/SDN displayed the highest transfer factor. Cu, on the other hand, showed the lowest transfer factors between sites and seasons. The transfer factors were in the following order: Mn>Zn>Fe>Cu for VDS/SDS, Fe > Mn >Zn >Cu for VRS/SRS, Fe>Mn>Zn> Cu for VDN/SDN and Fe>Zn> Mn> Cu for VRN/SRN.

Table 3. Mean micronutrient contents and standard deviation (in brackets) in *S. scabrum* from Bafut Sub-Division and predefined levels for growth and consumption (n =4)

Vegetable code \ Vegetable micronutrient (mg kg^{-1} in dry weight)	Fe	Cu	Mn	Zn
VDS	103.00^a (25.75)	1.99^b (0.50)	72.80^a (18.20)	32.50^a (8.13)
VRS	192.00^b (48.00)	1.46^b (0.37)	67.10^a (16.78)	43.60^a (10.90)
VDN	202.30^b (50.58)	0.78^a (0.20)	81.30^b (20.33)	33.70^a (8.43)
VRN	149.70^b (37.43)	1.01^a (0.25)	58.30^a (14.58)	28.40^a (7.10)
Normal levels in plant [48]	50- 500	5-20	20-1000	1-400
sufficiency levels in plants [48]	50-250	5-20	20-500	25-150
Critical levels for plant growth [48]	50-150	20-100	300-500	100-400
Toxicity levels in plants [46]	>500	20-30	>500	>400
Permissible levels in food [53]	-	<200	-	<500

VDS: Dry season S. scabrum from South Bafut; VRS: Rainy season S. scabrum from South Bafut; VDN: Dry season S. scabrum from North Bafut; VRN: Rainy season S. scabrum from North Bafut. Values in the same column followed by different letters are significantly different at P < .05; standard deviation in brackets

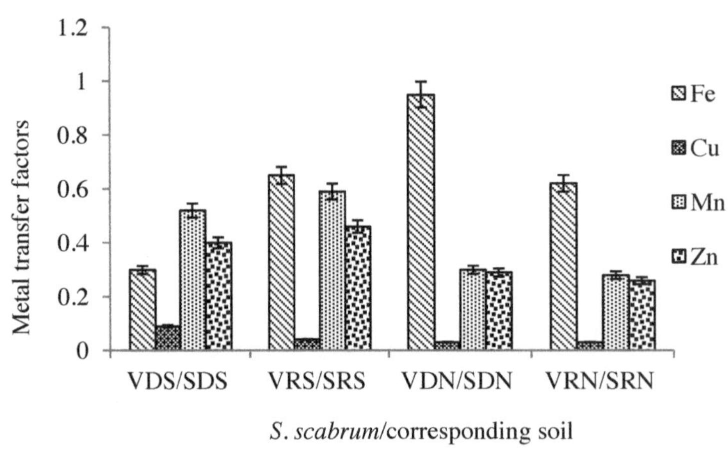

Fig. 4. Transfer factors (± standard error) of micronutrients

VDS: Dry season S. scabrum from South Bafut; SDS: Dry season soil sample from South Bafut; VRS: Rainy season S. scabrum from South Bafut; SRS: Rainy season soil from South Bafut; VDN: Dry season S. scabrum from North Bafut; SDN: Dry season soil from North Bafut; VRN: Rainy season S. scabrum from North Bafut; SRN: Rainy season soil from North Bafut

Table 4. Relationship between micronutrients in soils and vegetables from Bafut sub-Division (n=4)

Regression relationship	Correlation coefficient (r)	Probability level (p)
VRS versus SRS	0.99**	0.001
VDS versus SDS	0.94**	0.003
VDN versus SDN	0.87*	0.01
VRN versus SRN	0.69	0.02

***Significant at the 0.01 level; *Significant at the 0.05 level; VRS: Rainy season S. scabrum from South Bafut; SRS: Rainy season soil from South Bafut; VDS: Dry season S. scabrum from South Bafut; SDS: Dry season soil sample from South Bafut; VDN: Dry season S. scabrum from North Bafut; SDN: Dry season soil from North Bafut; VRN: Rainy season S. scabrum from North Bafut; SRN: Rainy season soil from North Bafut)*

Table 5. Mean soil-to-plant metal transfer factors and standard deviation (in brackets) of metals from the different sites and seasons (n=4)

Metal transfer factors / Vegetable/corresponding soil	Fe	Cu	Mn	Zn
VDS/SDS	0.30^a	0.09^b	0.52^b	0.40^b
	(0.075)	(0.023)	(0.13)	(0.1)
VRS/SRS	0.65^a	0.04^a	0.59^b	0.46^b
	(0.1625)	(0.01)	(0.1475)	(0.115)
VDN/SDN	0.95^a	0.03^a	0.30^a	0.29^a
	(0.2375)	(0.0075)	(0.075)	(0.073)
VRN/SRN	0.62^a	0.03^a	0.28^a	0.26^a
	(0.155)	(0.0075)	(0.07)	(0.065)

VDS: Dry season S. scabrum from South Bafut; SDS: Dry season soil sample from South Bafut; VRS: Rainy season S. scabrum from South Bafut; SRS: Rainy season soil from South Bafut; VDN: Dry season S. scabrum from North Bafut; SDN: Dry season soil from North Bafut; VRN: Rainy season S. scabrum from North Bafut; SRN: Rainy season soil from North Bafut

4. DISCUSSION

4.1 Particularities of Micronutrients in Bafut Soils and Vegetables

Fe was the most abundant micronutrient in all the soil samples and its contents varied between 213.0 and 342.1 mg kg^{-1}. These values fell between the permissible range for agricultural soils [39]. These soil iron levels were higher than the 21.4-58.9 mg kg^{-1} reported for some contaminated soils in Jos (Nigeria) by Pasquini [40] but lower than Fe contents documented for vegetable soils of the Cameroonian inland valleys of the Nkoup River in Foumbot [18] and Mezam River in Bamenda [20,11] (Table 6). In some environments, Fe availability under stress conditions is enhanced through the production of organic chelating agents by microorganisms [41]. The high Fe concentration of the soil might therefore be the result of the slightly acidic nature of the Bafut soils, conditions favorable to Fe accumulation. The main sources of Fe are probably the parent rocks, iron pipes, waste can containers and rusted iron rods commonly dumped into nature following rapid population growth in the Bafut sub-division as well as Fe present in irrigation water [42].

The Mn concentrations in Bafut soils, quite close to those of Fe and ranged from 113 to 264.1 mg kg^{-1}, were considered as moderate to high. These values fell within the permissible levels for crop cultivation [39]. Generally, the concentration of Mn in soil covers a particularly wide range, from as little as 0.20 mg kg^{-1} to over 300 mg kg^{-1} [43]. The lower values are typical of severely leached acid soils. In contrast, Mn toxicities are common in unleached acid soils and waterlogged conditions. Mn deficiency is most common in alkaline and calcareous soils. Soil pH is one of the main factors determining Mn availability in soils. A pH value close to or below 6.0 could favor Mn reduction and the formation of the more available divalent form (Mn^{2+}); higher pH favors Mn oxidation to Mn^{4+} ion, forming the insoluble oxides like MnO_2, Mn_2O_3 and Mn_3O_4 [43]. The moderate to high Mn contents recorded in the present study could be related to the slightly acidic pH of the studied soils [44].

Table 6. Comparison of micronutrient concentrations in soils and *S. scabrum* from Bafut with those of some areas in Africa

Micronutrient (mg kg^{-1})	Present study (Bafut)	Cameroon (Foumbot) (a)	Cameroon (Yaoundé) (b)	Cameroon (Bamenda) (c)	Tanzania (b)	Kenya (b)	Nigeria (d)
Soil							
Fe	213.0-342.1	NA	NA	297.0-507.0	NA	NA	NA
Cu	23.0-39.0	NA	NA	1.30-4.10	NA	NA	NA
Mn	113.0-264.0	NA	NA	49.60-84.10	NA	NA	NA
Zn	82.0-118.0	NA	NA	2.10-12.50	NA	NA	NA
S. scabrum							
Fe	103.0-202.3	115-882	182.00-387.93	297-619	184.45-266.52	147.37	280.5
Cu	0.78-1.99	0.013-0.097	NA	1.30-4.10	NA	NA	22.40
Mn	58.3-81.30	29.1-61.2	NA	49.6-84.10	NA	NA	359.40
Zn	28.40-43.60	14.7-47.2	39.70-38.33	2.10-12.50	38.90-39.21	41.80	52.99
Transfer factors							
Fe	0.30-0.95	NA	NA	0.92-1.20	NA	NA	NA
Cu	0.03-0.09	NA	NA	0.69-1.62	NA	NA	NA
Mn	0.28-0.59	NA	NA	0.56-0.72	NA	NA	NA
Zn	0.26-0.49	NA	NA	4.58-7.09	NA	NA	NA

(a) [9]; b) [20]; (c) [9]; (d) [14]; NA: not available

Zn contents in soils of the two studied sites globally ranged from 82.0 to 118.0 mg kg^{-1} and fell below the permissible limits for agricultural soils [39]. Zn is deficient in a wide range of soil types; soils with less than 0.30 mg kg^{-1} of zinc are likely to require zinc addition for optimum crop production. With highly alkaline soils, the critical level may attain 0.80 mg kg^{-1}. Certain soil conditions reduce Zn availability, notably high pH as the solubility of Zn decreases with pH [45].

The soil Cu contents of the studied Bafut soils fell between the permissible levels for optimum plant growth [39]. Most plants like *S. scabrum* (as well as cereals, oats, wheat, barley, maize, etc) are particularly sensitive to Cu deficiencies [45,46]. Soils derived from basic volcanic rocks and metamorphic rocks usually show low Cu contents whereas acidic conditions favor the availability of Cu [47]. Copper deficiency is most common on highly calcareous soils as well as in sandy soils with low organic matter contents [46]. High concentrations of other metals such as Fe, Mg and Al in soil can induce Cu deficiency due to "ionic balance effect" [48].

The concentrations of the various metals in the Bafut soils fell below the maximum permissible levels for agricultural soils [39] and consequently could not be expected to present any toxicity risks for the supported vegetables. There could instead be a potential risk of Zn and Cu deficiency.

Fe was the most represented element in all the vegetable samples from the studied sites. The vegetable Fe contents fell within the adequate levels and sufficiency levels for optimum plant growth, and were below the toxicity levels [49]. The high Fe contents of *S. scabrum* could be due to the naturally higher occurrence of Fe relative to the other metals in the soil environment and the slightly acidic pH which favors availability and uptake by plants [50]. These metal levels are similar to those already documented in Ngaoundéré (North Cameroon) [19] and Yaoundé [20], but far below those reported for vegetables grown along the Mezam River in Bamenda [9]. These metal values are also in the same order of magnitude as those reported in Kenya and Tanzania [20] and in Nigeria [14].

The Mn contents of the Bafut *S. scabrum* ranged between 58.3 and 81.3 mg kg^{-1}. These values were significantly ($P<.05$) higher than those (0.02 to 14.0 mg kg^{-1}) reported for some Russian vegetables [51] and the 0.18- 2.8 mg kg^{-1} range documented for some Indian leafy vegetables [52]. Globally, the dry season *S. scabrum* showed higher Mn contents than those grown in the rainy season. The vegetables metal contents in this study fell within the sufficiency levels and below the toxicity levels for optimum plant growth [48] as well as within the permissible values in food [53]. Mn and Fe are essential mineral elements for both plants and animals, and the recorded levels in this study indicate that the vegetables constitute very good sources of nutrients in the diet of the population.

The Zn contents of *S. scabrum* varied from 28.4 to 43.6 mg kg^{-1}. These values fell below the critical range for optimum plant growth [48] as well as the permissible level in food [53]. All the

S. scabrum samples contained inadequate Zn levels [48]. These observed values were in agreement with those published by Kamga et al. [20] and Idowu et al. [5], but higher than values reported for vegetables grown along the Mezam River valley in Bamenda [9]. These Zn levels were also analogous to the 25.2-50.0 mg kg^{-1} reported for *S. scabrum* in India [52], but higher than values recorded for vegetables in Nigeria [54], Egypt [12] and Latin America [55].

The Cu contents in *S. scabrum* were significantly low (0.78 to 1.99 mg kg^{-1}) compared to the other micronutrients in the vegetables. The Cu concentrations fell below the normal range and were inadequate for optimum plant growth [48] as well as below the permissible values in food [53]. These values were close to those already reported in Ngaoundéré (North Cameroon) [19], but higher than those for vegetables grown along the Nkoup inland River valley in Foumbot [18]. This range is within values reported for most vegetables irrigated with waste water [56-59].

The levels of Fe and Mn in *S. scabrum* from Bafut Sub-division fell within the sufficiency range. Cu and Zn contents suggested a dietary intake deficiency for the inhabitants of the study area, probably due to low soil concentrations and /or non-bioavailability of these metals. These deficient metals could be compensated through the consumption of other foodstuffs. Nevertheless, all the micronutrients were within the permissible range in food suggesting that *S. scabrum* could serve as cheaper source of micronutrients for the inhabitants of Bafut and beyond.

4.2 Soil-to-plant Metal Transfer Rates

The soil-to-plant metal transfer factors were globally low and less than 1. Although the highest correlation coefficients existed between VRS and SRS, and between VDN and SDN, one would have expected the highest metal transfer factors to correspond to soils with the highest metal concentration, which is not the case. This could imply that factors other than soil micronutrient concentrations are inducing metal uptake by plants [9,59]. The availability of metals to plants in the soils could be influenced by soil intrinsic factors (clayey to heavy clayey texture, humidity, mediocre porosity, acidic pH, high organic matter content, etc) and extrinsic factors (precipitation, temperature, gradient of slope, etc). These factors control solubility and mobility of soil metals and subsequent availability for plants absorption [36,60]. Similar findings [9] in Bamenda town revealed very high metal transfer factors for Zn (4.58-7.09 mg kg^{-1} dry weight) and Cu (0.69-1.62 mg kg^{-1} dry weight) compared to Bafut vegetables, while the Fe and Mn factors are comparable (Table 6). The differences results from vegetable irrigation with contaminated water from River Mezam in Bamenda which leads to the deposition of excess plant available metal in the soils.

5. CONCLUSION

The aim of the present study was to investigate the composition of Zn, Cu, Mn and Fe in soils and grown *Solanum scabrum* in the Bafut Sub-division (North-West Cameroon). The results showed that the soil metal contents were within the permissible level for agricultural soils. Also, Fe and Mn, most represented metals in vegetables, were within the adequate and normal levels for optimum plant growth and human consumption, while Cu and Zn were deficient suggesting compensation through the consumption of other food types. Metal transfer factors were not directly proportional to soil metal concentrations suggesting that factors other than soil metal concentrations could induce metal uptake by the plants.

Overall, the present study provides valuable information on the relationship between metal contents of soils and *S. scabrum* and suggests that *S. scabrum* could serve as cheaper source of micronutrients especially Fe and Mn for the population of Bafut and beyond.

ACKNOWLEDGEMENTS

The authors duly thank the anonymous Reviewers for their helpful suggestions and critical comments that greatly improved the quality of the manuscript.

COMPETING INTERESTS

All authors have declared that no competing interests exist.

REFERENCES

1. Fontem DA, Berinyuy JE, Schippers RR. Selecting promising varieties from farmer's landraces-an experience from Cameroon. FASA/CUDs, Dschang, Cameroon. 2002; 11.

2. Reilly C. Metal contamination of food. 2nd ed. Elsevier Applied Science: London; 1991.
3. Skurinkhin IM. Method of analysis for toxic elements in food. Part IV. General method of ashing for the determination of toxic elements. Journal of AOAC international. 1993;6:257-262.
4. Itanna F. Metals in leafy vegetables grown in Addis Ababa and toxicology implications. Ethiopian J. Health Dev. 2002;16:295-302.
5. Odhav B, Beekrumb S, Akulaa US, Baijnath H. Preliminary assessment of nutritional value of traditional leafy vegetables in KwaZulu-Natal, South Africa. J. Food Comp. Anal. 2007;20:430-435.
6. Pennington JT, Calloway DH. Copper contents of foods. J. Am. Diet Association. 1988;63:143.
7. Eisler R. Zinc hazards to fish, Wildlife and Invertebrates: A synoptic review. US fish Wildlife Service Biology for Reproduction. 1988;7:85.
8. Qui XX, Huang DF, Cai SX, Chen F, Ren ZG, Cai YC. Investigations on vegetables pollution and pollution sources and its control in Fuzhou, Fujian Province. Fujian J. Agric. Sci. 2000;15:16-21.
9. Tita MA, Tsala GN, Kamgang KVB. Levels of metals in huckleberry (Solanum scabrum) grown along the Mezam river banks Bamenda, North-West Cameroon: Implications for crops growth and human consumption. Syllabus Review. 2011;2(2): 69-78.
10. Tindall HD. Vegetables in the tropics. Macmillan Education Limited: London; 1983.
11. Nesamvuni C, Steyn NP, Potgieter MJ. Nutritional value of wild, leafy plants consumed by the Vhavenda. S. Afr. J. Sci. 2001;97:51-54.
12. Salama AK, Radwan MA. Heavy metals (Cd, Pb) and trace elements (Cu, Zn) contents in some foodstuffs from Egyptian markets. Emir. J. Agric. Sci. 2005;17(1): 34-44.
13. Yang RY, Keding GB. Nutritional contributions of important African indigenous vegetables. In: Shackleton CM, Pasquini MW, Drescher AW, Editors. African indigenous vegetables in urban agriculture. London: Earthscan; 2009.
14. Idowu MK, Oyedele DJ, Adekunle OK, Akinremi OOl, Eilers B. Effects of planting methods and seed density on vegetable yield and nutrient composition of *Solanum macrocarpon* and *Solanum scabrum* in Southwest Nigeria. Food and Nutrition Sciences. Scientific Research. 2014;5: 1185-95.
15. Schippers R. Notes on huckleberry, Solanum scabrum and related black nightshade species. Natural resources institute University of Greenwich, Greenwich, UK. 1998;17.
16. Stevels JMC. Légumes traditionnels du Cameroun, une étude Agro-Botanique. Wageningen Agricultural University paper, Wageningen, Netherlands. 2001;262. French.
17. Njofang C, Matschullat J, Amougou A, Tchouakoue JP, Heilmeier H. Soil and plant composition in the Noun River catchment basin, Western Cameroon. A contribution to the development of a biogeochemical baseline. Environment geology. 2006;56(7):1427-1436.
18. Tita MA, Tsala GN, Kamgang KVB. Levels of heavy metals in vegetables grown along the Nkoup River banks in Foumbot, Western Cameroon: Implications for crop growth and human consumption. In: Proceedings of the 5th Conference of the Africa Soil Science Society. Soils and new challenges for Sustainable Development in Africa. ASSS: Yaoundé, Cameroon. 2009; 57.
19. Adjia R, Fezue WML, Tchatchueng JB, Sorho S, Ngassoum MB. Heavy metals in five leafy vegetables from urban and peri-urban sites in Ngaoundéré, Cameroon. Iranica Journal of Energy and Environment. 2010;1(2):124-131.
20. Kamga TR, Kouamé C, Atangana AR, Chagomoka T, Ndango R. Nutritional evaluation of five African indigenous vegetables. Journal of Horticultural Research. 2013;21(1):99-106.
21. Anonymous. Presentation of the Bafut Village Community Project. VCP: Bafut, Cameroon; 2005.
22. Etia PM. Climate and climatic zones of Cameroon. In: JA, editor. Les Atlas Jeune Afrique: Paris, France; 1980.
23. Letouzey R. Phytogeographic map of Cameron. In: JA, editor. Les Atlas Jeune Afrique: Paris, France; 1980.
24. Acho-Chi C. Human interference and environmental instability addressing consequences of rapid urban growth in Bamenda. Environment and Urbanisation. 1998;10(2):161-174.

25. Morin S. Geomorphological map of Bafut. In: JA, editor. Les Atlas Jeune Afrique: Paris, France; 1980.
26. Nougier J. Geological map of Cameroon. In: JA, editor. Les Atlas Jeune Afrique: Paris, France; 1980.
27. Gavaud M, Muller JP. Soil map of Cameroon. In: JA, editor. Les Atlas Jeune Afrique: Paris, France; 1980.
28. Available:http://www.bafut.council.org (28/06/2015)
29. FAO. Guidelines for soil description, a framework for international classification, correlation and communication. 4th Ed.FAO: Rome, Italy; 2006.
30. McLean EO. Soil pH and lime requirement. In: Buxton DR, editor. Methods of soil analysis. Part 2. American Society of Agronomy Inc. and SSSA Inc.: Madison, USA; 1982.
31. Nelson DW, Sommers LE. Total carbon, organic carbon and organic matter. In: Buxton DR, editor. Methods of soil analysis. Part 2. American Society of Agronomy Inc. and Soil Science Society of America Inc.: Madison, USA; 1982. Walkey A, Black IA. Determination of organic matter in soil. Soil Science. 1934;37: 549-556.
32. Bremner JM, CS Mulvaney. Total nitrogen. In: Buxton DR, editor. Methods of soil analysis. Part 2. American Society of Agronomy Inc. and SSSA Inc.: Madison, USA; 1982.
33. US EPA (United States Environmental Protection Agency). SW-846 Method 3050B. Standard operating procedure for the digestion of soil/sediment using hotplate/beaker digestion technique. Chicago, USA; 1999.
34. Jones BJ, Case VW. Sampling, handling and analyzing plant tissue samples. In: R.L. Westerman editor. Soil Testing and Plant Analysis. 3rd ed. SSSA: Madison, USA; 1990.
35. Sauve S, Hendershot W, Allen HE. Solid solution partitioning of metals in contaminated soils: Dependence on pH, total metal burden and organic matter. Environmental Science and Technology. 2000;34:1125-1131.
36. Euroconsult. Agricultural compendium for rural development in the tropics and the subtropics. Elsevier: Amsterdam, Netherlands. 1989;740.
37. Tabi FO, Bitondo D, Yinda GS, Kengmegne SSA, Ngoucheme M. Effect of long term integrated soil fertility management by local farmers on nutrient status of a Typic Dystrandept under potato-based cropping system. International Research Journal of Agricultural Science and Soil Science. 2013;3(4):134-140.
38. MAAF. Fertilizer recommendation for agriculture and horticulture crops. 209th Ed, Ministry of Agriculture, Fisheries and Reference book Series: London, UK. 2000; 177.
39. Pesquini MW. The use of town refuse ash in urban agriculture around Jos, Nigeria: health and environmental risks. Science of the total environment. 2006;354:43-59.
40. Waheed A, Jaffar M, Masud K. Comparative study of selected essential and non essential metals in various canned and raw foodstuffs consumed in Pakistan. Nutr. And food Sci. 2003;33(6): 34-41.
41. Dara SS. Environmental chemistry and pollution control. 1st ed. Rajendra Ravindra Printer (PVT) Ltd: New Delhi, India; 2006.
42. Harmsen K, Vlek PLG. The Chemistry of micronutrients in soils. Fertilizer Research. 1985;7:56-67.
43. Ayangnigni MF. Effet de stress hydrique et de la fréquence d'irrigation sur la croissance et le rendement de la morelle noire (*Solanum scabrum*). Mémoire de fin d'étude, FASA/CUDs, Dschang, Cameroun; 2004. French
44. Murphy LS, Walsh LM. Correction of micronutrient deficiencies with fertilizers. In: Mortvedt JJ, et al. editors. Micronutrients in Agriculture. SSSA: Madison, Wisconsin; 1972.
45. Marschner H. Functions of Mineral Nutrients. In: Marschner H, editor. Mineral Nutrition of Higher Plant. 2nd Ed. Academic Press: New York, USA; 1998.
46. Stevenson FJ, Fitch A. Chemistry of complexation of metal ions with soil solution organics. In: Huang, PM, Schnitzer M, editors. Interactions of soil minerals with natural organic and microbes. SSSA: Madison, USA; 1986.
47. Idowu MK. Soil fertility management strategies for peri-urban vegetable production in the Southwestern Nigeria. In: Adeboye OC, Taiwo KA, Fatufe AA, editors. Food, health and environmental issues in developing countries: The Nigerian Situation. Alexander von

Humboldt Foundation: Cuvillier Verlag Gottinger; 2008.
48. Landon JR. Booker tropical soil manual: A handbook for soil survey and agriculture evaluation in the tropics and sub-tropics. Longman, Harlow, UK.1984;450.
49. Stolzfus RJ. Iron deficiency: Global prevalence and consequences. Food Nutr. Bull. 2003;24(4):99-103.
50. Gorbunov SM, Lyapunov Okina OI, Frontasyeva MV, Gundorina SF. Assessment of human organism's intake of trace elements from stapled foodstuffs in Central Region of Russia. Journal of Environmental Chemistry. 2004;14:1-16.
51. Singh BK, Taneja SK. Concentration of Zn, Cu, and Mn in vegetables and meat foodstuffs commonly available in Manipur: a North Eastern state of India. Electronic Journal of Environmental, Agricultural and Food Chemistry. 2010;9(3):610-616.
52. FAO/WHO. Joint FAO/WHO Food Standards Programme. 24th session. Codex Alimentarius Commission: Geneva, Switzerland. 1993;391.
53. Iyaka YA. Concentration of Cu and Zn in some fruits and vegetables commonly available in North-Central Zone of Nigeria. EJEAFCHE. 2007;6(6):2150-2154.
54. Oliveres M, Pizarro F, De Pablo S, Araya M, Uauy R. Iron, zinc and copper: Contents in common Chilean foods and daily intakes in Santiago, Chile. Nutrition. 2004;2:205-212.
55. Somer E. Toxic potential of trace metals in foods. Journal of Food Science. 1974;39: 215-217.
56. Bahemuka TE, Mubofu EB. Heavy metals in edible green vegetables grown along the sites of the Sinza and Msimbazi rivers in Dar-es-Salaam. Tanzania Food Chem. 1999;66:63-68.
57. Mapanda M, Mangwayana EN, Nyamangara J, Giller KE. Uptake of metals by vegetables irrigated using wastewater and the subsequent risks in Harare, Physics and Chemistry of the Earth. 2007; 32:1399-1405.
58. Lucas RE, Davis JF. Relationships between pH values of organic soils and availabilities of 12 plant nutrients. Soil Science. 1961;92:177-182.
59. Chaney RL. Toxic element accumulation in soils and crops: protecting soil fertility and agricultural food-chains. In: Bar-Yosef B, Barrow NJ, Goldshmid J, editors. Inorganic Contaminants in the Vadose Zone. Springer-Verlag: Berlin, Germany; 1989.
60. Memon AR, Aktoprakligil D, Ozdemir A, Vertii A. Heavy metal accumulation and detoxification mechanisms in plants. Turkish Journal of Botany 2001;25: 111-122.

Permissions

All chapters in this book were first published in IJPSS, by SCIENCE DOMAIN international; hereby published with permission under the Creative Commons Attribution License or equivalent. Every chapter published in this book has been scrutinized by our experts. Their significance has been extensively debated. The topics covered herein carry significant findings which will fuel the growth of the discipline. They may even be implemented as practical applications or may be referred to as a beginning point for another development.

The contributors of this book come from diverse backgrounds, making this book a truly international effort. This book will bring forth new frontiers with its revolutionizing research information and detailed analysis of the nascent developments around the world.

We would like to thank all the contributing authors for lending their expertise to make the book truly unique. They have played a crucial role in the development of this book. Without their invaluable contributions this book wouldn't have been possible. They have made vital efforts to compile up to date information on the varied aspects of this subject to make this book a valuable addition to the collection of many professionals and students.

This book was conceptualized with the vision of imparting up-to-date information and advanced data in this field. To ensure the same, a matchless editorial board was set up. Every individual on the board went through rigorous rounds of assessment to prove their worth. After which they invested a large part of their time researching and compiling the most relevant data for our readers.

The editorial board has been involved in producing this book since its inception. They have spent rigorous hours researching and exploring the diverse topics which have resulted in the successful publishing of this book. They have passed on their knowledge of decades through this book. To expedite this challenging task, the publisher supported the team at every step. A small team of assistant editors was also appointed to further simplify the editing procedure and attain best results for the readers.

Apart from the editorial board, the designing team has also invested a significant amount of their time in understanding the subject and creating the most relevant covers. They scrutinized every image to scout for the most suitable representation of the subject and create an appropriate cover for the book.

The publishing team has been an ardent support to the editorial, designing and production team. Their endless efforts to recruit the best for this project, has resulted in the accomplishment of this book. They are a veteran in the field of academics and their pool of knowledge is as vast as their experience in printing. Their expertise and guidance has proved useful at every step. Their uncompromising quality standards have made this book an exceptional effort. Their encouragement from time to time has been an inspiration for everyone.

The publisher and the editorial board hope that this book will prove to be a valuable piece of knowledge for researchers, students, practitioners and scholars across the globe.

List of Contributors

A. A. Farag
Central Laboratory for Agricultural Climate, Agricultural Research Center, Egypt

Shaimaa H. Abd-Elrahman
Department of Soils and Water, Faculty of Agriculture, Ain Shams University, Egypt

Bienvenue Nawan Sourou, Christine Ajokè Ifetayo Nougbodé Ouinsavi and Nestor Sokpon
Faculty of Agronomy, Laboratory of Forestry Studies and Research, University of Parakou,
BP 123 Parakou, Benin

Anuj Kumar Garg and A. K. Gupta
Department of Chemistry, S.S.V. P.G. College, Hapur, U.P, India

Ashu Rani
Department of Pure and Applied Chemistry, University of Kota, Kota, Rajasthan, India

Harinder Singh, Madhu Sharma and Monika Bansal
School of Agriculture, Lovely Professional University, Phagwara, Punjab, India

Aakash Goyal
Biodiversity and Integrated Gene Management Program, International Center for Agriculture
Research in the Dry Areas (ICARDA), P.O.Box 6299, Rabat-Institutes Rabat, Morocco

J. B. Ndiso and H. M. Saha
Department of Crop Sciences, Pwani University, P.O.Box 195 - 80108, Kilifi, Kenya

G. N. Chemining'wa and F. M. Olubayo
Department of Plant Science and Crop Protection, University of Nairobi, P.O.Box 30197 00100 GPO, Nairobi, Kenya

Buri Mohammed Moro, Issaka Roland Nuhu and Essien Ato Martin
CSIR-Soil Research Institute (SRI), Academy Post Office, Private Mail Bag, Kwadaso, Kumasi,
Ashanti Region, Ghana

O. A. Agba
Department of Agronomy, Cross River State University of Science and Technology, Obubra, Cross River State, Nigeria

E. E. Ikenganyia and J. E. Asiegbu
Department of Crop Science, University of Nigeria, Nsukka, Enugu State, Nigeria

Martin Maluki, Joshua Ogweno and Robert Morwani Gesimba
Department of Crops and CHS, Horticulture and Soils, Egerton University, P.O.Box 536 - 20115, Egerton, Kenya

B. Ndiso and H. M. Saha
Department of Crop Sciences, Pwani University, P.O.Box 195 - 80108, Kilifi, Kenya

G. N. Chemining'wa and F. M. Olubayo
Department of Plant Science and Crop Protection, University of Nairobi, P.O.Box 30197 00100 GPO, Nairobi, Kenya

Onesmo F. Assenga, Tileye Feyissa and Patrick A. Ndakidemi
School of Life Sciences and Bioengineering, The Nelson Mandela African Institution of Science and Technology, P.O.Box 447, Arusha, Tanzania

M. E. Essilfie, H. K. Dapaah, J. W. Tevor and K. Darkwa
College of Agriculture Education, University of Education, Winneba P.O.Box 40, Mampong-Ashanti, Ghana

Namugwanya Margaret, John Stephen Tenywa and Twaha Ali Basamba
Department of Agricultural Production, Makerere University, P.O.Box 7062, Kampala, Uganda

Namugwanya Margaret
Department of Agriculture, Kyambogo University, P.O.Box 1, Kyambogo, Kampala, Uganda

Erasmus Otabbong
Departmentof Soil and Environment, Swedish University of Agricultural Sciences, Ornskoldsvik, Sweden

Douglas B. Sims and Christopher J. Collumb
Department of Physical Sciences, College of Southern Nevada, North Las Vegas, NV, USA

Vernon F. Hodge
Department of Chemistry, University of Nevada, Las Vegas, NV, USA

Konstantinos I. Vatalis
Environmental Engineering Department, Technological Institution, Western Macedonia, Kozani, Greece

George Valakas
School of Mining and Metallurgical Engineering, NTUA, Athens, Greece

Alaaldin Idris H. Khairy and Kwang Soo Roh
Department of Biology, College of Life Sciences, Keimyung University, Daegu 42601, Korea

J. Jeyapraba, S. Mahendran and N. Sujirtha
Department of Agricultural Biology, Faculty of Agriculture, Eastern University, Sri Lanka

M. A. Abul-Soud
Central Laboratory for Agricultural Climate, Agricultural Research Centre, Giza, Egypt

Shaimaa H. Abd-Elrahman
Soils and Water Department, Faculty of Agriculture, Ain Shams University, Cairo, Egypt

TAŞ Birol
Department of Park and Horticulture, Vocational School of Technical Science, Uludag University, Bursa, Turkey

M. S. Chernyshenko and L. I. Musatenko
Department of Phytohormonology, MG Kholodny Institute of Botany, NAS of Ukraine, Ukraine

V. O. Chernyshenko
Department of Protein Structure and Function, Palladin Institute of Biochemistry, NAS of Ukraine, Ukraine

Sauban Musa Jibril and Bello Hassan Jakada
Faculty of Applied Science, Jodhpur National University, Jhanwar Road Village Narnadi, Near Boranada, Jodhpur, Rajasthan, India

G. M. B. Bohm
Departamento de Gestão Ambiental, Instituto Federal Sul-rio-grandense- IFSul, Praça 20 De Setembro 455, Pelotas, 96015-360, RS, Brazil

T. L. Oliveira, E. M. Bohm and E. B. Bohm
Graduando no Curso Superior de Gestão ambiental, Instituto Federal Sul-rio-grandense- IFSul, Praça 20 De Setembro 455, Pelotas, 96015-360, RS, Brazil

P. J. Sanches
Departamento de Química, Instituto Federal Sul-rio-grandense- IFSul, Praça 20 De Setembro 455, Pelotas, 96015-360, RS, Brazil

Yaşar Özyigit
Department of Horticulture, Akdeniz University, Korkuteli Vocational High School, Antalya, Turkey

Sadik Çakmakci
Department of Field Crops, Faculty of Agriculture, Akdeniz University, Antalya, Turkey

J. C. Nwite
Department of Crop Production Technology, Federal College of Agriculture, Ishiagu, Ebonyi State, Nigeria

A. O. Onunwa
Department of Soil Science and Land Resources Management, Nnamdi Azikiwe University, Awka, Anambra State, Nigeria

C. A. Igwe
Department of Soil Science, University of Nigeria, Nsukka, Enugu State, Nigeria

T. Wakatsuki
Faculty of Agriculture, Shimane University, Nara, Japan

P. J. Brown, L. B. McCarty and V. L. Quisenberry
Clemson University, Clemson, SC, 29634, USA

J. L. Atkinson
11550 North Meridian Street, Suite 600, Carmel, IN 46032, USA

A. Manikandan
ICAR - Central Institute for Cotton Research, Nagpur, India

K. S. Subramanian
Tamil Nadu Agricultural University, Coimbatore, India

Most. Fatema Kaosar Mishu, Md. Waliur Rahman, Mohammad Abul Kalam Azad, Bhabendra Kumar Biswas, Md. Omar Kayess, Md. Rafiqul Islam and Md. Rakibul Alam
Department of Genetics and Plant Breeding, Faculty of Agriculture, Hajee Mohammad Danesh
Science and Technology University, Dinajpur, Bangladesh

Md. Aminul Islam Talukder
Department of Agricultural Chemistry, Faculty of Agriculture, Hajee Mohammad Danesh Science and Technology University, Dinajpur, Bangladesh

Olaoye J. Olawale
Department of Soil Science, Ahmadu Bello University, P.M.B 1044, Zaria, Nigeria

S. T. Abu and Ogunwole O. Dorcas
Department of Crop Science, Ahmadu Bello University, P.M.B 1044, Zaria, Nigeria

E. Indira
Manonmaniam Sundaranar University, Tirunelveli-627 012, Tamilnadu, India

B. Annadurai
Department of Physics, Aditanar College of Arts and Science, Tiruchendur-628216, Tamilnadu, India

P. Latha
Department of Botany, Govt. Degree College, Armoor, Nizamabad, India

B. Vidya Vardhini
Department of Botany, Telangana University, Dichpally, Nizamabad -503322, India

Margaret Awah Tita, Pierre Wotchoko and Alice Magha
Higher Teachers' Training College, University of Bamenda, P.O.Box 20, Bamenda, Cameroon

Primus Azinwi Tamfuh and Rebecca Bienkaa Azinwi
Faculty of Science, University of Yaoundé 1, P.O.Box 812, Yaoundé, Cameroon